Computational Science, Engineering and Technology Series: 28

Civil and Structural Engineering Computational Technology

Saxe-Coburg Publications on Computational Engineering

Trends in Parallel, Distributed and Grid Computing for Engineering
Edited by: P. Iványi and B.H.V. Topping

Developments and Applications in Engineering Computational Technology
Edited by: B.H.V. Topping, J.M. Adam, F.J. Pallarés, R. Bru and M.L. Romero

Developments and Applications in Computational Structures Technology
Edited by: B.H.V. Topping, J.M. Adam, F.J. Pallarés, R. Bru and M.L. Romero

Substructuring Techniques and Domain Decomposition Methods
Edited by: F. Magoulès

Soft Computing in Civil and Structural Engineering
Edited by: B.H.V. Topping and Y. Tsompanakis

Trends in Civil and Structural Engineering Computing
Edited by: B.H.V. Topping, L.F. Costa Neves and R.C. Barros

Parallel, Distributed and Grid Computing for Engineering
Edited by: B.H.V. Topping and P. Iványi

Trends in Computational Structures Technology
Edited by: B.H.V. Topping and M. Papadrakakis

Trends in Engineering Computational Technology
Edited by: M. Papadrakakis and B.H.V. Topping

Computational Methods for Acoustics Problems
Edited by: F. Magoulès

Mesh Partitioning Techniques and Domain Decomposition Methods
Edited by: F. Magoulès

Object Oriented Methods and Finite Element Analysis
R.I. Mackie

Programming Distributed Finite Element Analysis
R.I. Mackie

Computer Aided Design of Cable-Membrane Structures
B.H.V. Topping and P. Iványi

Domain Decomposition Methods for Distributed Computing
J. Kruis

Civil and Structural Engineering Computational Technology

Edited by
B.H.V. Topping and Y. Tsompanakis

SAXE-COBURG
PUBLICATIONS

© Saxe-Coburg Publications, Stirlingshire, Scotland

published 2011 by
Saxe-Coburg Publications
Dun Eaglais
Station Brae, Kippen
Stirlingshire, FK8 3DY, UK

Saxe-Coburg Publications is an imprint of Civil-Comp Ltd

Computational Science, Engineering and Technology Series: 28
ISSN 1759-3158
ISBN 978-1-874672-55-5

British Library Cataloguing in Publication Data
A catalogue record for this book is available from the British Library

Cover image: Design created, with permission, using Yuri Vishnevsky's "Silk".
http://weavesilk.com

Printed in Great Britain by Bell & Bain Ltd, Glasgow

Contents

Preface		iii
1	Role and Perspective of Computational Structural Analysis for Sustainable Reconstruction and Seismic Risk Mitigation after an Earthquake V. Gattulli	1
2	Concrete as the Multiphase Material in Biological Shields against Nuclear Radiation C.E. Majorana, B. Pomaro, V.A. Salomoni, F. Gramegna and G. Prete	35
3	Development of Realistic Three-Dimensional Track Models for Railway Vehicle Dynamic Analyses J. Pombo and J. Ambrósio	65
4	Dynamic Analysis of Beam Structures under Moving Loads: A Review of the Modal Expansion Method Z. Dimitrovová	99
5	Virtual Experiments and a Statistically Equivalent Representative Volume Element for Macroscopic Constitutive Laws M. Šejnoha, J. Vorel, R. Valenta and J. Zeman	131
6	Applications of Graph Products and Canonical Forms in Structural Mechanics: A Review A. Kaveh and H. Rahami	157
7	Numerical and Experimental Assessment of Stainless and Carbon Bolted Tensioned Members P.C.G. da S. Vellasco, L.R.O. de Lima, J. de J. dos Santos, A.T. da Silva, S.A.L. de Andrade and J.G.S. da Silva	187
8	Numerical Evaluation of the Ultimate Bearing Capacity of Steel Structures C.J. Gantes	219
9	Recent Developments in the Analysis of Stiffened Plates E.J. Sapountzakis	243

| 10 | Numerical Modelling and Analysis of Composite Structures subject to Extreme Loading
R.Y. Xiao, Z.W. Gong, C.S. Chin and C.G. Bailey | 279 |

Author Index   **301**

Keyword Index   **303**

Preface

This volume comprises the Invited Lectures presented at "The Thirteenth International Conference on Civil, Structural and Environmental Engineering Computing" (CC 2011) held in Chania, Crete, Greece from 6-9 September 2011. The CC conference series began in London in 1983. The CC 2011 conference was held concurrently with "The Second International Conference on Soft Computing Technology in Civil, Structural and Environmental Engineering" (CSC 2011). We are grateful to the authors and co-authors of the Invited Lectures included in this volume. Their contribution both to these conferences and to this book is greatly appreciated.

Other papers presented at this conference are published as follows:

- *The Invited Review Lectures from CC 2011 and CSC 2011 are published in:*
  Computational Technology Reviews, Volume 4, Saxe-Coburg Publications, Stirlingshire, Scotland, 2011.

- *The Invited Lectures from CSC 2011 are published in:*
  Soft Computing Methods for Civil and Structural Engineering, Y. Tsompanakis and B.H.V. Topping, (Editors), Saxe-Coburg Publications, Stirlingshire, Scotland, 2011.

- *The Contributed Papers from CC 2011 are published in:*
  Proceedings of the Thirteenth International Conference on Civil, Structural and Environmental Engineering Computing, B.H.V. Topping and Y. Tsompanakis, (Editors), (Book of Summaries with a CD-ROM of full-text papers), Civil-Comp Press, Stirlingshire, Scotland, 2011.

- *The Contributed Papers from CSC 2011 are published in:*
  Proceedings of the Second International Conference on Soft Computing Technology in Civil, Structural and Environmental Engineering, Y. Tsompanakis and B.H.V. Topping, (Editors), (Book of Summaries with a CD-ROM of full-text papers), Civil-Comp Press, Stirlingshire, Scotland, 2011.

We would like to thank the members of the CC 2011 Conference Editorial Board for their help before and during the conference: Dr J.M. Adam, Spain; Prof. H. Adeli, USA; Prof. Dr S.M.L. Adriaenssens, Belgium; Dr N.A. Alexander, UK; Prof. A.H. Al-Gadhib, Saudi Arabia; Dr S.A. Alghamdi, Saudi Arabia; Prof. H.M. Al-Humaidi, Kuwait; Dr N.S. Al-Kaabi, United Arab Emirates; Mr J.M. Alonso, Spain; Prof. D. Ambrosini, Argentina; Assoc Prof. R. Amor, New Zealand; Dr L. Andersen, Denmark; Prof. C.J. Anumba, USA; Dr Y. Arayici, England; Prof. A.J. Aref, USA; Dr

A.F. Ashour, UK; Dr I.M. Asi, Jordan; Dr C.E. Augarde, UK; Prof. Y. Ayvaz, Turkey; Prof. J.R. Banerjee, UK; Prof. C.C. Baniotopoulos, Greece; Prof. A. Baratta, Italy; Dr M.A. Barkhordari, Iran; Dr B. Basu, Ireland; Prof. J.W. Baugh, USA; Dr R.G. Beale, UK; Prof. I. Becchi, Italy; Prof. A.A. Becker, UK; Prof. J. Bencat, Slovakia; Prof. A. Benjeddou, France; PD Dr V. Berkhahn, Germany; Dr A. Bjelanovic, Croatia; Prof. J. Blachut, UK; Prof. F. Bontempi, Italy; Dr D. Borza, France; Prof. M.A. Bradford, Australia; Prof. F.A. Branco, Portugal; Dr J. Bridgeman, UK; Dr J. Brozovsky, Czech Republic; Prof. J.W. Bull, UK; Prof. F. Buyle-Bodin, France; Dr P. Cacciola, UK; Prof. R. Cajka, Czech Republic; Prof. D. Camotim, Portugal; Prof. F. Casciati, Italy; Dr L.M.Santos Castro, Portugal; Dr S. Chandra, India; Dr A.P. Chassiakos, Greece; Dr H.-P. Cheng, USA; Prof. Y.K. Chow, Singapore; Prof. I. Colominas, Spain; Prof. A. Combescure, France; Assoc. Prof. L. Contrafatto, Italy; Prof. J.G.S. da Silva, Brazil; Prof. L. Damkilde, Denmark; Assoc. Prof. J. de Brito, Portugal; Prof. G. De Matteis, Italy; Prof. G. De Roeck, Belgium; Prof. P.R.B. Devloo, Brazil; Prof. J. Dias, Germany; Prof. Z. Dimitrovová, Portugal; Prof. L.M.J.S. Dinis, Portugal; Dr M. Dolenc, Slovenia; Dr M. Domaneschi, Italy; Prof. L. Dunai, Hungary; Dr A. Dusi, Italy; Prof. A. Ebrahimpour, USA; Dr D.J. Edwards, UK; Dr-Ing. A. Elenas, Greece; Prof. A. Elhami, France; Prof. J. Ermopoulos, Greece; Dr D. Eyheramendy, France; Dr I. Faraj, UK; Prof. F. Farinha, Portugal; Prof. I. Flood, USA; Prof. L. Fryba, Czech Republic; Prof. M.B. Fuchs, Israel; Dr E. Gal, Israel; Prof. Ch. Gantes, Greece; Dr P. Gardner, UK; Prof. V. Gattulli, Italy; Prof. R.I. Gilbert, Australia; Prof. F. Gonzalez Vidosa, Spain; Prof. D.E. Grierson, Canada; Prof. N. Grujovic, Serbia; Dr Z.W. Guan, UK; Prof. M.N.S. Hadi, Australia; Prof. F. Hadipriono, USA; Dr M. Hartnett, Ireland; Assoc. Prof. O. Hededal, Denmark; Dr M. Hirokane, Japan; Prof. G. Hofstetter, Austria; Prof. C.A. Issa, Lebanon; Dr P. Ivanyi, Hungary; Prof. B.A. Izzuddin, UK; Prof. P. Jayachandran, USA; Dr L. Jendele, Czech Republic; Prof. D.S. Jeng, UK; Dr T. Ji, UK; Prof. X. Jia, USA; Dr X. Jiang, USA; Dr O. Jirousek, Czech Republic; Dr E.S. Kameshki, Kingdom of Bahrain; Prof. M. Kaminski, Poland; Prof. A. Kaveh, Iran; Prof. T. Kerh, Taiwan; Dr A.I. Khan, Australia; Prof. J.T. Kim, Korea; Prof. N. Kirac, Turkey; Prof. S. Kmet, Slovakia; Dr P. Komodromos, Cyprus; Prof. J. Kosmol, Poland; Dr T. Koudelka, Czech Republic; Prof. V.K. Koumousis, Greece; Dr D. Kovacevic, Serbia; Dr K. Krabbenhoft, Australia; Dr T. Krejci, Czech Republic; Dr J. Kruis, Czech Republic; Prof. J.S. Kuang, Hong Kong; Prof. H.G. Kwak, Korea; Dr J.E. Laier, Brazil; Prof. R. Landolfo, Italy; Prof. F. Lebon, France; Dr C.K. Lee, Singapore; Prof. P. Leger, Canada; Dr V.M A. Leitão, Portugal; Dr C.J. Leo, Australia; Prof. A.Y.T. Leung, Hong Kong; Prof. R. Levy, Israel; Prof. L. Lima, Brazil; Prof. S.Y.Liong, Singapore; Prof. J. Logo, Hungary; Prof. P.B. Lourenco, Portugal; Prof. P.E.D. Love, Australia; Dr J. Wei-Zhen Lu, Hong Kong; Dr R.I. Mackie, UK; Prof. C.E. Majorana, Italy; Prof. M. Malafaya-Baptista, Portugal; Prof. Dr K. Marti, Germany; Dr M.D. Martinez-Rodrigo, Spain; Prof. I. May, UK; Prof. F.M. Mazzolani, Italy; Prof. T.J. McCarthy, Australia; Dr G. Milani, Italy; Prof. S.A. Mirza, Canada; Assoc. Prof. E.S. Mistakidis, Greece; Prof. A. Miyamoto, Japan; Prof. R. Montenegro Armas, Spain; Prof. G. Montero, Spain; Prof. C.A. Mota Soares, Portugal; Dr I. Mura, Italy; Prof. G. Muscolino, Italy; Prof. T. Nakai, Japan; Dr J. Naprstek, Czech

Republic; Prof. L.F. Costa Neves, Portugal; Dr S.T. Ng, Hong Kong; Dr J. Novak, Czech Republic; Prof. E. OBrien, Ireland; Prof. H.R. Ovesy, Iran; Dr F.J. Pallares, Spain; Dr A. Palmeri, UK; Assist. Prof. P.G. Papadopoulos, Greece; Prof. M. Papadrakakis, Greece; Dr B. Patzak, Czech Republic; Prof. O. A. Pekau, Canada; Dr C. Pellegrino, Italy; Dr M. Phiri, UK; Dr Y.-L. Pi, Australia; Dr B. Picoux, France; Dr J. Pombo, Portugal; Ms V. Pomezanski, Hungary; Prof. C.P. Providakis, Greece; Dr P.N. Psarropoulos, Greece; Dr M.Y. Rafiq, UK; Prof. A.A. Ramezanianpour, Iran; Prof. M. Raoof, UK; Dr Y. Ribakov, Israel; Dr A. Riccio, Italy; Prof. N. Rizzi, Italy; Dr C. Romanel, Brazil; Prof. M.L. Romero, Spain; Dr M. Rouainia, UK; Dr D. Rypl, Czech Republic; Prof. R.L. Sack, USA; Assoc. Prof. F. Saje, Slovenia; Prof. M.P. Saka, Turkey; Dr M. Salehi, Iran; Prof. L.M. Santos Castro, Portugal; Assoc. Prof. E.J. Sapountzakis, Greece; Prof. T. Schanz, Germany; Prof. K. Schilling, Germany; Prof. G.I. Schuëller, Austria; Dr W. Sebastian, UK; Dr M. Scjnoha, Czech Republic; Dr A.G. Sextos, Greece; Prof. K.-J. Shou, Taiwan; Dr R.C. Silva, Brazil; Prof. L. Simoni, Italy; Prof. S. Singh, USA; Prof. M.J. Skibniewski, USA; Prof. R.M. Skitmore, Australia; Prof. J. Sladek, Slovakia; Prof. I.F.C. Smith, Switzerland; Prof. J.E. Souza de Cursi, France; Dr R. Spallino, Denmark; Dr M.C. Suarez Arriaga, Mexico; Prof. I. Takahashi, Japan; Dr T.T. Tanyimboh, UK; Prof. P.F. Teixeira, Portugal; Prof. R.A. Tenenbaum, Brazil; Prof. A. Tesar, Slovakia; Prof. D. Thambiratnam, Australia; Dr D. Theodossopoulos, UK; Prof. G. Thierauf, Germany; Prof. H.R. Thomas, UK; Dr W. Tizani, UK; Prof. B.H.V. Topping, UK; Prof. V.V. Toropov, UK; Prof. C.P. Tsai, Taiwan; Prof. T.-K. Tsay, Taiwan; Prof. G. Turk, Slovenia; Prof. Z. Turk, Slovenia; Dr H.S. Turkmen, Turkey; Prof. J. Turmo, Spain; Dr G.J. Turvey, UK; Prof. Y. Uematsu, Japan; Dr C. Vale, Portugal; Dr L.S. Vamvakeridou-Lyroudia, UK; Prof. M.M.B.R. Vellasco, Brazil; Prof. P.C.G. da S. Vellasco, Brazil; Prof. Dr-Ing. T. Vietor, Germany; Prof. W. Wagner, Germany; Prof. C.M. Wang, Singapore; Dr A.S. Watson, UK; Dr M.C. Weng, Taiwan; Prof. T.P. Williams, USA; Prof. R.Y. Xiao, UK; Prof. Y.-B. Yang, Taiwan; Dr J.Q. Ye, UK; Dr J. Zeman, Czech Republic; Prof. C. Zeris, Greece; Prof. Q.L. Zhang, China; Prof. A. Zingoni, South Africa; and Prof. P.P. Zouein, Lebanon.

Finally, we are grateful to Jelle Muylle (Saxe-Coburg Publications) for his help in coordinating the publication of this book and for all his administrative and organisational skills in organising these conferences. We also wish to thank Dawn Sewell (Civil-Comp Press) for her administrative support.

Professor B.H.V. Topping
University of Pécs, Hungary
& Heriot-Watt University, Edinburgh, UK

Professor Y. Tsompanakis
Technical University of Crete
Chania, Greece

Chapter 1

©Saxe-Coburg Publications, 2011.
Civil and Structural Engineering Computational Technology
B.H.V. Topping and Y. Tsompanakis, (Editors)
Saxe-Coburg Publications, Stirlingshire, Scotland, 1-34.

Role and Perspective of Computational Structural Analysis for Sustainable Reconstruction and Seismic Risk Mitigation after an Earthquake

V. Gattulli
Dipartimento di Ingegneria delle Strutture, delle Acque e del Terreno (DISAT)
CEntro di Ricerca e Formazione in Ingegneria Sismica (CERFIS)
University of L'Aquila, Italy

Abstract

The catastrophic earthquake which struck the city of L'Aquila at the beginning of April 2009, caused several failures in buildings. A serious level of damage occurred in masonry structures, including partial collapses; however the reinforced concrete structures generally performed satisfactorily despite large deformations. After the immediate recovery process, a long reconstruction period is expected, during which the issue of seismic mitigation is going to play a fundamental role. The benefits achievable with innovative techniques at any level of application, such as new materials to be integrated into the old masonry structures, reinforcement with different fibers for both structural and non structural elements, innovative protective systems increasing the dissipation characteristics or isolating the structures at their base, need to be supported by evidence of reliable modelling in order to be confident on the achievable benefits. The chapter illustrates the role and the perspective of computational mechanics in different areas currently under investigation at the new earthquake research centre of the University of L'Aquila (CERFIS www.cerfis.it). In particular a synthesis of the lessons learned through comparing the immediate damage scenario and its evolution during the aftershocks with its modelling reproduction through computational nonlinear structural analysis for both masonry and reinforced concrete is presented. Different methodologies for in-situ data acquisition are compared in order to assess their contribution towards reliable modelling. Different techniques and methods are used to calibrate modelling procedures able to reproduce the observable post-earthquake scenario in detail in a set of structures. The knowledge acquired in accurate modelling has been used to evaluate the efficiency of specific techniques for seismic mitigation. In the framework presented two case studies are discussed: the masonry structure of the Palazzo Camponeschi and the reinforced concrete structure of Building A of the Engineering Faculty of the University of L'Aquila.

Keywords: structural modelling, earthquake engineering, existing buildings, damage assessment.

1 Introduction

At 01:32:39 UTC (03:32:39 at epicentre), on the 6th of April 2009, a devastating earthquake struck the city of L'Aquila and surrounding villages in the Abruzzo Region of Central Italy. The magnitude of the main shock was estimated to be Mw=6.3 (moment magnitude scale) by the UGS US Geological Survey, or Mw=6.2 and ML=5.8 (Richter magnitude scale) by the Italian INGV - Earthquake Remote Sensing Group. The epicentre was located at depth of 8.8km at 5km SW of the city of L'Aquila (42.423°N, 13.395°E) [1].

The earthquake caused extensive damage to buildings and other engineering structures; partial collapses or severe damage affected virtually all of the constructions in the city center, which consisted primarily of masonry-type buildings, including traditional palaces and high-density residential quarters. Nonetheless, many of these buildings survived, though heavily damaged, thanks to the earthquake-resistant countermeasures which characterize the local traditional construction practices, schooled by a long history of recurrent seismic events. The principal residential areas, which mostly have emerged in recent decades to satisfy the needs of the growing population and are mainly located in the western and eastern peripheries of the city, are by contrast characterized by the extensive use of reinforced concrete technology. The concrete structures have shown a high variability of response to the seismic activity, with a generally good or optimal performance in most cases, considering the exceptionality of the event and despite a diffuse and significant scenario of non-structural damage. The scattered distribution of the heaviest damage reveals that site-specific conditions could have played a crucial role in the amplification of the seismic action. Isolated structural failures, or partial collapses, occurred in a few buildings; although their causes are still under investigation.

An extensive investigation activity is involving different research fields in order to characterize the main shock [1–3] and the seismic sequence [4, 5], based on the available National Accelerometric Network (RAN) registered data. As a consequence of the geological and topographic characteristics of the L'Aquila region, specific microzonation studies have been developed, based on the aftershock activity, to understand the polarization and amplification effects of the seismic shock [6–11]. Preliminary efforts to correlate the anomalous seismic intensity with the damage scenario on the buildings, have been done [12].

In respect of the structural engineering, continuous activity is developing the level of understanding of the structural behaviour during the earthquake. In the very short initial period, qualitative deductions have been done based the observation of the damage scenario on both masonry and reinforced concrete bindings [13–18]. These report activities had the specific remit to characterize, the scenario immediately after the earthquake for the forthcoming debate in the scientific community. Subsequently, a wider investigation activity to improve the level of knowledge about the public and private facilities have included both punctual tests to determine the material properties [19], and structural monitoring to achieve an overall comprehension of building modal characteristics [20–22]. During this investigation, a simultaneously strong and, in some areas novel, modelling effort has been conducted to understand the seismic behaviour of damaged buildings [23, 24], based on the survey testing and the preliminary consideration on the seismic input, and to verify the validity of the design procedures for both provisional [25] and rehabilitation interventions [19].

Several independent research groups have developed and used different levels of computational complexity to investigate the seismic structural behaviour of monuments and buildings. Particular attention has been devoted to the masonry buildings, because of their high artistic and economic importance. Starting from the analysis of the behaviour of specific portions of the masonry walls modelled as rigid body kinematic chains, which represent a useful and simplified approach to obtain a first indicative evaluation on the level of adequacy of each wall, many models have been developed; they aim to assess the damage occurring [26, 27] or to define the capacity of the structures against the seismic forces when a global behaviour is established. These activities represent the results of a enduring collaboration among many Italian researchers motivated by the examination of the damage level reached in several representative edifices and churches in L'Aquila region, aiming to research and install innovative techniques for a sustainable rehabilitation plan. As a result of the complexity in geometry, material distribution and damage scenario shown in many edifices, only the preliminary results of such efforts are currently available in the literature.

The results available in the literature evidence the high complexity in the masonry modelling arising from the non homogeneous material characteristics and the particular construction methods used in L'Aquila, where the poor quality of the mortar and the irregular stones cause the crumbling of the masonry. The current practice models, where homogenized material is considered, do not allow advanced and realistic analysis of the effective behaviour during the main shock. For this reason several approaches were proposed, either considering the overturning of a monolithic wall, or considering the incoherent nature of the masonry at L'Aquila.

The present chapter summarizes the efforts made to model the seismic structural behaviour of a selection of the historic and modern buildings belonging to the University of L'Aquila. In particular, the main emphasis is placed on the comparison between the immediate damage scenario and its evolution during the aftershocks with the results obtained by nonlinear structural analysis. The case studies of the Palazzo Camponeschi and the Building A of the Engineering Faculty of L'Aquila represent a particular choice arising from the special and thorough investigation activities done by different research groups from the Training and Research in Earthquake Engineering Centre (CERFIS). The macro element and the finite element models, used to examine the masonry edifice and the reinforced concrete structure, are fully described, highlighting the role of nonlinear modelling in the final design of the seismic retrofitting of the two buildings.

2 Seismic structural analyses: research versus current practice

The history of the last fifty years clearly shows that earthquake engineering plays a significant role in the civil construction area thanks to continuous efforts made in different research fields related not only to civil engineering. The inherent earthquake unpredictability, modelled often through the use of stochastic variables such as the return period, the intensity and the frequency content of the seismic action, makes it difficult to have a simple path for the prediction of the seismic induced effects on structures. Moreover, as a result of the wide availability of computers with high computational potential together with the ease of developing and using numerical tools, it has become relatively effortless to model the time dependence of the seismic action and the possible nonlinear structural behaviour. The great development of research in the field of structural mechanics and especially in computational mechanics has modified considerably the way of approaching structural design. Consequently, the technical codes which rule and drive the practice activity of civil engineers have been modified to take into account innovations in structural analysis.

In Italy, the first rules for structural design in seismic area, were proposed in the nineteenth century, as a consequence of the Messina (1783), Ancona (1857) and Norcia (1859) earthquakes. Immediately after the disastrous earthquake of Messina (1908), the technical codes suggested the use of wood and steel elements to reinforce the structures. The above mentioned suggestions did not furnish any values for the seismic action and did not request any numerical verification of these elements. This approach, which could have appeared inappropriate, was consistent with the building codes of those years. Only in the *Regio Decreto* of 10th January 1907, was the requirement to include consideration of tensional limits in the structural components introduced in the deign of reinforced concrete structures.

In that period, both in Italy and in the USA (after the San Francisco earthquake in 1906), an understanding of modern seismic engineering, from which the national seismic codes were developed, made great advances. In particular, it was recognized that the effects of dynamic actions on the structures could be represented by horizontal and mass-proportioned forces; their values, in the absence of direct registrations, were assumed to be reasonable values for designing structures that would remain undamaged after an earthquake; it was conceptually allowed that for earthquakes with high intensity, structural damage can occur, but it was not economically sustainable to prescribe a structural design under seismic actions with a high period of return, and consequently a scarce probability of occuring. In this respect, the horizontal seismic forces were reduced by one third (which is similar to the concept of the ductility factor). Finally, it was immediately recognized that the specific aim for the structural design in seismic areas is primarily the safeguarding of human lives and consequently the admission that the possible occurrence of damage should avoid the structural collapse.

Subsequently, in the USA, the first "formulae" for seismic calculation of the structures were introduced. The damage observation on einforced concrete structures, struck by earthquakes between the 1925 and 1935, showed that the structures designed to resist the wind performed sufficiently well during the seismic events without particular damage. Therefore, the guidelines defined how to take into account the lateral inertia forces as horizontal static forces considered unidirectional and with constant distribution in height. Linear models were used to establish the effects of the horizontal forces, which were considered dependent on permanent weight distribution.

It was only from the nineteen forties the importance of structural dynamics in the evaluation of seismic action and behaviour, was evidenced. In this period, the first digital registrations, on the El Centro (M_w=6.99) seismic event, were obtained. Based on these findings, still used nowadays, the response spectrum, which allows the definition of the linear structural response under the seismic input, based on basic concepts of modal decomposition [28], was developed.

The American Uniform Building Code (UBC) was proposed in 1961 with the aim of avoiding serious structural damage and the loss of human lives. The seismic action was modelled through a monodirectional constant and static force, independent from the height variation. The response spectrum approach for seismic action evaluation was unchanged, but a factor which represents the seismic zone was included in the intensity definition.

In Italy, this improvement was implemented only in 1971, with the national law n.64/74, issued after the Ancona earthquake, which occurred in the same year; in the Italian situation, this law marked an important change in the seismic conception of the structural design procedure. Until that period, the seismic area was located only where an earthquake had occurred in a limited interval of knowledge. Moreover, the previous code, law n.1684 of 1962 did not furnish any specific information either on the construction detail (the minimum area of longitudinal and transversal steel reinforcements, *etc.*) and the regulatory requirements to guarantee good behaviour under seismic forces. The seismic action intensity was calculated through the following expression, appearing similar to those cited in the American code, such as:

$$F_i = K_i W_i \qquad (1)$$

where F_i represents the horizontal force at the i-th floor, K_{hi} the seismic coefficient, W_i the total permanent weight. For the first time, the seismic action was evaluated by taking into account the dynamic characteristics of the structure and the zone where the structure was located. Indeed, the seismic coefficient K_{hi} was the product of different coefficients, as

$$K_i = CR\epsilon\beta\gamma_i \qquad (2)$$

where C represents the seismic intensity, equal to $(S-2)/100$, which depends on the seismic grade S of the seismic zone considered ; R is the response coefficient which depends on the natural period of the structure T_0 defined with simple formulas the height of the structure and their maximum dimension in plant; ϵ is the foundation coefficient and, finally, γ_i is the distribution coefficient of the horizontal forces. Generally, the distribution of static forces increase with the construction height.

In 1971, the earthquake on the Pacoima Dam (San Fernando), drastically changed the philosophy on which the seismic codes were based. The records of ground acceleration above $1g$ and the damage observed in some structures, such as the Olive View UCLA Medical Center, highlighted clearly that strong seismic accelerations made it easy to overcome the elastic range threshold for a large number of structures, but even more, very often the interstorey drift is too large to be tolerated by non structural elements and equipment that can be very expensive. Therefore, it was admitted that structures under intense earthquake can dissipate energy in the plastic range given that enough ductile behaviour is assured. Besides that, the economic loss arising from the impossibility of immediately using the production facilities, in a short time can overtake the cost of the damage which occured. Consequently, the idea of designing and constructing protective systems able to avoid the input of seismic energy into the structure or to increase the dissipation of energy, become motivated by economics [29].

The Friuli earthquake, in 1976, speeded up Italian research activities and, looking at the current progress in the USA, emphasised the need to thoroughly analyze the structural behaviour beyond the linear-elastic range. The studies on the fundamentals of dynamics and plasticity became central to the Italian studies on structural analysis and seismic engineering [30]. The use of the ductility factor, expressed by the ratio between the ultimate displacement over the yield displacement, in seismic design permitted referal to linear structural models subjected to horizontal actions evaluated through a design spectrum reduced by the ductility factor. Dynamics and ductility produced a great impact on design and construction practice.

The Italian law n.219 in 1981, enacted immediately for the emergency situation arising from the earthquake which struck the Irpinia region in 1980, was substituted in 1984 by a national code, which considered for the first time all the Italian territory subjected to seismic risk. Confirming the historical path, after a series of strong earthquakes occured in an interval between 1989 (Loma Prieta) and 1995 (Kobe), subsequence observations, investigations and developments permitted the itroduction of new concepts in the design codes. In this phase the probabilistic studies, especially devoted to identifying the return period and the intensity of earthquake, were used to improve calibration of the design action for the different performance required. Indeed, a differentiation in the design spectra was introduced to define the seismic action for several limit states of the construction, which prevented the structural and non structural components from different failures or damage levels. Moreover, the strategic importance of the construction or the owner's request can drive the definition of the design structural performance requirements clearly introducing the concept of performance based design. The objective is to abandon the design approach in which the code is prescribing the performance indirectly, through the use of technical prescription, to move towards new codes which define different level of performances at the local and the global level, which should be reached through the ability of the designers.

This new conceptual design viewpoint, reasonable from a theoretical point of view, has modified substantially the approach to the problem which began to be introduced ten years later in Italy. Indeed, in Italy, enhancement of the national seismic code was introduced in 1986 and in 1996 (DM1996), based on the risk seismic map introduced in 1984. In this period, the classical allowable stress design (ASD) method is accompanied by the new limit state verifications which use the semi-probabilistic approach. The main objectives of these new codes were the design and construction of structures which can sustain low intensity earthquakes without either the structure or the equipment being damaged, medium intensity earthquakes sustaining only non structural damage, and finally, intense ones producing no collapse or human loss. The strategic social importance of the structure is also evidenced, increasing the design horizontal forces, with the aim, immediately after the earthquake, of preserving the full effectiveness. In this period, the novelty in comparison to the previous standard is the use of the modal analysis and the consequent use of the design response spectrum to evaluate the seismic structural response which the code allowed as an alternative to the static analysis.

The apparent complication permits the dynamic amplification factor being taken into consideration directly on the simple oscillators in which the structure is decomposed (characterized by the natural period and the modal shapes). Consequently, the seismic actions are dependent on the modal characteristics by means of the participation factors, making the distribution of the inertia forces in the elastic range more realistic, especially in irregular three-dimensional frames, with respect to the distribution considered in the static analysis. In order to make use only of the linear elastic analysis the ductility factor concept is still used to define the design response spectrum.

The lastest generation of seismic codes in Italy were introduced in 2003, with the OPCM3431, and its integration, the OPCM3274, which have been consolidated by the national code DM2005.

The OPCM3274 is the first Italian code which introduced the possibility of specifically evaluating the structural ductility. Different from the previous codes, the ductility factor, previously implicitly considered, was determined either by the assumption of specified values, which depend on the structure's characteristics (material, geometry, structural scheme), or by more advanced nonlinear analyses at the local or the global level.

The seismic action was defined in probabilistic terms and analyses based on both the elastic and design response spectra were admitted. Different design levels were related to the expected structural performance. For the first time, the employment of nonlinear models in structural analysis was possible and the performance based design philosophy was introduced.

A comparison uniquely in terms of forces is not allowed; the idea that yielding in each structural element induces large displacements against low reactive forces variation produces great emphasis on displacements in the parameters to be controlled.

The force pattern related to an increasing level of seismic intensity (associated with specific structural performance: operational, immediate occupancy, life safety, collapse prevention) is difficult to evelute in an elasto-plastic analysis taking into account only the reactive forces. In terms of displacements, the different performance levels are clearly separated. Therefore, the description of the expected damage results easily can be described in terms of displacements and rotations in the structural elements.

Three different limit states were introduced in the seismic structural design: limited damage (SLDL), severe damage (SLDS) and collapse (SLCO). In particular, the first one aimed to reduce the damage in the structural and non structural elements, and is controlled by the limitation of the interstory drifts. The limit states of severe damage and collapse, by contrast, were related to the ultimate capacity, and were evaluated mainly in terms of resistance in the structural elements.

To maximize the structural capacity to dissipate the seismic energy, overcoming the elastic field, high ductility was requested and the capacity design approach, widely used in the USA [31], was introduced. The new approach aimed to hierarchize the succession of the ductility mechanisms while avoiding any fragile rupture in the structural elements and to achieve the maximum level of global ductility. A lower resistance against the ductile mechanisms than the fragile ones were requested as a design objective. The plastic hinges have to be developed in the beam elements first, then in the column, avoiding fragile ruptures in the nodes, slabs and foundations.

For the first time, the OPCM3274 defined specific criteria to evaluate the seismic adequacy of existing structures.

The actual code, the NTC08, became mandatory immediately after the L'Aquila earthquake and is a slight adaption of OPC3274 and the subsequent NT2005. In addition to taking into account the collapse, the necessity to minimize the damage on the structure and the associated economic losses, is emphasised. Four different limit states were defined (operative, immediate occupancy, life safety, collapse prevention limit states) introducing a further limit state for strategic buildings, assuring their efficiency under low intensity seismic events. The capacity design methodology, adopted by the OPCM3274, is extended to ordinary structures where a low level of ductility is requested. This historical review of the seismic codes has evidenced the significant changes which have occured in recent decades to structural design, due to a continuous improvement of the understanding of the physical phenomenon. Structural engineers are allowed, in theory, through the new concept introduced in the recent codes, to design civil structures with an ultimate capacity adequate to the local seismic demand, and to manage a correct sizing in order to ensure the structure will work optimally during its service life.

Important advances also concern the operations carried out to verify the seismic adequacy of existing structures, as actually defined in the codes. Particular attention was given by the NTC08, in a specific section, to the achievement of the acceptable knowledge and methods to analyze the existing structures, with specific paragraphs devoted to each material (reinforced concrete, steel and masonry). Nowadays, after the L'Aquila earthquake, the civil engineering community is mainly involved in applying the new code, evaluating the adequacy of the existing structures in a post-earthquake scenario. The specific complexity which characterizes these analyses is represented by the necessity to achieve an overall knowledge of the structure examined (such as the material characteristics, the damage sustained, the ageing effects, *etc.*) and reproduce them in reliable models. The modelling effort, which implies the adoption of advanced tools, aims to reproduce the seismic action, the actual structural configuration, including the damage and the deteriorated material characteristics.

This phase, which can be termed "diagnosis", as reported in Figure 1, runs parallel to the design idea which is the result of the structural designer's capacity and experience and is not characterized by an unique solution. A continuous comparison between the two phases aims to adapt the design idea on the modelling deductions or testing results, and, at the same time, manage the computational effort in order to allow the evaluation of seismic performance on the rehabilitated structure.

The central role of the structural modelling, in order to analyze the existing structures, induces the necessity to achieve an overall comprehension of the nonlinear behaviour of the different structural typologies, based on the material used. These requests require a calibration of the model complexity to reach a necessary and comprehensible output. In the following paragraphs, specific consideration of the seismic characteristics of the L'Aquila earthquake are summarized, then a review of the available computational models for both reinforced concrete and masonry structural elements is reported. Finally, examples of the complexity of modelling existing structures in their nonlinear behaviour under seismic actions are given through two case studies selected from among the activities developed by CERFIS.

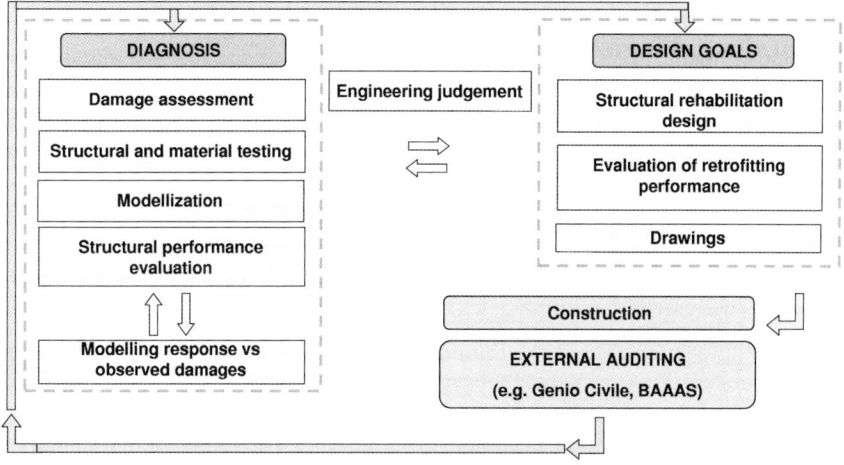

Figure 1: Coordinated actions for seismic retrofitting of existing building

3 Seismic action modelling

Thanks to a wide coverage by the National Accelerometric Network (55 strong motion stations) great quantities of data in acceleration, velocity and displacements are available to well describe the earthquake sequence. Significant data were produced by 21 stations of the National Accelerometric Network (RAN), situated less than 100km from the epicentre, fourteen being in the Abruzzo Region; four of them are placed less than 10km from the main shock epicentre.

In particular, nine stations, reported in Table 1, were situated less than 30km from the epicentre, located on different soil-types and altitudes. The higher values of ground acceleration were registered by the station AQV, with a maximum of 0.67g in an EW direction, 0.56g in a NS direction and 0.52g vertically. The other stations, including the closest AQG station, at 4.3 km from the epicentre, show lower values of PGA, in both horizontal and vertical directions. A fruitful correlation, through a standardized mitigation curve [32] between the RAN station registrations, in terms of maximum horizontal PGA, was done. In Figure 3, the curve, defined through a soft type-soil, was compared by the PGA of the closest accelerometer stations based on their distance from the epicentre. An overall coherence between the registration and the theoretical curve was demonstrated, showing an anomalous position of AQV being the PGA around two times what was predicted by the curve. Moreover, the registration around 20km distance, showed lower values of PGA in respect of that predicted probably arising from the presence of soil A which expedites seismic mitigation. The

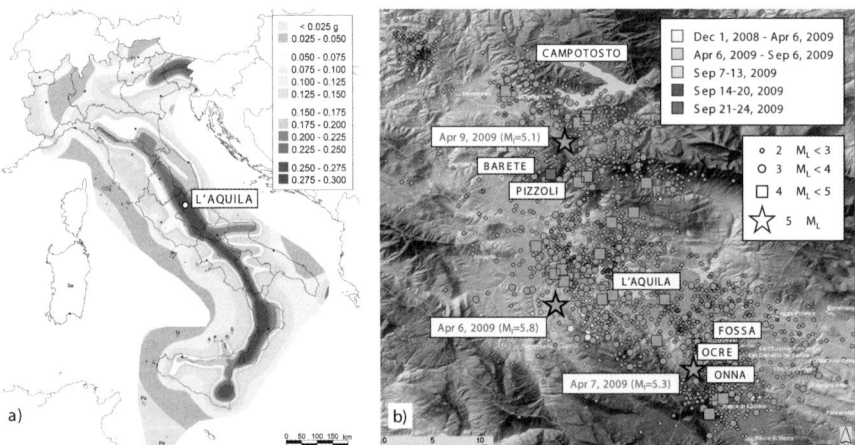

Figure 2: The earthquake of April 6, 2009: (a) localization of L'Aquila on the Italian map of the seismic hazard (ground acceleration with 10% probability of occurrence in a return period $T_r = 50$years on a A-type soil), (b) map of instrumental seismicity around the city of L'Aquila for the period December 2008-September 2009

stations AQG, AQK and AQA, placed on soils B or C, show good accordance with the curve.

However, despite particular attention being devoted to the analyses of these data, further information could be necessary to analyze the earthquake waves development in depth. In particular, as previously noted, the observation of the damage scenario distribution on the L'Aquila region, produced some anomalous effects on the civil structures. Contradicting consideration of the mitigation effects of the seismic waves, some particularly damaged villages were found far from the epicentre. The effects could be attributed both to the poor quality of the material and to site effects arising from the geological stratifications and the different altitude. An extensive microzonation study was developed by the Department of Civil Protection on the geological region's characteristics [33]. A more thorough inspection of the damaged buildings and other engineering structures allowed the authors to show a map of intensity distribution (I_s, MCS scale) of the region in question [34]. The highest level of I_s (= 9.0) were distributed along the NW-SE direction, according to the fault orientation. The devastating effects on the city of L'Aquila and in many villages on the Roio hill (I_s=8.0) were documented. It recognized that probable site effects contributed to amplify the shock waves in many villages situated on the top of hills, depending also on the geological stratifications.

Taking into account the structures of the University buildings, some conclusions were drawn on the site effects at Roio hill, seat of the Engineering Faculty Campus, at S-E of L'Aquila. A preliminary evaluation of the effective PGA perceived at the Roio hill (R point), taking into account the value of peak acceleration determined on the mitigation curve at around 2km, furnishing a value of 0.41g. An overall correspondence with the horizontal PGA at AQG allows the development of a reasonable estimation based on this registration.

The integration of motion equation under both the AQG horizontal accelerograms furnished the response spectra, in terms of acceleration and displacements (Figure 4).

Record Id	Station Id	Latitude [o]	Longitude [o]	Altitude [m]	Soil	Epic. dist. [km]	PGA_{EW} [g]	PGA_{NS} [g]	PGA_Z [g]
FA030	AQG	42.37	13.34	721	B	4.3	0.42	0.43	0.22
GX066	AQV	42.38	13.34	692	B	4.8	**0.67**	**0.56**	**0.51**
AM043	AQK	42.34	13.40	726	B	5.6	0.34	0.34	0.35
CU104	AQA	42.38	13.34	693	C	5.8	0.39	0.45	0.38
EF021	GSA	42.42	13.52	1062	A	18.0	0.15	0.15	0.11
BX007	FMG	42.27	13.12	1071	A	19.3	0.02	0.03	0.02
BY048	MTR	42.52	13.24	975	A	22.4	0.04	0.06	0.02
GE1463	GSG	42.46	13.55	1200	A	22.6	0.02	0.03	0.02
DF006	ANT	42.42	13.08	568	B	23.1	0.02	0.03	0.01

Table 1: Accelerometric National Network records of stations situated less than 100km from the epicenter

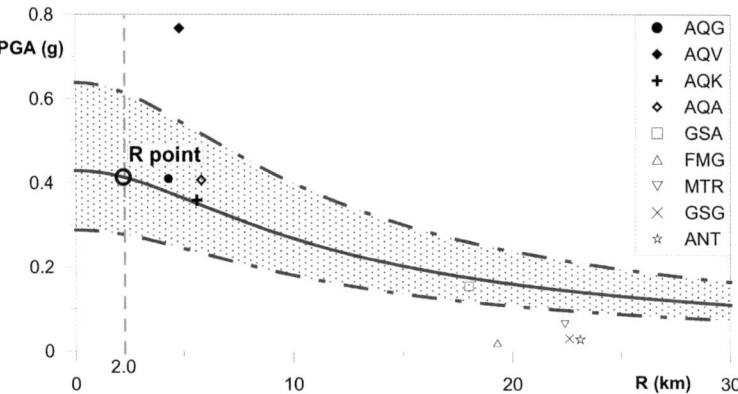

Figure 3: Seismic mainshock mitigation curve with the accelerometer stations closer than 30km. The Engineering Faculty is about 2km from the epicentre (R point, Roio)

Different values of damping (2-5-7%) were considered. A significant reduction effect on higher frequencies was evidenced up to 0.6s, from $\xi = 2\%$ to $\xi = 5\%$. In both directions, the damping ratio equal to 2%, furnishes acceleration values higher than 1g, with a maximum of 1.44g and 1.28g in N-S and E-W directions respectively. The N-S spectrum presents its maximum at 0.22s, while the E-W spectrum is at 0.51s. In both cases, the spectra several times overcome the value of 1g. Increasing ξ to 5%, the reduction induced only two peaks, in the N-S direction, around 1g, at periods lower than 0.3s. In terms of displacements, considering the lower damping ratio, the peak values both are placed at around 0.9s, being 14.8cm and 16.5cm in the N-S and E-W direction, respectively. At higher period values, the displacements decrease, quicker in E-W direction.

A standardized spectrum, which could be valid in both directions, is proposed (the dashed line in Figure 4). The criteria used to realize the spectrum include the parameters defined by the Italian code. Starting from a PGA of 0.41g, the new spectrum was calibrated considering a damping ratio equal to 5%. The fundamental periods, T_B and T_C, were defined with the NTC08 rules considering the location on a B-type soil, being 0.16s and 0.48s, respectively. The principal parameters are reported in Table 2. The plateau, placed at 0.75g, calculates the average of acceleration values in both directions. The spectrum aims mainly to approximate the AQG spectra, in the lower frequencies range. An overall under evaluation of the acceleration spectra is shown for a period higher than 0.6s. In terms of displacements, the standardized spectrum shows a good approximation until 0.4s; after a range of underestimation until around 1.0s, the regularized spectrum overtakes the others.

Moreover, one of the aspect of the 2009 seismic sequence discussed, was represented by the influences of site effects on the observed damage. The microzonation study carried out by the Italian Department of Civil Protection (DPC), dedicates a

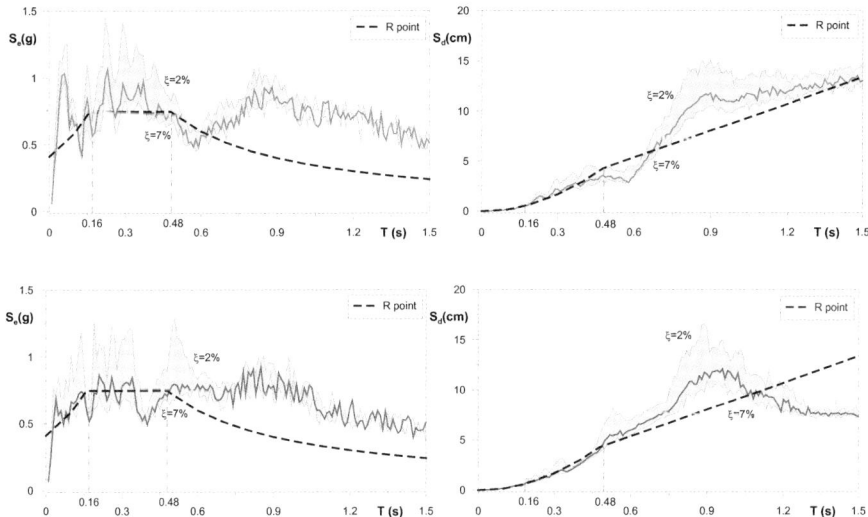

Figure 4: The response spectra of the AQG main shock (ξ=5%) compared with the elastic spectrum relative to the target PGA (R point): (a) N-S component, (b) E-W component

significant effort to analyzing the local effects both in the L'Aquila center, and in the surrounding hills, such as Roio hill. In particular the parameters needed to define the local spectra are furnished. However the L'Aquila region is divided into different districts, and a set of parameters is defined. Focusing attention on the areas where the cases studied are placed, an overall understanding of the local effects is achievable. The microzonation of the L'Aquila center furnishes only the value of the amplification factor for the acceleration which ranges between 1.2 and 2.0. Overall amplification effects are noted at the high frequencies range.

A more thorough investigation and data treatment was developed to define the microzonation of Roio hill. The results of different numerical models were illustrated, furnishing values of amplification factors and the fundamental periods to define the local spectra. No information from the top of Roio hill is available; several surrounding villages were examined, and a high variability of parameters is shown. In particular, the values of amplification factor ranges between 0.40 and 3.04, while the values of T_B and T_C range around 0.10s-0.24s and 0.29s-0.71s, respectively.

In addition, an extensive microzonation campaign, based on the analysis of after-

	a_g	F_O	T_C^*	Soil	S_t	ξ
R point	0.41g	1.28	0.356	B	1	5%

Table 2: Parameters for the definition of the spectra in Roio hill (R point)

shock-swarm, started a few days after the main shock [9], developed on Roio hill. Comparison between the measured accelerations and the observed damage evidenced the possibility of local amplification of the ground motion [1, 9, 35]. Focusing attention on Roio hill, the site amplification effects were analyzed taking into account 152 aftershocks from a network of six stations [9]. A summary of the results obtained is presented here in view of the discussion of the damage scenario relating to the estimate of the main shock. Colle di Roio is characterized by the overthrust of the Mesozoic carbonatic deposits above a Miocenic clay-marls succession, with a 130°N trending asymmetric anticline. The tests conducted examined the ground motion at the Engineering Faculty Campus, Poggio di Roio (two stations), Colle di Roio, Roio Piano and San Ruffina.

Signals were processed numerically to obtain the transfer functions between different nodes of the network. The results suggest an overall amplification effect on both alluvium-based and rock-based soils at the edges of the plain stations. The resonance frequency is characterized by a high variation from 1.3Hz to 4.00Hz. The Engineering Faculty station exhibits a high seismic amplification of the horizontal component, especially around 3.6 Hz, with an amplification factor equal to 2 in the range 2-10Hz. Here, the results of the site characterization are roughly synthesized in terms of elastic response spectra in Figure 5. In particular, firstly the amplification from soil B-type to soil C-type provided by the National code (NTC08) is taken into consideration, then the amplification evaluated from the study of the site effects is taken into account in the spectrum with local effect (dot-dashed line in Figure 5). Moreover, a three-dimensional effect which polarized the seismic waves was found. In particular, the maximum amplification denoted in the station closer to the Faculty of Engineering, was obtained for azimuth close to 120°N, as reported in the frame of Figure 5. The absence of a direct transfer function between the AQG station and the Engineering Faculty node, for the previously presented network, does not allow a complete reconstruction of the main shock characteristics felt by the structures of the Building A. However, the previously discussed results permit the determination of a reasonable range in which the elastic response spectrum of the main shock of April 6

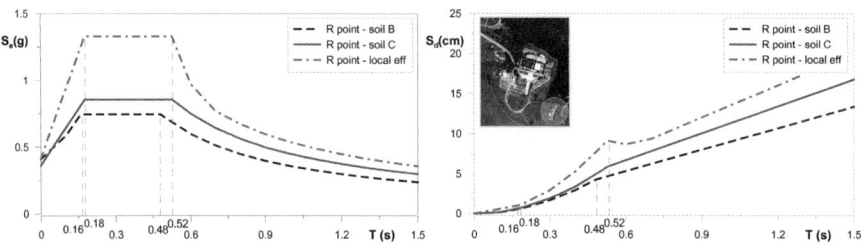

Figure 5: Comparison between the AQG average spectrum, the Roio spectrum and the hypothesized spectrum due to the local effects. In the frame the polarization (120°N) of seismic waves due to three-dimensional local effects

at the Engineering Faculty site is contained. Furthermore, a discussion on the damage which occurred with respect to the acceleration estimate can be developed [35, 36], as presented in the following sections.

4 Structural modelling

The progress in computational mechanics and the need to describe the structural behaviour beyond the elastic range have permitted the development of several models able to reproduce the nonlinear behaviour of the structural elements realized by different materials. The proposed models aim to achieve the correct combination between the specific level of detailed description requested and the computational effort. An extensive scientific literature is available on nonlinear modelling of both reinforced concrete and masonry elements.

4.1 Reinforced concrete structures

The modelling of nonlinear behaviour of reinforced concrete elements, characterized by the interaction of two different materials, represents currently a challenging research field. Most of the reinforced concrete elements, as beams and columns, due to their geometry properties, allow the assumption of the Euler-Bernoulli beam bringing back the behaviour description on the axis. Therefore, the overcoming of elastic range is described considering the plan section hypothesis and the uniform plasticization in determined point along the axis. This approach concentrates the plasticity in a determined cross-section, based on the axial force and the bending moment distribution, left in elastic range the remaining element. Based on the nonlinear constitutive laws of each material and the plane section hypothesis, the capacity curve of each section is modelled.

The definition of the so-called plastic hinges generally implies a two-component model acting in parallel: one which is linear elastic and one which is elastic-perfectly plastic with the plastic deformations concentrated in plastic hinges at the ends of the element [37]. The elastic modulus of the first component is equal to the strain hardening modulus $p \cdot EI$ of the moment-curvature relation, where EI is the pre-yield section stiffness. The elastic modulus of the elasto-plastic component is equal to $q \cdot EI$ where $q = 1 - p$. The nonlinear behaviour under both static and cyclic loads can be analyzed defining an hysteretic law.

To reduce the computational effort, the plastic hinges are defined where the bending moment development suggests the achievement of the yield limit. The concentration of all the plasticization in one point, overestimates the plastic rotation of the node, causing, for example in a column, an higher inter-storey drift. To reduce this effect, a refinement, which spreads the plasticity on a determined length, allows a more realistic rotation on the yielding tip.

However, when the ratio between the section dimension results are significantly different from the unit (out of the range 0.25-4), or the section shape is not regular, the plane section hypothesis loses validity and overcoming the elastic range shows a more diffusive crack pattern. In this case, the concentration of the plasticity at a determined point is not advised; many models proposed follow two different approaches, with different levels of computational effort. In the first case, the macroelements models aim to achieve an overall description of the element behaviour; many models, generally based on the strut-and-tie theory, are composed by a reticular-placed frame with an axial nonlinear behaviour defined. The behaviour under both static and cyclic load histories is well reproduced.

A second group of models, which require high computational effort and complexity, allow a more detailed description of the nonlinear behaviour with information at the continuum level. A discretized continuum model allows a complete knowledge of the stress and deformation scenario in the reinforced concrete element. The refinements consist on either a better definition of concrete micromechanics [38–40], or the interaction between the concrete and steel reinforcement [41]. The fragile and anisotropic nature of the concrete, represents a highly investigated aspect, taking into account the fractures development and their influence on the reinforced concrete stiffness and ultimate capacity. The finite element fracture analysis has been based principally on two-dimensional theory, considering either a discrete fracture or a smeared crack concept. The presence of single or multiple cracks patterns, in a defined area, are analyzed applying an incremental stress vector, updating the stiffness matrix, and, as requested, the tensor of principal direction (in rotating crack approaches). A reduced computational effort is required by this second approach [38, 42] allowing the development of more useful models. The majority of these models are based on linear elastic material constitutive laws coupled in series with a softening crack-bridging law, where tensile fracture occurs. More advanced constitutive laws were defined considering the consequence of shear-tensile loading to predict the three-dimensional material capacity [43].

Finally a novel numerical model has been proposed for the analysis of arbitrary shape cross-sections under three-dimensional loading, considering the interaction between all internal forces, which arise from anisotropic characteristics of constitutive materials [44]. The derived model is suitable for isotropic or anisotropic materials and nonlinear analysis of an arbitrary shaped cross section. The full three-dimensional state of a given cross-section is reproduced by means of the superposition of a distortion-warping displacement field ($\mathbf{u}_w$) with the traditional plan-section (from EB theory) displacement field ($\mathbf{u}_{ps}$).

4.2 Masonry structures

The modelling of masonry structures appears, generally, more complex due to the geometry of the walls, and the material characteristics. The fragility, heterogeneity and anisotropic characteristics of the masonry elements, as well as the typological variety, compromise the possibility of describing univocally the mechanics behaviour. Moreover, the evaluation of the seismic performance of the existing masonry structure is affected by the difficult to achieve complete consciousness of the entire storey of the structure, including, for example, the different construction periods, methodologies, any damage which has occurred and the subsequent interventions. The presence of bi-dimensional and three-dimensional elements, irregularly assembled, does not allow a simplified approach similar to that adopted by the reinforced concrete structures. For this reason, the construction of a reliable model also depends on the structure examined. Consequently, a preliminary intensive investigation effort, involving different professional skills, is needed to furnish a quantitative evaluation of the seismic capacity.

The first approach used to analyze the masonry structures consists of the definition of local kinematic chains to evaluate the overturning of each wall or portion of them. The capacity of the monolithic element, in terms of PGA, against the overturning, is calculated and compared with the demand. This approach represents a useful method to analyze the out-of-plane behaviour of the most representative portions of the building, also through the continuous updating of the wall configuration due to its rotation, without considering the presence of any contribution to the stability by the orthogonal walls. The material is considered only in terms of weight without any influence on the wall resistance or the kinematic chains shape.

Beyond the local analyses, the necessity to evaluate the global structural capacity under horizontal forces encouraged the investigation toward suitable models to capture the in-plane behaviour of the masonry walls. The progress of computational mechanics allows development of advanced models to analyze the in-plane capacity of the wall, and moreover, the interaction between orthogonally-oriented walls.

Finite element models, with either membrane or plate-elements allowed, added an overall understanding of global behaviour to the definition of the masonry nonlinear constitutive law, until achievement of the ultimate capacity. This approach frequently used in practice synthesizes the material in a equivalent continuum.

Simple models that consider each wall as a single structural element characterized by a nonlinear response in terms of applied shear and lateral drift have been proposed [46]. Brick units and mortar joints were considered as continuum element. The constitutive equations for in-plane loaded brick masonry are obtained on the hypothesis of a plane stress condition and by considering the brick masonry as a stratified medium made up of two typical layers. The masonry walls, modelled as an equivalent frame, and connected with the floors, composed the earthquake resistant structure. The floors share vertical loads with the walls and are planar stiffening elements, with an orthotropic membrane behaviour. The piers and lintels elements are joined in rigid nodes. The model is based on the hypothesis that the walls out-of-plane behaviour and flexural floors response are negligible with respect to global ones. The model allows both static and dynamical analyses.

Several models for in-plane analyses, based on a multilevel approach, were presented. An iterative scheme which uses two different, local and global, modellings of the masonry mechanics is developed. The significant influence of shape, dimension and disposition of the regular stones in the capacity prediction, is taken into account, as a unilateral frictional problem between rigid bodies. Therefore, the mechanical behaviour of the masonry is modelled as the interaction between each two bricks by means of the nonlinear response of the mortar joints modeled as a springs at contact [47].

More complex multiscale models, to represent the behaviour in the case of irregular stones masonry, were still under investigation. The greater difficulty is represented by the definition of an interaction law which depends on the shape and size of the stone and the thickness of the mortar. Some valid example were used to define the out-of-plane capacity [27].

5 Reinforced concrete buildings: the case of the Engineering Faculty Building A

As a result of the extensive attention devoted by a research group of CERFIS, the Building A of the Engineering Faculty, is considered as case study. The Building A represents one of the recently-built structures of the Engineering Faculty Campus (Figure 6), sited at the top of Roio hill, at the south-east of the city. The building hosts many classrooms, the Dean's Faculty Office, the student secretary and the laboratories of structures and materials and water engineering. Because of the ground slope, the building having four levels, sites two of these partially underground. The external covering is realized by glass curtain and split-face bricks. The principal entry, placed at the third floor, which is hidden by a semicircular bleacher, called an "amphitheatre", is over topped by a 4.50 m height glass curtain, and surrounded by 9.00 m height split-face bricks walls. At the opposite side, the last two levels are tapered, and are covered solely by wide windows. Transversally, perpendicular to the principal facade, some partition walls divide the different classrooms. Finally a large glass structure supported by a metallic frame was installed longitudinally, to cover the gap between the facade and the remained structures. Even though it appears as a single building, the Building A is composed of seven independent reinforced concrete substructures (A1-A7), with adequate seismic joints. Each substructure has significant irregularities in the geometry characteristics, mass and stiffness distribution. The principal facade covering is supported by planar reinforced concrete frames which are connected with the principal substructures through reinforced concrete slabs and horizontal metallic tubes. Between the substructures A3 and A4, simply supported reinforced concrete slabs are placed at each floor. The resistant structures are formed by shear walls positioned mainly at the low levels and along the perimeter, while internally and at the top two floors, a series of rectangular and circular columns are distributed. Wide beams in the slab heights represent the majority of horizontal elements in the intermediate floors. Two elevator cores, and the staircases are located in the substructures A2 and A6; while in A6 a passageway connects the Building B. The roof of the substructure A7 supports the passageway to the main entrance.

Immediately after the main shock, the damage occurring at Building A was exam-

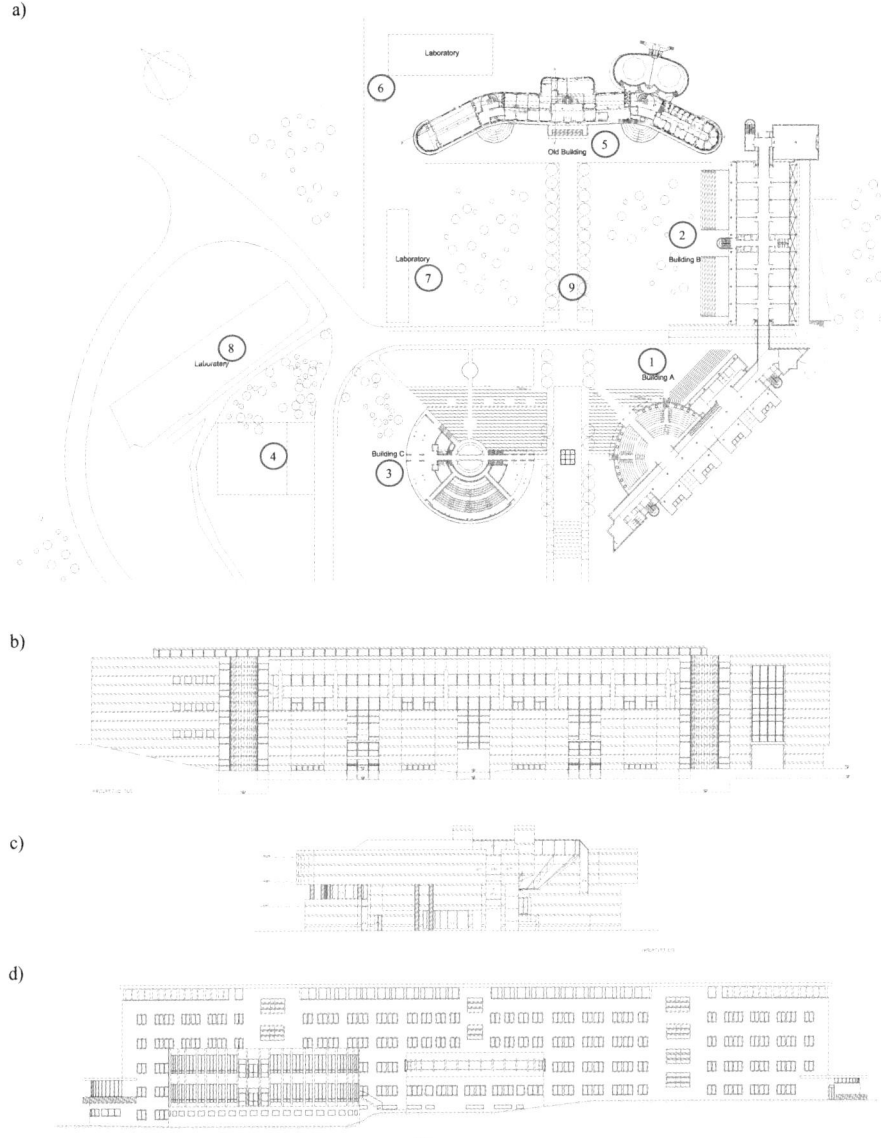

Figure 6: Faculty of Engineering: (a) Roof plan; (b) Building A, South view; (c) Building C, East view; (d) Old building, Nord view

ined both to understand the structural deficiencies under seismic loads and to register the damage evolution during the aftershocks. The main emblematic scenario was represented by the collapse of portions of split-face bricks walls on the main facade (Figure 7). Different profiles of damage were visible in the external and the internal

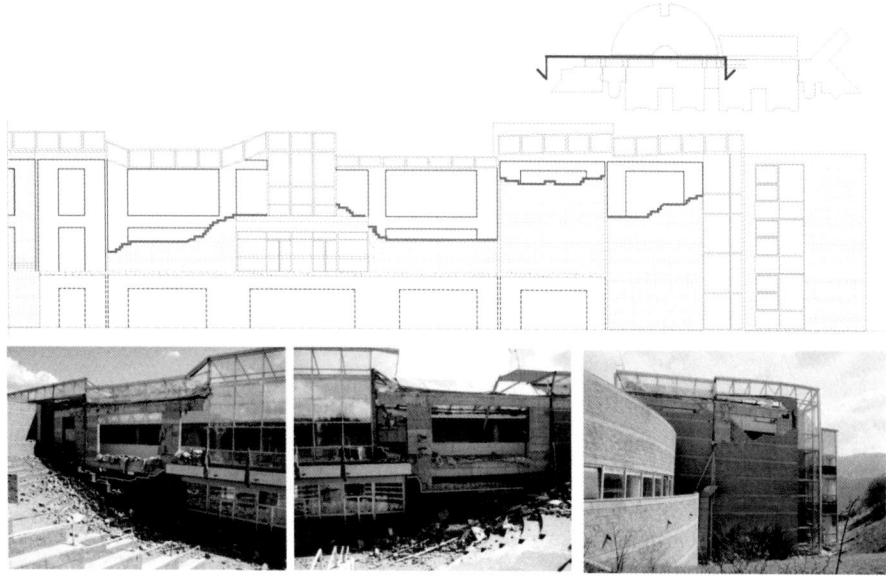

Figure 7: The facade after the main shock. The damage profile is shown with a thick line

layers. Due to the collapse of the split-face bricks the metallic tubes were seriously damaged, along the element and at the anchorages. The cover glass sheets and the metallic frames collapsed, probably arising from both the relative motion between the planar frames and the principal substructures and the damage to the walls which supported the structure on the facade-side. While the external substructures (A1, A2 and A6) were slightly damaged, the highest level of damage was visible on A3 and A4; in particular on the last two floors, starting from the entrance, where the transverse length of the building was reduced. Internally, in A3 and A4, the transversal partition walls, which did not provide significant resistance to the drift motion, presented X-shape cracks. The longitudinal partition, which divided the classrooms from the common areas, collapsed. The partially underground floors, due to the presence of wide shear walls appeared safe without significant non-structural damages. In all substructures, the reinforced concrete structural elements appear undamaged. The first surveys put into evidence the exceptionality of the seismic action which caused important relative displacements between the floors, especially where the shear walls were almost completely interrupted. In the lower floor, where wide shear walls surround all the substructures, the interstorey experienced drift was small and less damage occurred. The most vulnerable aspect was represented by the brick walls of the facade, which disaggregated as a result of the large accelerations and displacements. Also the internal partitions, due to their thinness and their length did not furnish resistance during the seismic shock.

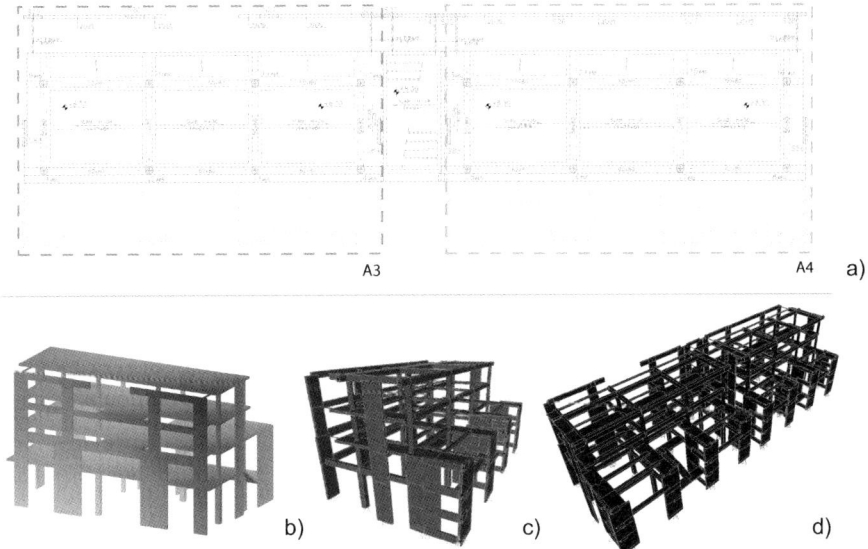

Figure 8: Structural modelling of substructure A3 of Building A: (a) the structural plan at the roof level, (b) three-dimensional reproduction, (c) finite element model of substructure A3, and (d) finite element model of substructure A3 and A4

5.1 Structural modelling

Several different finite element models were tested to reproduce the structural behaviour of the Building A. The finite element models were continuously improved thanks to the information derived firstly from the visual surveys, and subsequently by the on-site testing. A fruitful updating were conducted based on the identified modal properties. Soil-structure interaction at the base foundation has been also considered. The models represent only the resistant frames without considering the possible increase of stiffness and damping arising from the non-structural elements. This choice should be considered strongly realistic, in the case studied, arising from the particular constructive method used for the face-split bricks, which are mounted externally to the reinforced concrete frame. Moreover, the limited number of partition walls, used to divided the classrooms, together with small thickness and small connection furnishes a negligible augment of stiffness. The modelling effort aims to understand the seismic behaviour of each substructure, to verify the level of adequacy with respect to the Italian Code, and, finally, to design the retrofitting. To achieve reliable knowledge of the dynamic behaviour of each substructure a series of dynamic measurements under ambient vibrations were registered on the A1 and A3 substructures [21]. Attention was focused mainly on analyzing the structures behaviour in the out-of-plane facade direction and the latero-torsional behaviour.

The modal characteristics of each substructure were identified through output-only procedures, already investigated in previous studies [48, 49]. The measurement testing, which was conducted on the damaged structures, provided valuable information on the modal characteristics of the substructures; in particular the role of coupling, considering the reduced number of metallic tubulars, was examined. The lower natural frequency of substructure A1 is equal to 3.86 Hz. A roto-translation shape characterizes the main mode, with prevalent translation in the y transversal direction, and in-phase motion of the facade and the structure. All the modes identified are characterized by a roto-translational shape. Substructure A3 is characterized by the lower natural frequency of 2.77 Hz. The first natural mode is characterized by a y transversal translational shape in the structure and a symmetrical roto-translational shape in the two sides of the facade with a prevalent y-direction. Where the interconnections are lacking, out-of phase motion is possible. The second and third modes are characterized by roto-translational shape both in the structure and in the facade. As a result of the position of accelerometers, the different shape of facade are localized at the last floor, over the level of the principal entry.

An exhaustive and coordinated on-site testing campaign was conducted on the structural elements of the Building A to improve the level of knowledge. A series of inspections *in situ* was carried out to assure the correspondence of the design geometric characteristics with the real structures. Moreover, a campaign of material tests, through non destructive and destructive methods, aimed to define the mechanical characteristics of the concrete, the presence of steel reinforcements, coherently with the design plans, and their mechanical characteristics. The testing locations were carefully selected at different points of each floor level to represent the spatial distribution of the material characteristics on each substructure. The results show relatively high values of the concrete compressive strength [19]. A core drilling specimen on one of the transversal shear walls which forms the top of the substructure A3, showed cracked reinforced concrete enabling the possibility of correctly performing the compression tests. This occurrence, which could be caused by a cyclic exciting of the elastic behaviour of the structural element, represents useful information for understanding the level of capacity achieved by the substructure during the seismic event. Finally, the testing on steel specimens assigned the steel reinforcements in FE44k class, according to the Italian Standard. The effects of the values assumed for the material resistance on the nonlinear behaviour of the substructure A3 (Figure 8) were examined through a pushover analysis. The capacity curves in the out-of-plane facade direction considering the mean and the characteristic values are reported in Figure 9. The analysis shows a significant increase of resistance arising from higher values of material strengths which does not correspond to an increase in ductility, with the ultimate displacement capacity remaining constant. An examination of the seismic performance has been developed, focusing attention on both structural elements and split-face brick walls.

First, the linear models were used to examine the modal characteristics of each substructure. The analysis of modal properties of the substructures, some of which are characterized by coupling with the planar frames, show different shape of displacement arising from the translational and torsional component of motion. In particular, an important latero-torsional coupling is detected in substructures A1 and A6, due to their irregularity. The substructures A3 and A4 are characterized by regular motion in the facade out-of-plane transversal direction. Finally, for the high irregular geometry of A6, the first natural mode results in having a translation prevalent in the x longitudinal direction. An improvement of in finite element models, through the definition of nonlinear material constitutive laws, allows us to conduct a series of nonlinear static analyses to define the capacity curve of each substructure in both the principal directions. Due to the severe damage which occurred, and based on inadequacy detected during preliminary linear analyses with the design spectra, in terms of lateral drift at different levels, as requested by the National Code, the nonlinear behaviour of substructure A3 was thoroughly examined. The most vulnerable aspect was represented by the interruption of the shear walls at the entry level, which left the circular columns characterized by an insufficient level of stiffness. Two different finite element models were developed: one of which represents the whole substructure, and the second one that neglects the stiffer part, partially at the underground level, considering only the 3rd (the level of the principal entry) and 4th floor. Moreover, the non linear model of A3 was utilized to better understand the role of the tubes and the reinforced concrete slabs in the coupling between the principal structure and the partially connected planar frame of the facade. Different hypotheses on the tubular connection were done: no connection, rod connection and beam connection. While in the first hypothesis the planar frame is connected with the principal structure uniquely by the reinforced concrete slabs, in the other cases the presence of tubes, considering all the elements as in the undamaged structure, were taken into account either as rod connections or adding the bending moment resistance (beam connections). In Figure 10, the capacity curves for the three different hypotheses are reported. The incremental forces applied in the facade out-of-plane direction, are proportional to the principal mode shape in that direction. A significant difference between the cases where the facade is connected with the tubes or only with a reinforced concrete slab coupling can be appreciated. Indeed, the tubes sustain the shear walls of the planar frame, transferring axially the tension in the three-dimensional structure, assigning more ductility to the whole substructure.

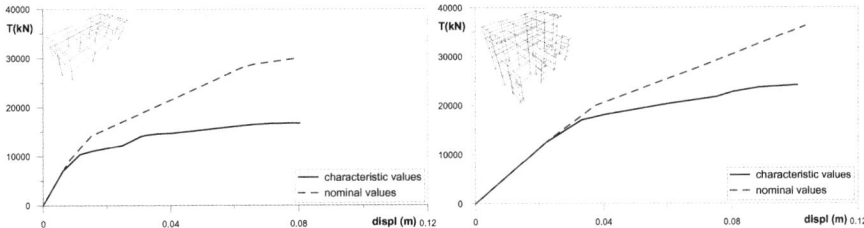

Figure 9: Structural capacity in out-of-plane facade direction using the material resistance characteristic values and nominal values: (a) finite element reduced model (only 3rd and 4th floor); (b) complete finite element model.

However, some differences in the global behaviour are visible between the rod connection and the beam connection. These small differences are only due to the low rotational stiffness of the tubular unit modelled as a rod, which leaves unchanged the stiffness ratio between the planar frame and the main system. In the complete model, there are larger differences between models using a rod connection or a beam connection. Taking into account the finite element models with rod connections, the capacity curves are compared with the demand, in terms of maximum displacement, considering both the spectrum at Roio with soil type-C and the spectrum which included the local effects (Figure 11). In both cases the demand is quite low with respect to the ultimate displacement capacity. Moreover, the demand is lower than the displacement required to achieve the first yielding in one element (represented as the blue point on the graphs), which occurred at the two external shear walls transversally oriented, the only ones that arrive at the top of the structure. Considering the complete model, the structure results appear more flexible and the demand, in terms of maximum displacement, increases. Taking into account the spectrum with the included local effects, the demands result become rather close to the first yielding point. This consideration can explain the anomalous cracked concrete found during the material testing, only in the shear walls. The results are coherent with the investigation tests and the *in situ* surveys, confirming that the substructure suffered a single large displacement which damaged the non-structural elements but left undamaged the structural elements. Preliminary analyses were carried out on the out-of-plane capacity of the face-split brick walls, against the seismic action, in respect of the monolithic overturning behaviour. The particular construction method, which was realized through two external layers in respect of the planar frame, allows the wall to be considered as an independent element, horizontally supported in a monolateral direction by the planar frame. Except for the substructure A1, the damages profile appears to be essentially horizontal suggesting the monolithic collapse of the non structural elements. The values of capacity in terms of PGA and maximum displacement, through, linear and nonlinear analyses respectively, and the respective adequacy levels, were calculated for each wall belonging to different substructures [50]. An overall inadequacy of the collapsed walls was demonstrated. Subsequently, the minimum value of horizontal force which the con-

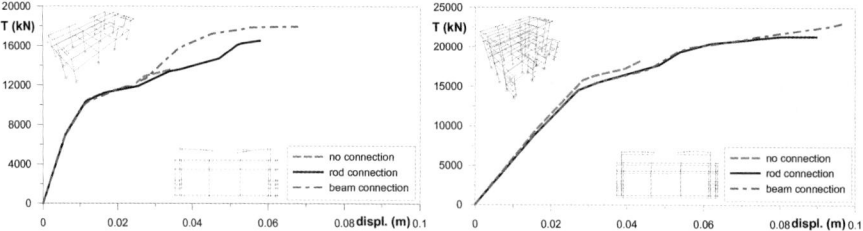

Figure 10: Structural capacity in out-of-plane facade direction with different connections between the facade and the main structure: (a) finite element reduced model (only 3rd and 4th floors); (b) complete finite element model

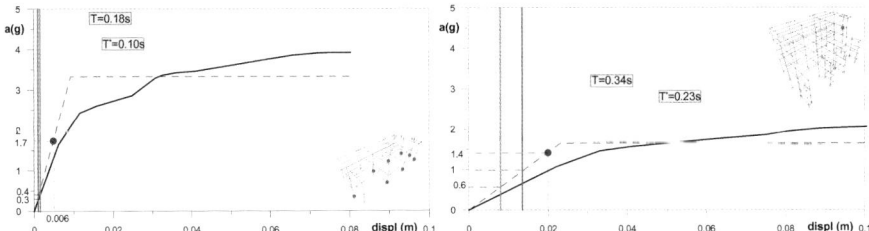

Figure 11: Structural capacity in out-of-plan direction, compared with the displacement demand of main shock, taking into account the (a) finite element reduced model (only 3rd and 4th floor); (b) complete finite element model. The green line represents the demand of Roio soil-C spectrum, the red line takes into account the local effects.

nections between the wall and the planar frame will be able to provide to achieve the full adequacy against the seismic action, was calculated. The analysis aims to high-

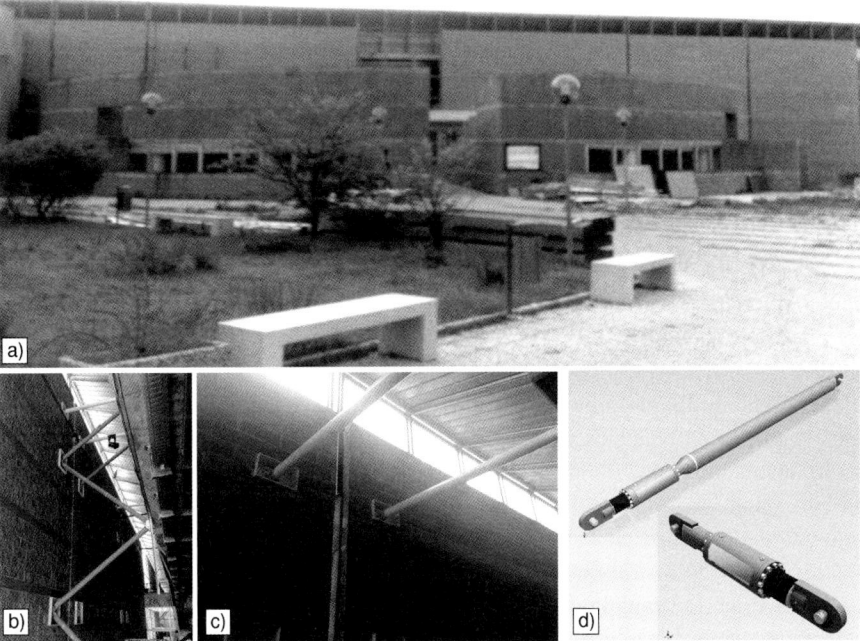

Figure 12: Retrofitting of the Engineering Faculty Building A: (a) reconstructed main facade, (b) coupling truss structures, (c) rod connections during construction, (d) transversal and longitudinal nonlinear viscous dampers

light the importance of a proper connection between the planar frame and both layers of the facade wall to avoid disastrous collapses.

Within the level of knowledge of the Building A behaviour system reached, the optimal design of retrofitting interventions on structural and non-structural components has been pursued based on energy dissipation concepts and new technologies [19]. The process has been supported by the free-of-charge cooperation of the academic staff and the resources of the university laboratories; the action has been supported by a voluntary non-profit organization which encourages high ethical standards. The purpose was to improve the seismic performance of the Building A through slightly invasive interventions. The dissipative coupling of adjacent structures with nonlinear viscous dampers was adopted [51]. An optimization of the damping features was done to maximize the dissipation effects. Moreover, strengthening of some shear walls by bonding FRP sheets was designed to increase the flexural and shear stress capacity. The works at the Engineering Faculty Building A are currently reaching an advanced level of development (see Figure 12).

6 Masonry buildings: the case of Palazzo Camponeschi

The Palazzo Camponeschi, seat of the Faculty of Letters and Philosophy, represents along with the Palazzo Carli, one of the historical building of the University of L'Aquila, located in the city centre. The construction and its development started in 1592 when the Jesuit order established the Aquilanum Collegium there. The Building was seriously damaged during the earthquake of 1703, and re-built in the following decades. In Figure 13, the three-dimensional geometry of Palazzo Camponeschi, joined to the church of S. Margherita, is depicted. Its irregular geometry, the different building periods of the structural components, the adjacency of other masonry structures, and the serious level of damage, represent a valid example of the high complexity involved in tackling the analysis of masonry structures.

The L-shaped masonry structure is based on a sloping site, being partially underground. The building is composed of three storeys with the last topped by a pitched roof. The main entrance, formed by an arched portal on the right side on the southeast facade, is located at the first floor. The external facades present a regular vertical and horizontal distribution of two orders of rectangular windows. The basement of the internal fronts, facing the internal courtyard, is characterized by regular arches supported by masonry columns. An embankment forms the internal courtyard.

From a structural viewpoint, the original resistant system is constructed by vertical masonry elements composed of irregular stone units. Again large inclusions of bricks or other low-quality materials reduce the regularity of the masonry framework. Additionally, the particular L-shape of the plan certainly increases its seismic vulnerability. Two wide passageways, at each level, connect all the rooms, along the wings. The floor slabs are supported by barrel and cross vaults, built of bricks and regular stones. A number of reinforcements, especially tie rods and concrete riddles, have been in-

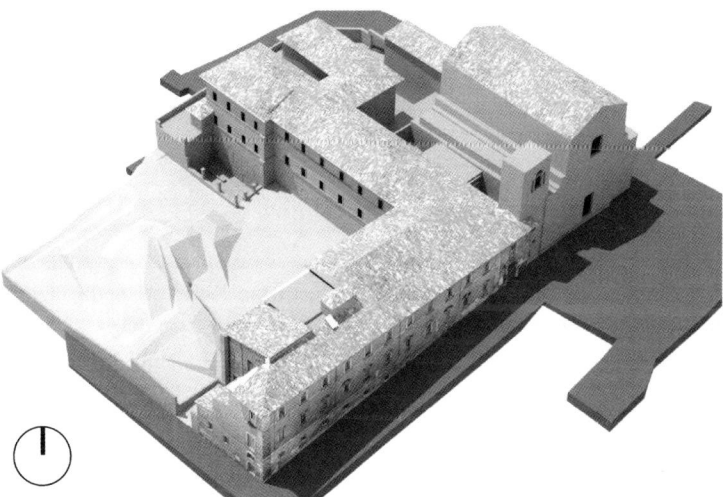

Figure 13: Architectural 3D model of Palazzo Camponeschi and S. Margherita church

serted over the years to fortify the overall strength of the structure. Wooden trusses support the roof. The west side, characterized by important interventions around the 1950s, consists of reinforced concrete slabs and beams.

The overall performance of Palazzo Camponeschi during the seismic event can be considered satisfactory without disastrous collapses. The high complexity of the geometry, the heterogeneous nature of the structural materials and, finally, the lack of adequate construction details, affected the seismic behaviour of the entire structure. Two regions of incipient overturning are clearly detected (Figure 14). The first one occurs on the internal facade on the N-E arm, where two monolithic portions rotate around a cylindrical hinge at the second floor (the first from the courtyard). At the end of the other arm the wall orthogonal to the principal facade presents an incipient overturning mechanism around a ground-level cylindrical hinge; a diagonal crack on the main facade wall suggests the movement of an additional lateral wedge, provoked by some internal tie rods, which guarantee a good interconnection.

The minor damage detected on the external walls demonstrates the important role played by the iron cross ties, effectively supplying a good connection between the orthogonal walls, thus increasing the global stiffness and ensuring the "box" behaviour of the whole structure. Nonetheless, shear-induced cracks of varying width and depth formed on the external walls, with losses of several architectural elements.

The overall scenario appears compatible with the development of a non-negligible torsional component in the seismic response of the building. The important crack pattern localized in the inner corner, at the connection between the two walls facing the internal courtyard, can be interpreted as an indirect confirmation of this conclusion. The indoor damage is severe but localized, as a large portion of the building appears

Figure 14: Palazzo Camponeschi kinematic mechanisms: (a) three hinges mechanism, (b) overturning mechanism.

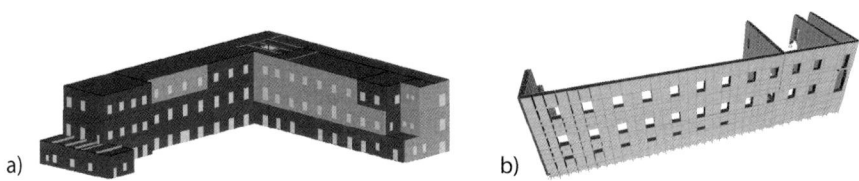

Figure 15: Palazzo Camponeschi models: (a) FME-B, (b) main facade finite element model

almost undamaged. In the zones where the damage is concentrated, a few collapses of the vaulted structures can be reported, and many internal vaults and arches present a serious crack pattern, including in particular those supporting the stairs. The ground of the internal courtyard was supported on the west side by a masonry retaining wall, which underwent a complete collapse under the seismically-induced pushing action.

6.1 Structural modelling

Extensive investigative testing allowed us to define almost four different masonry typologies, arising from the different phases of reconstruction and development of the structural scheme; medium dimension irregular stones, a double layer of regular stone, and inclusion of regular bricks characterized the masonry [52]. The experimental tests demonstrated the low quality of the mortar mechanics characteristics, while locally the complete lack of connection between each of the stones was detected.

Based on the extensive information, a modelling activity was carried out to analyze the structural vulnerabilities against the seismic action. The computational effort aimed to achieve an overall comprehension of the damage scenario, and the global seismic capacity. Moreover, a detailed analysis was performed uniquely on the main facade to investigate the diffuse damage scenario through including different hypotheses on the effective boundary condition. The frame by the macro element (FME) approach [45, 46] was used to realize a nonlinear model of the entire edifice. The heterogeneity in the masonry properties justified the necessity to maintain two different hypotheses; the first one (FME-A) considered only one uniform material, the second one (ME-B) used two materials; both the models were calibrated to reduce the differences between the effective mechanical properties. Figure 15a evidences with different colors the portions belonging to the two materials for the FME-B model.

The regular openings distribution, both vertically and horizontally, allows the structural modelling through an equivalent three-dimensional frame, with a juxtaposition of piers, lintels, rigid nodes, and horizontal diaphragms closing the "box". The staircases were not modelled, while the vaults were considered as horizontal slabs. Two different structures were jointly modelled, based on the survey considerations, the first one realized through regular stone blocks, the other concerning a reinforced concrete structure. The analysis consisted of several nonlinear static analyses developed by varying both the load distribution and their eccentricities. The results in terms of seismic demand and structural capacity, reported in Table 3, furnished an overall understanding of the adequacy of the resistant structure for serviceability (s) and ultimate limit states (u). The level of adequacy α is expressed as the ratio between the capacity and the demand displacement, in each direction.

The model is oriented to have the principal facade in the y-z plane, while the facade which runs along the adjacent church, is in the x-z plane. Both the models are characterized by a satisfying level of adequacy at the serviceability limit states with α values higher than 1.60. The structure results are inadequate at the ultimate limit states, with a worst performance in the principal facade out-of-plane direction. The FME-B model furnishes better results, due to the presence of the second material with better mechanical characteristics.

A more detailed model, which needs a large computing effort, was developed with the finite element approach (Figure 15(b)). To improve the understanding of the nonlinear behaviour of the main facade wall, under horizontal forces, with different constraint conditions, the nonlinear constitutive laws of two different materials were defined. The masonry wall was modelled through triangular and quadrilateral shell ele-

	$D_{c,s}$ (cm)	$D_{d,s}$ (cm)	$D_{c,u}$ (cm)	$D_{d,u}$ (cm)	α_u	α_s
FME-A						
dir x	2.55	1.57	2.92	4.38	**1.62**	**0.67**
dir y	3.00	1.08	3.56	3.93	**2.11**	**0.91**
FME-B						
dir x	2.68	1.59	3.20	4.40	**1.68**	**0.74**
dir y	1.99	0.87	2.53	2.51	**2.13**	**1.01**

Table 3: Displacement evaluated by nonlinear static analysis: D_c structural capacity, D_d seismic demand at s and u limit states

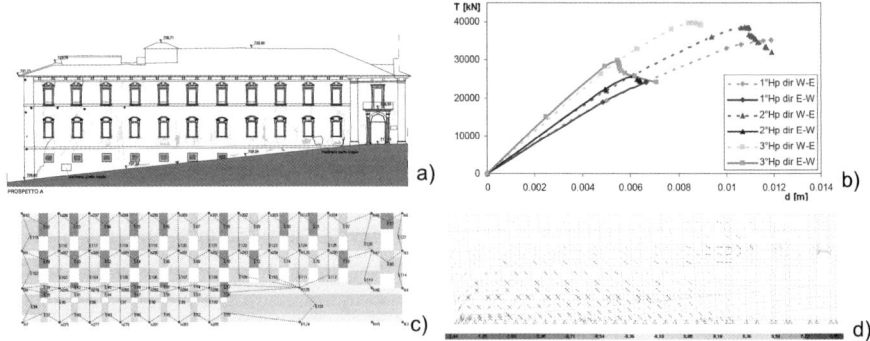

Figure 16: Damage on the main facade of Palazzo Camponeschi: (a) damage scenario, (b) capacity curves by the main facade model (c) damage mechanisms in the FME model (light/dark element in elastic/plastic range), (d) stress behaviour through static nonlinear analysis in the main facade model

ments. Two values of thickness, 1.0m and 0.9m, were defined and include the portion of wall under the entrance level, and all the superior portion, respectively. The orthogonal walls were modelled.

A sensitive analysis, varying the boundary conditions, furnishes useful information on the influence of the in-plane structural capacity. In particular, the effect of the sloping soil and the adjacent church were considered. The facade is constrained: at the bottom, at the right side to simulate the presence of the adjacent church and, finally, at the sloping ground level. The pushover analysis in both directions was developed and the results were reported in Figure 16. A fruitful comparison between the damage scenario and the modelling results is reported in Figure 16. High levels of stress, directed towards the left-lower side, confirmed by both models may correspond to the effective damage scenario.

7 Conclusion

The paper illustrates some of the activities conducted by several researchers and young fellows at the University of L'Aquila after the dramatic 2009 earthquake. The first part presents a brief critical historical review of the impact of research on the practical activities in the field of earthquake engineering. The second part, summarizing a series of relevant results, to the author's knowledge, describes the role played by computational mechanics on different engineering aspects of the reconstruction phase currently in progress. Two case studies are used to evidence the need for a balance between sophisticated modelling and realistic and practical tools usable by many practitioners continuously working on the decision-making process.

Acknowledgments

The author expresses deep gratitude to all the colleagues and young fellows, coauthors in many of the cited references, who have made possible the development of research in spite of the tremendous difficulties that the University of L'Aquila has faced after the April, 6, 2009 earthquake.

References

[1] A. Akinci, L. Malagnini, F. Sabetta, "Characteristics of the strong ground motions from the 6 April 2009 L'Aquila earthquake, Italy", Soil Dynamics and Earthquake Eng., 30, 320-335, 2010.

[2] R. Puglia, R. Ditommaso, F. Pacor, M. Mucciarelli, L. Luzi, M. Bianca, "Frequency variation in site response as observed from strong motion data of the L'Aquila (2009) seismic sequence", Bull Earthquake Eng, 9, 869-892, 2011.

[3] A. Masi, L. Chiauzzi, F. Braga, M. Mucciarelli, M. Vona, R. Ditommaso, "Peak and integral seismic parameters of L'Aquila 2009 ground motions: observed versus code provision values", Bull Earth Eng, 9, 139-156, 2011.

[4] B. Lolli, P. Gasperini, E. Boschi, "Time variations of aftershock decay parameters of the 2009 April 6 L'Aquila (central Italy) earthquake: evidence of the emergence of a negative exponential regime superimposed to the power law", Geophys J Int, 185, 764-774, 2011.

[5] G.A. Papadopoulos, M. Charalampakis, A. Fokaefs, G. Minadakis, "Strong foreshock signal preceding the L'Aquila (Italy) earthquake (M_w6.3) of 6 April 2009", Nat Hazards Earth Syst Sci, 10, 19-24, 2010.

[6] M. Compagnoni, F. Pergalani, P. Boncio, "Microzonation study in the Paganica-San Gregorio area affected by the April 6, 2009 L'Aquila earthquake (central Italy) and implications for the reconstruction", Bull Earthquake Eng, 9, 181-198, 2010.

[7] B. Pace, D. Albarello, P. Boncio, M. Dolce, P. Galli, P. Messina, L. Peruzza, F. Sabetta, T. Sanò, F. Visini, "Predicted ground motion after the L'Aquila 2009 earthquake (Italy, M_w6.3): input spectra for seismic microzoning", Bull Earthquake Eng, 9, 199-230, 2011.

[8] M. Maugeri, A.L. Simonelli, A. Ferraro, S. Grasso, A. Penna, "Recorded ground motion and site effects evaluation for the April 6, 2009 L'Aquila earthquake", Bull Earthquake Eng, 9, 157-179, 2011.

[9] E. Bertrand, A.-M. Duval, J. Régnier, R.M. Azzara, F. Bergamashi, P. Bordoni, F. Cara, G. Cultrera, G. Di Giulio, G. Milana, J. Salichon, "Site effects of the Roio basin, L'Aquila", Bull Earthquake Eng, 9, 809-823, 2011.

[10] S. Marzorati, C. Ladina, E. Falcucci, S. Gori, M. Saroli, G. Ameri, F. Galadini, "Site effects "on the rock": the case of Castelvecchio Subequo (L'Aquila, centrla Italy)", Bull Earthquake Eng, 9, 841-868, 2011.

[11] G. Cultrera, M. Mucciarelli, S. Parolai, "The L'Aquila earthquake - A view of site effects and building behavior from temporary networks", Bull Earthquake Eng, 9, 691-695, 2011.

[12] M. Mucciarelli, M. Bianca, R. Ditommaso, M. Vona, M.R. Gallipoli, A. Giocoli, S. Piscitelli, E. Rizzo, M. Picozzi, "Peculiar earthquake damage on a reinforced concrete building in San Gregorio (L'Aquila, Italy): site effects or building defects?", Bull Earthquake Eng, 9, 825-840, 2011.

[13] A.M. Ceci, A. Contento, L. Fanale, D. Galeota, V. Gattulli, M. Lepidi, F. Potenza, "Structural performance of the historic and modern buildings of the University of L'Aquila during the seismic events of April 2009", Engineering Structures, 32, 1899-1924, 2010.

[14] N. Augenti, F. Parisi, "Learning from construction failures due to the 2009 L'Aquila, Italy, earthquake", J of Perf of Constr Facilities, 24(6), 536-555, 2010.

[15] P. Ricci, F. De Luca, G.M. Verderame, "6th April 2009 L'Aquila earthquake, Italy: reinforced concrete building performance", Bull Earthquake Eng, 9, 285-305, 2011.

[16] H. Kaplan, H. Bilgin, S. Yilmaz, H. Binici, A. Oztas, "Structural damages of L'Aquila (Italy) earthquake", Nat Hazards Earth Syst Sci, 10, 499-507, 2010.

[17] F. Braga, V. Manfredi, A. Masi, A. Salvatori, M. Vona, "Performance of non-structural elements in RC buildings during the L'Aquila, 2009 earthquake", Bull Earthquake Eng, 9, 307-324, 2011.

[18] D. D'Ayala, S. Paganoni, "Assessment and analysis of damage in L'Aquila historic city centre after 6th April 2009", Bull Earthquake Eng, 9, 81-104, 2011.

[19] A.M. Ceci, L. Fanale, D. Galeota, V. Gattulli, M. Lepidi, F. Potenza, "Seismic retrofitting of the recently-built edifices of the Engineering Faculty of L'Aquila", Proc. of Sustainable development strategies for constructions in Europe and China, Rome, Italy, 6 April, 2010.

[20] D. Spina, B.G. Lamonaca, M. Nicoletti, M. Dolce, "Structural monitoring by the Italian Department of Civil Protection and the case of 2009 Abruzzo seismic sequence", Bull Earthquake Eng, 9, 325-346, 2011.

[21] D. Foti, M. Mongelli, V. Gattulli, F. Potenza, A.M. Ceci, "Output-only structural identification of the Engineering Faculty Edifice A at L'Aquila", Proc. of Fourth IOMAC Conference, 9-11 May, Istanbul, Turkey, 2011.

[22] C. Modena, F. Da Porto, F. Casarin, M. Munari, E. Simonato, "Cultural heritage buildings and the Abruzzo earthquake: performance and post-earthquake actions", Advanced Materials Research, 133-134, 3-17, 2010.

[23] G. Lucibello, G. Brandonisio, E. Mele, A. De Luca, "Seismic behavior of some Basilica churches after L'Aquila 2009 earthquake", Advanced Materials Research, 133-134, 801-806, 2010.

[24] L. Di Sarno, A.S. Elnashai, G. Manfredi, "Assessment of RC columns subjected to horizontal and vertical ground motions recorded during the 2009 L'Aquila (Italy) earthquake", Engineering Structures, 33, 1514-1535, 2010.

[25] M. Carraro, T. Ghedina, A. De Sabbata, C. Modena, F. Casarin, M.D. Benetta, "The S. Marco church in L'Aquila: Provisional interventions after the 2009

Abruzzo earthquake", Advanced Materials Research, 133-134, 953-958, 2010.
[26] C. Carocci, S. Cattari, S. Lagomarsino, C. Tocci, "The case study of Santa Maria Paganica church damaged by 2009 L'Aquila earthquake", Advanced Materials Research, 133-134, 163-168, 2010.
[27] G. De Felice, A. Mauro, "On overturning of the facade in churches with single nave: some cases studies from L'Aquila, Italy, 2009 earthquake", Advanced Materials Research, 133-134, 807-812, 2010.
[28] R.W. Clough, J. Penzien, Dynamics of structures, McGraw-Hill, NY, 1993.
[29] J.T.P. Yao, "Concept of structural control", ASCE Journal of the Structural Division, 98, 1567-1574, 1972.
[30] R. Baldacci, G. Ceradini, E. Giangreco, "Dinamica e stabilitá", Colonna Italiana, Tamburini, Milano, 1971.
[31] M.J.N. Priestley, G.M. Calvi, M.J. Kowalsky, "Displacement-based seismic design of structures", IUSS Press, 2007.
[32] F. Sabetta, A. Pugliese, "Attenuation of peak horizontal acceleration and velocity from Italian strong-motion records", Bull Seism Soc Am, 77(5), 1491-1513, 1987.
[33] Work Group MS-AQ, "Microzonazione sismica per la ricostruzione dell'area aquilana", Report Civil Protection Department, 2010. (in Italian)
[34] P. Galli, R. Camassi, "Rapporto sugli effetti del terremoto aquilano del 6 Aprile 2009", Joined Report DPC-INGV, 2009.
[35] S. Marzorati, L. Luzi, V. Petrini, F. Meroni, F. Pergalani, "Detection of local site effects through the estimation of building damages", Soil Dyn and Earth Eng, 23, 497-511, 2003.
[36] E. Irtem, K. Turker, U. Hasgul, "Causes of collapse and damage to low-rise RC buildings in recent Turkish earthquakes", J of Performance of Constructed Facilities, 21(5), 351-360, 2007.
[37] F.C. Filippou, A. Issa, "Nonlinear analysis of reinforced concrete frames under cyclic load reversals", Earthquake Engineering Research Center, Report No. EERC88/12, University of California, Berkeley, 1988.
[38] M. Petrangeli, J. Ozbolt, "Smeared crack approaches - material modeling", J. of Eng. Mechanics, 122(6), 545-554, 1996.
[39] P. Pivonka, J. Ožbolt, R. Lackner, H.A. Mang, "Comparative studies of 3D-constitutive models for concrete: application to mixed-mode fracture", Int. J. Numerical Methods in Engineering, 60, 549-570, 2004.
[40] Z.P. Bažant, "Concrete fracture models: testing and practice", Eng. Fracture Mechanics, 69, 165-205, 2002.
[41] G. Ranzo, M. Petrangeli, "A fibre finite beam element with section shear modelling for seismic analysis of RC structures", J. of Earthquake Engineering, 2(3), 443-473, 1998.
[42] M.P. Collins, D. Mitchell, P. Adebar, F.J. Vecchio, "A general shear design method", ACI Structural Journal, 93(1), 36-60, 1996.
[43] J. Cervenka, V.K. Papanikolaou, "Three dimensional combined fracture plastic material model for concrete", Int. J. of Plasticity, 24, 2192-2220, 2008.

[44] J.M. Bairan, "A non-linear coupled model for the analysis of reinforced concrete sections under bending, shear, torsion and axial forces", PhD thesis, Universitat Politècnica de Catalunya, Barcelona, Spain, 2005.
[45] S. Lagomarsino, L. Gambarotta, "Damage models for the seismic response of brick masonry shear wall. Part I: the mortar joint model and its application", Earthquake Eng and Struct Dyn, 26, 423-439, 1997.
[46] S. Lagomarsino, L. Gambarotta, "Damage models for the seismic response of brick masonry shear wall. Part II: the continuum models and its applications", Earthqiake Eng and Struct Dyn, 26, 441-462, 1997.
[47] S. Brasile, R. Casciaro, G. Formica, "Multilevel approach for brick masonry walls - Part I: A numerical strategy for the nonlinear analysis", Comput Methods Appl Mech Eng, 196, 4934-4951, 2007.
[48] E. Antonacci, A. De Stefano, V. Gattulli, M. Lepidi, E. Matta, "Comparative study of vibration-based parametric identification techniques for a three-dimensional frame structure", Structural Control and Health Monitoring, 2011. doi:10.1002/stc.449
[49] M. Lepidi, V. Gattulli, D. Foti, "Swinging-bell resonances and their cancellation identified by dynamical testing in a modern bell tower", Engineering Structures, 31, 1486-1500, 2010.
[50] A.M. Ceci, V. Gattulli, F. Potenza, "Serviceability and damage scenario in RC irregular structures: post-earthquake observations and modelling predictions", J of Performance of Constructed Facilities, 2011. (submitted)
[51] V. Gattulli, M. Lepidi, F. Potenza, A.M. Ceci, "Nonlinear viscous dampers interconnecting adjacent structures for seismic retrofitting", 8th Int. Conf. on Structural Dynamics EURODYN, 4-6 July, Leuven, Belgium, 2011.
[52] A.M. Ceci, L. Fanale, D. Galeota, V. Gattulli, "Analisi del comportamento sismico di Palazzo Camponeschi: modelli rappresentativi dello scenario di danno", XIV ANIDIS Conference, 18-22 September, Bari, Italy, 2011.

Chapter 2

©Saxe-Coburg Publications, 2011.
Civil and Structural Engineering Computational Technology
B.H.V. Topping and Y. Tsompanakis, (Editors)
Saxe-Coburg Publications, Stirlingshire, Scotland, 35-64.

Concrete as the Multiphase Material in Biological Shields against Nuclear Radiation

C.E. Majorana[1], B. Pomaro[1], V.A. Salomoni[1], F. Gramegna[2] and G. Prete[2]
[1] Department of Structural and Transportation Engineering
  University of Padua, Italy
[2] National Laboratories of Legnaro
  National Institute of Nuclear Physics, Legnaro, Padua, Italy

Abstract

Concrete is commonly used as a biological shield against nuclear radiation. As long as, in the design of the nuclear facility, its load carrying capacity is known together with its shielding properties, changes in the mechanical properties due to nuclear radiation are of particular significance and may have to be taken into account in such circumstances. The study presented here shows the first evidence of the behavior of concrete when exposed to nuclear radiation in order to evaluate the consequent effect on the mechanical field, by means of a proper definition of the radiation damage, strictly connected with the strength properties of the building material. Experimental evidence on the decay of the mechanical modulus of concrete has allowed implementation of the required damage law within a three-dimensional finite element (FE) research code which accounts for the coupling among moisture, heat transfer and the mechanical field in concrete treated as a fully coupled porous medium. The upgrade of the numerical model allows for assessment of the durability of concrete under the effects of a radioactive environment; considerations on the ultimate strength resource of concrete cannot neglect the temperature rise due to heat produced by radiation which, in fact, is proved to represent the most serious source of damage for concrete. The case study is represented by a next generation nuclear facility (currently under design) for the National Institute of Nuclear Physics (INFN) at the National Laboratories of Legnaro (LNL) in Padua, Italy: the SPES Project; the research structure is expected to produce neutron-rich unstable nuclei, called "exotic beams", by fission reactions of a primary radioactive proton beam on an uranium-carbonium target, in a dedicated underground bunker. Contour diagrams and time evolution graphs of the most significant variables (temperature, humidity, displacements, damage parameter) are reported for the scenario of the SPES Project, under an irradiation profile lasting six months, *i.e.* a working rate of the facility for the production of the exotic species of approximately 4500 hours per year, in order to limit temperatures to admissible exercise ranges for concrete.

Keywords: shielding, radioactive ion beams, radiation damage, nuclear facility, concrete durability, radiation heat, coupled problem.

1 Introduction

A recently developed extension of a finite element code, NEWCON3D, that can perform fully coupled hygro-thermo-mechanical three-dimensional analyses for cementitious materials, is outlined here; the research code is adopted to study the effects of nuclear radiation on a concrete shielding for the specific neutron source of the study case, in conjunction with a Monte Carlo code developed by CERN and INFN of Milan, Fluka [4], used to describe the radiation field (neutron fluence and deposited energy) which the mechanical field is dependant on. In fact, radiation in the form of either fast and thermal neutrons, primary gamma rays or gamma rays produced as a result of neutron capture, can affect concrete. Changes in the properties of concrete appear to depend primarily on the behavior of concrete aggregates that can undergo a volume change when exposed to radiation. Radiation damage in concrete aggregates is caused by changes in the lattice structure of the minerals in the aggregates. Fast neutrons are mainly responsible for the considerable growth, caused by atomic displacements, that has been measured in certain aggregates (*e.g.* flint). Quartz aggregates, made of crystals with covalent bonding, seem to be more affected by radiation than calcareous aggregates that contain a weaker ionic bonding. Neutron fluences of the order of 10^{19} n/cm^2 and gamma radiation doses of 10^{10} rad seem to become critical for concrete strength. The physical problem additionally requires that the collateral effect represented by the development of heat within the shielding, as a consequence of absorbed radiation, as long as the power density is not negligible, *i.e.* for values of energy flux density below 10^{10} MeV/(cm^2 s).

In Section 2 the main features of NEWCON3D are described, moving from a thermodynamic formulation which leads to recover the constitutive laws applied in the model; this Section conveys some theoretical aspects of the physical phenomenon of radiation/matter interaction, pointing out that two alternative solution approaches can be envisaged: an analytical one at the macro-scale and a stochastic one at the micro-scale, the latter being of interest for the numerical application with Fluka.

Section 3 sets out the specific upgrade of the FE code, introduced to take into account the effects of nuclear radiation on the material via the definition of a new damage variable; specifically, radiation damage is defined based on experimental evidence and its supposed acting in conjunction with the already implemented chemo-thermo-mechanical damage.

Numerical results are collected in Section 4: a proper interface between Fluka outputs, for the main physical quantities, and NEWCON3D has allowed the generation of coupled thermo-hygro-mechanical analyses for a slice of the directly impinged wall of the concrete shielding, under a design neutron exposure spectrum. What is relevant is that thermal effects due to radiation heat development are envisaged to be the most restrictive ones for finding a neutron source duration compatible with the durability of the shielding material (namely, six months).

2 The mathematical model

The FE model adopted to perform radiation-induced damage analyses is the three-dimensional rescrach code NEWCON3D [2, 19-22, 29-33]. Concrete is treated here as a multiphase system where the micropores of the skeleton are partially filled with liquid water, both in the form of bound or absorbed water and free or capillary water, and partially filled with a gas mixture composed of dry air (a non-condensable constituent) and water vapour (condensable), supposed to behave like an ideal gas. When higher than standard temperatures are taken into account, several phenomena are considered within the code: heat conduction, vapour diffusion and liquid water flow in the voids. In respect of the mechanical field, the model couples shrinkage, creep, damage and plasticity effects within the constitutive law of the material, here developed starting from a thermodynamic approach.

2.1. Thermodynamic formulation

By using the irreversible thermodynamic theory, a rational framework is developed to deal with the formulation of the constitutive equations of elastic-plastic materials, with damage, autogenous and viscous effects included, with the assumptions of infinitesimal deformations and coupling between thermal, hygral and mechanical fields, in the local state approach. Under the first hypothesis, the total strain ε can be split into the components given by the elastic strain supposed to be affected by damage ε^{ed}, the plastic strain ε^p, the autogenous strain ε^a and the viscous strain ε^c.

Within the local state approach, thermodynamics of a solid continuum, in a point and at a given instant of time, is defined as long as the value of a certain number of state variables is known, depending on the space coordinates only and not time-varying [18]. This assumption implies that the evolution of the thermodynamic state can be considered as a sequence of equilibrium states. We distinguish between observable variables and internal variables. The physical phenomenon here analysed, i.e. the considered constitutive law for the material under thermo-hygro-mechanical loads, requires the former, being ε the total strain tensor, T temperature and m the moisture mass per unit volume of the solid (moisture concentration), for which the conservation law gives:

$$\frac{\partial m}{\partial t} = -\nabla \mathbf{f} \tag{1}$$

where $\mathbf{f}$ is the flux of moisture mass; these are quantities strictly necessary to describe the reversible part of the process. The other variables are required when dissipative phenomena occur and, in our problem, they include the cumulated plastic strain p, defined as:

$$p = \int_0^t \left[\frac{2}{3} \dot{\varepsilon}^p(\tau) : \dot{\varepsilon}^p(\tau) \right]^{1/2} d\tau \tag{2}$$

(the symbol : denoting tensor product), a hardening variable $\boldsymbol{\alpha}$, a scalar variable taking into account damage effects due to irreversible changes on the internal structure D, a scalar variable β providing the current state of damage -introduced to prescribe the effects of a current damage state on the further development of damage, in agreement with [7] which investigates thermodynamics for elastic-plastic materials with damage; in addition the amount of total strain accounting for autogenous effects ε^a and the one accounting for creep/relaxation effects ε^c are introduced.

The physical process is said to be thermodynamically admissible if it satisfies the Clausius-Duhem inequality at each instant of the evolution in time of the process. The state variables being identified, the Helmholtz free energy of the material per unit volume ψ can be written assuming the decomposition of the total free energy ψ into four parts: the fraction coming from the elastic behaviour, affected by damage $\psi_e(\varepsilon^{ed}, D, m, T)$, the one due to the development of microscopic defects $\psi_d(\beta)$, the one given by plastic deformation, i.e. the hardening process $\psi_p(\varepsilon^p, p, \alpha, m, T)$, the one due to autogenous effects (e.g. given by chemical reactions due to carbonation or to the action of chlorides or sulphates) $\psi_a(\varepsilon^a, m, T)$, the one linked to relaxation effects, i.e. creep (at low to medium temperatures), which is supposed to evolve into load-induced thermal effects (at high temperatures) $\psi_c(\varepsilon^c, m, T)$:

$$\rho\psi = \rho\psi_e\left(\varepsilon^{ed}, D, m, T\right) + \rho\psi_d(\beta) + \rho\psi_p\left(\varepsilon^p, p, \alpha, m, T\right) + \rho\psi_a\left(\varepsilon^a, m, T\right) + \rho\psi_c\left(\varepsilon^c, m, T\right) \quad (3)$$

It is to be noticed that internal structural changes arising from damage are supposed to influence the elastic deformation only, under the assumption that the free energy arising from plastic, autogenous and viscous deformation is small in comparison with the elastic one, so that the effects of damage on plasticity, autogenous fields and creep can be neglected.

Acknowledged analytical expressions for the introduced free energy components are available in literature for ψ_e and ψ_c. Focusing on the component of the total free energy due to creep effects ψ_c, its definition implies a requirement to analyse the class of thermodynamically admissible creep and relaxation functions, **c** and **r** respectively, in the classic visco-elastic constitutive equations, relating the deformation and stress tensor ε and σ:

$$\varepsilon = \mathbf{c} \circ \sigma \qquad \sigma = \mathbf{r} \circ \varepsilon \qquad (4)$$

where the operator $\circ$ is the Stieltjes convolution. It is in fact shown that the field of thermodynamically admissible creep-relaxation functions is larger than the traditional set of monotone positive definite functions used in practice (exponential or power functions) and includes several non-monotone and even locally negative functions [9, 10]. Particularly, the use of exponentials is antecedent to power functions mainly due to historical reasons, i.e. linear viscosity was born in connection with the study of polymer behaviour and, only later on, with creep on porous materials, where the power integral kernels K, required by the heredity theory, were introduced in the Volterra integral form of the constitutive law (one-dimensional condition):

$$\sigma(t) = \int_{-\infty}^{t} K(0, t-v)\, d\varepsilon(v) \qquad K(0, t) = r(t) \qquad (5)$$

in which *r(t)* is the relaxation function and the integral is a Stieltjes integral.

One of the expressions for the density of the Helmholtz free energy, still most frequently applied, comes from Staverman-Schwarzl [35, 36], established for exponential relaxation functions for polymers in isothermal conditions:

$$\psi_c(t) = \frac{1}{2} \int_{0^-}^{t} \int_{0^-}^{t} r(2t - u - v) d\varepsilon(v) d\varepsilon(u) \tag{6}$$

where the lower limits of the integrals are justified when loading starts at time *t=0* with a jump; Equation (6) represents a particular case of the more general expression:

$$\psi_c(t) = \frac{1}{2} \int_{-\infty}^{t} \int_{-\infty}^{t} K(t - u, t - v) d\varepsilon(v) d\varepsilon(u) \tag{7}$$

where *K* is a kernel such that functional ψ_c in Equation (7) is non-negative definite. It is in fact to be noticed that (6) comes from (7) when the following holds:

$$K(x, y) = K(x + y) \tag{8}$$

and in (6) *r(t)* is a function of the relaxation spectrum $\rho(\mu)$:

$$r(t) = \int_0^\infty \rho(\mu) e^{-\frac{t}{\mu}} d\mu + r_\infty \tag{9}$$

r_∞ being the long-term modulus.

Non isothermal conditions were successively taken into account by Staverman assuming small changes in temperature, leading to a similar expression, though in incremental form [35]. As shown by Equation (6), the Staverman-Schwarzl's expression is a non-aging-like model for visco-elastic materials, since only heredity is taken into account *i.e.* the relaxation function, within the Volterra integral, is dependant just on the time lag *t-u-v* (duration of unit constant stress), and not separately on *t, u, v*. This second scenario would account for the description of the typical time-hardening process exhibited by concrete under solidification, mainly arising fom the mass growth of cement hydration products per unit volume. A thermodynamic formulation for ageing visco-elasticity recently has been suggested by Bazant and Huet [3], by generalization of the Staverman-Schwarzl's formula; after consideration of the integration domain, an expression for ψ_c is given that involves double Stieltjes integration, over a square domain for the strain history, of a quadratic expression depending only on a symmetrised form of the relaxation function, according to the following:

$$\psi_c(t) = \frac{1}{2} \int_0^t \int_0^t f(t, u, v) d\varepsilon(v) d\varepsilon(u) \tag{10}$$

where *f(t, u, v)* is found to be symmetric with respect to *u* and *v*:

$$f(t, u, v) = f(t, v, u) \tag{11}$$

and:
$$f(t,u,v) = \min[R(2t-v,u), R(2t-u,v)] \quad (12)$$

with $R(t,t')$ being a general relaxation function for an aging material.

Expression (10) suggested by Bazant and Huet for the free energy is shown to be a potential for stress, i.e. $\sigma = \rho \dfrac{\partial \psi_c}{\partial \varepsilon}$, and allows, conversely, one to obtain a constitutive equation for ageing visco-elasticity consistent with continuum thermodynamics.

As regards the free energy in the elastic field ψ_e, where damage is supposed to occur, the requirement of its convexity with respect to all the state variables (in order to verify *a priori* the second inequality law) leads to an expression for ψ_e, in the presence of damage, which is still quadratic in ε^e and T, as when damage is not accounted for [18]. Generalizing the thermo-elastic approach provided by Lemaitre to coupled thermo-hygro-mechanics, the following expression for the elastic contribution to the total free energy is obtained:

$$\rho\psi_e = \frac{1}{2}(1-D)\mathbf{a}:(\varepsilon^e - \mathbf{k}\Delta T - \boldsymbol{\mu}\Delta m):(\varepsilon^e - \mathbf{k}\Delta T - \boldsymbol{\mu}\Delta m) + C\Delta T^2 + M\Delta m^2 \quad (13)$$

in which ε^e is the elastic strain tensor of the undamaged material, $\mathbf{a}$ the fourth order tensor of the elastic components, $\mathbf{k}$ the tensor containing thermal expansion coefficients, $\boldsymbol{\mu}$ the tensor containing the coefficients of hygrometric expansion, ΔT and Δm the variations in temperature and moisture concentration, respectively.

Differentiating Equation (13) with respect to time and substituting the resulting expression into the Clausius-Duhem inequality, the requirement for entropy of being non-negative in isothermal conditions and at constant moisture content implies the state law reported below, which follows by generalizing to the thermo-hygro-mechanical case, the thermo-mechanical relationship defined by Lemaitre [18]:

$$\sigma = \rho \frac{\partial \psi}{\partial \varepsilon^e} = (1-D)\mathbf{a}:(\varepsilon^e - \mathbf{k}\Delta T - \boldsymbol{\mu}\Delta m) = \bar{\mathbf{a}}:(\varepsilon^e - \mathbf{k}\Delta T - \boldsymbol{\mu}\Delta m) \quad (14)$$

under the hypothesis that the change in mass density ρ is neglected.

Equation (14) shows that the free energy is a stress potential, as previously stated for ψ_c, or, in other words, that the stress tensor σ is the conjugate variable for the elastic strain tensor and can be considered as a modified stress tensor $\bar{\sigma}$ in the presence of damage [18]:

$$\bar{\sigma} = \frac{\sigma}{1-D} = \mathbf{a}:(\varepsilon^e - \mathbf{k}\Delta T - \boldsymbol{\mu}\Delta m) \quad (15)$$

The same differential approach for the other observable variables further gives the conjugate thermodynamic variables for temperature and moisture concentration:

$$s = -\rho\frac{\partial \psi}{\partial T} \qquad \bar{\mu} = \rho\frac{\partial \psi}{\partial m} \quad (16)$$

where s is the specific entropy, of which the specific Helmholtz free energy is a function, together with the specific internal energy e according to:

$$\psi = e - sT \tag{17}$$

and $\bar{\mu}$ is the moisture potential defined as:

$$\bar{\mu} = \bar{h} - T\bar{s} \tag{18}$$

in which $\bar{s}$ represents the amount of entropy brought about by the moisture flow and $\bar{h}$ is the enthalpy of moisture, function of pressure $\bar{p}$, mass density $\bar{\rho}$ and internal energy density $\bar{e}$ of moisture:

$$\bar{h} = \frac{\bar{p}}{\bar{\rho}} + \bar{e} \tag{19}$$

In analogy to what is stated for the stress tensor, temperature and moisture concentration, the conjugate forces corresponding to the internal state variables D, p, $\boldsymbol{\alpha}$, β and $\boldsymbol{\varepsilon}^i$ ($i=a, c, p$) can be found by taking into account the different components of the total free energy $\rho\psi$:

$$\boldsymbol{\sigma} = -\rho \frac{\partial \psi_p}{\partial \boldsymbol{\varepsilon}^p} \quad Y = \rho \frac{\partial \psi_e}{\partial D} = -\frac{1}{2} \mathbf{a} : \left(\boldsymbol{\varepsilon}^e - \mathbf{k}\Delta T - \boldsymbol{\mu}\Delta m \right) : \left(\boldsymbol{\varepsilon}^e - \mathbf{k}\Delta T - \boldsymbol{\mu}\Delta m \right)$$
$$R = \rho \frac{\partial \psi_p}{\partial p} \quad \mathbf{A} = \rho \frac{\partial \psi_p}{\partial \boldsymbol{\alpha}} \quad B = \rho \frac{\partial \psi_d}{\partial \beta} \quad \boldsymbol{\sigma} = -\rho \frac{\partial \psi_a}{\partial \boldsymbol{\varepsilon}^a} \quad \boldsymbol{\sigma} = -\rho \frac{\partial \psi_c}{\partial \boldsymbol{\varepsilon}^c} \tag{20}$$

In order to describe the inelastic dissipative process, as well as the evolution of the internal state variables, it is necessary to refer to the dissipation potential, *i.e.* a function continuous and convex with respect to the flux variables (*i.e.* time derivatives of the internal state variables and heat flux). By means of such a function the relationships between flux variables and dual variables given by the conjugate forces, *i.e.* the complementary laws, can be found.

If the expressions of the dissipation potential and its dual potential, function of the conjugate forces, are, respectively:

$$\varphi = \varphi\left(\dot{\boldsymbol{\varepsilon}}^p, \dot{D}, \dot{p}, \dot{\boldsymbol{\alpha}}, \dot{\beta}, \boldsymbol{\varepsilon}^a, \boldsymbol{\varepsilon}^c, \frac{\mathbf{q}}{T}, \mathbf{f} \right) \qquad \varphi^* = \varphi^*(\boldsymbol{\sigma}, Y, R, \mathbf{A}, B, \mathbf{g}, \mathbf{i}) \tag{21}$$

where $\mathbf{g}$ and $\mathbf{i}$ are the thermodynamic conjugate forces related to heat flux $\mathbf{q}$ and moisture flux $\mathbf{f}$, respectively:

$$\mathbf{g} = \nabla T \qquad \mathbf{i} = \nabla \bar{h} \tag{22}$$

then, the complementary laws are the following:

$$\dot{\boldsymbol{\varepsilon}}^p = \frac{\partial \varphi^*}{\partial \boldsymbol{\sigma}} \quad \dot{D} = -\frac{\partial \varphi^*}{\partial Y} \quad \dot{p} = -\frac{\partial \varphi^*}{\partial R} \quad \dot{\boldsymbol{\alpha}} = -\frac{\partial \varphi^*}{\partial \mathbf{A}}$$
$$\dot{\beta} = -\frac{\partial \varphi^*}{\partial B} \quad \boldsymbol{\varepsilon}^a = \frac{\partial \varphi^*}{\partial \boldsymbol{\sigma}} \quad \boldsymbol{\varepsilon}^c = \frac{\partial \varphi^*}{\partial \boldsymbol{\sigma}} \quad \frac{\mathbf{q}}{T} = -\frac{\partial \varphi^*}{\partial \mathbf{g}} \quad \mathbf{f} = -\frac{\partial \varphi^*}{\partial \mathbf{i}} \tag{23}$$

It is to be underlined that the first law leads to the plasticity law whereas the following four relationships represent the evolution laws for the internal state variables, particularly the second is the so-called property of normality for the damage variable D.

Now, if the flux vector **J** is defined as:

$$\mathbf{J} = \rho \{\dot{\varepsilon}^p, \dot{D}, \dot{p}, \dot{\alpha}, \dot{\beta}, \varepsilon^a, \varepsilon^c, \mathbf{q}, \mathbf{f}\}^T \qquad (24)$$

and its thermodynamic conjugate force vector **X** as follows:

$$\mathbf{X} = \left\{\boldsymbol{\sigma}, Y, R, \mathbf{A}, B, \frac{\mathbf{g}}{T}, \mathbf{i}\right\} \qquad (25)$$

then the rate of entropy production can be expressed as:

$$\mathbf{X} \cdot \mathbf{J} = \boldsymbol{\sigma} : \dot{\boldsymbol{\varepsilon}}^p - Y\dot{D} - R\dot{p} - \mathbf{A} : \dot{\boldsymbol{\alpha}} - B\dot{\beta} + \boldsymbol{\sigma} : \boldsymbol{\varepsilon}^a + \boldsymbol{\sigma} : \boldsymbol{\varepsilon}^c - \frac{1}{T}\mathbf{g} \cdot \mathbf{q} - \mathbf{i} \cdot \mathbf{f} \geq 0 \qquad (26)$$

If the uncoupling among intrinsic, thermal and hygral dissipation is conceived (which does not imply the uncoupling of effects) and the hardening effect is supposed to happen without damage and, *vice-versa*, damage occurs without a macroscopic plastic deformation, the following inequalities hold separately:

$$\boldsymbol{\sigma} : \dot{\boldsymbol{\varepsilon}}^p - R\dot{p} - \mathbf{A} : \dot{\boldsymbol{\alpha}} - B\dot{\beta} + \boldsymbol{\sigma} : \boldsymbol{\varepsilon}^a + \boldsymbol{\sigma} : \boldsymbol{\varepsilon}^c \geq 0 \qquad -Y\dot{D} \geq 0 \qquad (27)$$

which, Y being quadratic and positive definite according to Equation (20), states the increase of damage in time, in order to fulfil the second principle of thermodynamics.

Looking at Equation (26), it can be seen that the Clausius-Duhem inequality becomes an extended version of the classical formulation, where an extra entropy flux γ, directly connected with $\mathbf{X} \cdot \mathbf{J}$, is introduced, related to the considered dissipative processes -in addition to heat dissipation and fluid flow dissipation, *i.e.* the mechanical dissipative processes connected with damage, autogenous effects and creep.

With the introduction of the new terms accounting for damage, plasticity, autogenous effects and creep, as suggested in [14], the second principle in coupled thermo-hygro-mechanics can be written in the form below (we refer to [37] for how the hygral part enters into the inequality):

$$\boldsymbol{\sigma} : \dot{\boldsymbol{\varepsilon}} - \rho(\dot{\psi} + s\dot{T}) + div\boldsymbol{\gamma} - (\mathbf{q} + \boldsymbol{\gamma}) \cdot \frac{\nabla T}{T} - \nabla \bar{h} \cdot \mathbf{f} + \bar{h}\dot{m} \geq 0 \qquad (28)$$

where the superimposed dot denotes time derivatives. The proof that an expression analogous to Equation (28), though in thermo-mechanics, leads to the Clausius-Duhem formalism for the second law of thermodynamics, is provided in [14] and the essential conditions under which the inequality is recovered are shown there, starting from an analogous approach for the definition of the internal state variables of the problem.

2.2 The radiation field

In order to answer the specificity problem represented by the SPES Project [13], the main theoretical aspects of radiation by neutron particles are collected in the following. The core of the facility, in fact, is the target made of a fissionable compound of uranium and carbon, and the target room is, therefore, the most critical environment, mainly exposed to neutrons generated by high energy fission reactions from the primary proton beam impinging on the target.

According to the deterministic approach, the transport equation (so-called "Boltzmann equation") is understood to describe the attenuation of radiation in matter; the Boltzmann equation is an integro-differential equation which is obtained by considering the balance (production minus losses) of inflow and outflow particles through the surface of an elementary closed volume, in steady state conditions for the radiation field, and for this reason it is considered as the basis of the "transport theory"; it describes in the most general form the physical phenomenon of radiation attenuation in a shielding, distinguishing among the amount of absorbed, scattered, and secondarily produced radiation, that are the different possible forms of removal of the incident particles. Hence neutron transport calculations can be treated by numerically solving the Boltzmann equation; the unknown is the neutron flux, a spatially, directionally and energetically dependent quantity, so the numerical resolution implies discretization in space, angles of propagation and energy of radiation. During the early 1990s the Monte Carlo technique was improved to face the same problem and nowadays many physical phenomena have been implemented in Fluka or similar codes.

In both cases a high knowledge of the shielding properties of media is required, in terms of a so-called "cross-section" exhibited by the atoms of the medium towards the different removal mechanisms, listed above.

In the following two simplified approaches for dealing with the transport theory are recalled briefly that, though under some easier assumptions, are enlightening to comprehend the behavior of neutrons colliding in a shielding.

The two approaches are applied in a semi-infinite plane geometry and are described by: the diffusion theory, which applies to "thermal neutrons", *i.e.* relatively slow neutrons (with energy often referred to as below 0,025eV) and the two-group theory, which allows for getting the attenuation model for "fast neutrons" (neutrons with energy between 0,5 MeV and 10 MeV) [34]. They both developed from the Boltzmann equation but renounce solving the problem in all its complexity, neglecting several contributions in the balance, *e.g.* the photons produced during inelastic scattering of neutrons, the captured thermal neutrons, the secondary gamma photons and neutrons as a result of fission reactions; nevertheless they offer, in their simplified way, acceptable estimates in the expected flux density field [8, 16, 34].

The basic hypothesis of the diffusion theory is that the neutron current is isotropic and given by:

$$I = -D\nabla\Phi \qquad (29)$$

thus leading to an equation of the type:

$$-D\nabla^2\Phi + \Sigma_a\Phi - S = 0 \qquad (30)$$

where Φ is the flux density (for instance in [n/(cm^2 s)], S is the source term, Σ_a is the absorption cross-section and D a diffusion coefficient, defined in a simple diffusion theory by:

$$D = \frac{\lambda_s}{3} \qquad (31)$$

λ_s being the scattering mean free path.

The solution for a plane source in an infinite medium can be derived from this approximated theory by imposing proper boundary conditions, *e.g.*:

$$\Phi(x) = \frac{SL}{2D} e^{-\frac{x}{L}}, \qquad L = \sqrt{\frac{D}{\Sigma_a}} \qquad (32)$$

where L is known as the diffusion length.

This theory is valid deep in the medium, where the flux Φ is often given by multiplied scattered particles and hence it can be expected to be nearly equal in all directions, whereas near a free surface or a source the flux density is quite anisotropic and the diffusion approximation may be poor. Moreover the same hypothesis makes the theory suitable for thermal neutrons rather than fast ones [8].

The second approach allows for describing a simplified behavior for fast neutrons: according to the two-group theory, neutrons in the shield are divided into neutrons at thermal energies and those above. The group diffusion equation (30) is applied to each group, the source term in each group being the neutrons slowed down in the immediately higher energy group. This leads to two equations of the type:

$$\begin{aligned} -D_f \nabla^2 \Phi_f + \Sigma_f \Phi_f = 0 \\ -D_t \nabla^2 \Phi_t + \Sigma_t \Phi_t - \Sigma_f \Phi_f = 0 \end{aligned} \qquad (33)$$

for the fast and the thermal group, respectively. The fast source term in the shield is zero, whereas the thermal source term equals the fast neutrons scattered into thermal energies ($\Sigma_f \Phi_f$) and, in a similar manner to the diffusion equation (*i.e.* according to an exponential decay, see Equation (32)), the solution of the fast group equation in a semi-infinite plane geometry is given by:

$$\Phi_f(x) = \Phi_0(x) e^{-\Sigma_R x}, \qquad \Sigma_R = \frac{1}{L_s} \qquad (34)$$

where Σ_R is the macroscopic removal cross section, which is roughly equal to the reciprocal of the average relaxation length for fast neutrons in the shielding material, and Φ_0 is the fast neutron flux at $x=0$.

The theory in this case is valid until the attenuation of fast neutrons is dominated by a removal process, *i.e.* if concrete contains sufficient moderating material (for concrete, hydrogen, with its intrinsic water content) [16, 17].

2.3 The coupled thermo-hygro-mechanical problem

The NEWCON3D model consists of a series of balance equations, *i.e.* a mass balance equation of water (both liquid and vapour, taking into account phase changes and hydration/dehydration processes), an enthalpy balance equation of the whole multiphase medium (considering the latent heat of phase change and the hydration/dehydration process), a linear momentum balance equation of the fluid phases (Darcy's equation) and a linear momentum balance equation of the whole medium.

Appropriate constitutive equations and some thermodynamic relationships are included as well. The field equations of the model are briefly recalled below; for additional details the reader is referred to [2, 19-22, 29-33]. The continuity equation for non-isothermal flow is expressed in terms of relative humidity as:

$$\frac{\partial h}{\partial t} - \nabla^T \mathbf{C} \nabla h - \frac{\partial h_s}{\partial t} - k\frac{\partial T}{\partial t} + \chi \mathbf{m}^T \frac{\partial \mathbf{\varepsilon}}{\partial t} = 0 \qquad (35)$$

where h is the relative humidity, directly connected with the moisture content w, T is temperature, $k = \left(\frac{\partial h}{\partial T}\right)_{w,\varepsilon}$ is the hygrothermic coefficient, i.e. the change in h due to a one-degree change of T at constant w, ε and a fixed degree of saturation, dh_s is the self-desiccation, $\chi = \left(\frac{\partial h}{\partial \varepsilon_v}\right)_{T,w}$ represents the change in h due to the unit change of volumetric strain ε_v at constant w, T and a given degree of saturation, $\mathbf{C}$ is the (relative humidity) diffusivity diagonal matrix.

Heat balance requires:

$$\rho C_q \frac{\partial T}{\partial t} - C_a \frac{\partial w}{\partial t} - C_w \mathbf{J} \cdot \nabla T = -div\mathbf{q} \qquad (36)$$

where ρ is the mass density of concrete, C_q the isobaric heat capacity of concrete (per kilogram of concrete) including chemically bound water but excluding free water, C_a is the heat of sorption of free water (per kilogram of free water); C_w is the isobaric heat capacity of liquid water, $C_w \mathbf{J} \cdot \nabla T$ is the rate of heat supply arising from convection by moving water, $\mathbf{J}$ is the flux of humidity, depending on the gradient of h by means of the permeability c:

$$\mathbf{J} = -c\, \nabla h \qquad (37)$$

$\mathbf{q}$ is the heat flux which can be attributed to the temperature gradient or moisture concentration gradient. Combining the two cases, the constitutive law defining the heat flux comes from a Fourier's law (heat flux due to temperature gradient) and a Dufour's flux (heat flux due to moisture concentration gradient):

$$\mathbf{q} = -a_{Tw}\, \nabla w - a_{TT}\, \nabla T \qquad (38)$$

where the coefficients a_{Tw} and a_{TT} depend on w and T.

Finally, consider the macroscopic linear momentum balance equation for the whole medium:

$$div\boldsymbol{\sigma} + \rho \mathbf{g} = 0 \qquad (39)$$

where ρ is the density of the multiphase medium (concrete plus water species) and $\mathbf{g}$ an acceleration related to gravity.

As regards the mechanical field, the constitutive relationship for the solid skeleton in incremental form can be written as:

$$d\boldsymbol{\sigma}' = (1-D)\mathbf{D}_T \left(d\boldsymbol{\varepsilon} - d\boldsymbol{\varepsilon}_T - d\boldsymbol{\varepsilon}_c - d\boldsymbol{\varepsilon}_{lits} - d\boldsymbol{\varepsilon}_p - d\boldsymbol{\varepsilon}_{sh} - d\boldsymbol{\varepsilon}_0\right) \quad (40)$$

where $\boldsymbol{\sigma}'$ is the effective stress tensor ($\boldsymbol{\sigma}' = \boldsymbol{\sigma} + p\mathbf{I}$, p being the mean pore pressure), directly responsible for deformations of the solid skeleton, D the upgraded scalar radio-chemo-thermo-mechanical damage (see below), $\mathbf{D}_T$ is the tangent stiffness matrix, $d\boldsymbol{\varepsilon}_T$ is the strain rate caused by thermo-elastic expansion, $d\boldsymbol{\varepsilon}_c$, the strain rate accounting for creep, $d\boldsymbol{\varepsilon}_{lits}$, the load induced thermal strain rate, $d\boldsymbol{\varepsilon}_p$, the plastic strain rate, $d\boldsymbol{\varepsilon}_{sh}$ is due to shrinkage and $d\boldsymbol{\varepsilon}_0$ represents the autogenous strain increments, due e.g. to chemical variations in the mixture and including the irreversible part of the strain rates not contained in the previous terms. Particularly, D is the damage scalar variable, which must include mechanic, thermo-chemical as well as radiation effects.

According to the classical effective stress concept [15] the elastic modulus of the damaged material at a given temperature, $\overline{E}(T)$ can be obtained from that of the undamaged material at the same temperature $E(T)$ and a damage parameter D, which can assume values between 0 and 1, D being a measure of the cracks' volume density within the material:

$$\overline{E}(T) = (1-D)E(T) \quad (41)$$

The total stress tensor $\boldsymbol{\sigma}$ is, then, modified to take into account damage, measuring a reduction in the resistant area due to cracks beginning and spreading:

$$\overline{\boldsymbol{\sigma}}' = \boldsymbol{\sigma}'\frac{S}{\overline{S}} = \frac{\boldsymbol{\sigma}'}{1-D} \quad (42)$$

where S and $\overline{S}$ are the resistant area of the uncracked and cracked material, respectively.

As for the mechanical effects, since the damaging mechanisms are different in uniaxial tension and compression experiments, the damage parameter D_m (the subscript stands for mechanical contributions only) is decomposed according to the Mazars model [23-26] into two parts, d_t for traction and d_c for compression, which are functions of the equivalent strain:

$$\tilde{\varepsilon} = \sqrt{\sum_{i=1}^{3}\left(\langle\varepsilon_i\rangle_+\right)^2} \qquad \left(\langle\varepsilon_i\rangle_+ = \frac{|\varepsilon_i| + \varepsilon_i}{2}\right) \quad (43)$$

ε_i being the principal strains. Hence:

$$D_m = \alpha_t d_t + \alpha_c d_c \quad (44)$$

where α_t and α_c are weighting coefficients defined in [24].

Chemo-mechanical damage has been introduced for the first time in [19]; thermo-chemical effects also have been taken into account in multiplicative way, as proposed by Gerard et al. [5] and Nechnech et al. [28]: a damage parameter D_{tc}, $0 \leq D_{tc} \leq 1$, describes thermo-chemical material degradation at elevated temperatures (mainly due to micro-cracking and cement dehydration) resulting in a reduction of the material strength properties, so that Equation (41) becomes:

$$\overline{\sigma}' = \frac{\sigma'}{(1-D_m)(1-D_{tc})} \qquad (45)$$

As mentioned before, the total effect of the mechanical and thermo-chemical damages acting at the same time is multiplicative, *i.e.* the total damage D is defined by:

$$D = 1-(1-D_m)(1-D_{tc}) \qquad (46)$$

The upgrade of the model has been developed by assuming that the nuclear radiation can activate a damage process which combines with the mechanical and thermo-mechanical ones so that the above multiplicative relation is maintained and the total damage is redefined:

$$D = 1-(1-D_m)(1-D_{tc})(1-D_r) \qquad (47)$$

in which D_r accounts for radiation damage.

The application, within the numerical code NEWCON3D, of a standard finite element discretization in the space of Equations (35), (36) and (39) results in:

$$\begin{bmatrix} \mathbf{K} & \mathbf{HU} & \mathbf{TU} \\ \mathbf{L}^T & \mathbf{I} & \mathbf{TP} \\ \mathbf{0} & \mathbf{TH} & \mathbf{TS} \end{bmatrix} \begin{Bmatrix} \dot{\mathbf{u}} \\ \dot{\mathbf{h}} \\ \dot{\mathbf{T}} \end{Bmatrix} + \begin{bmatrix} 0 & 0 & 0 \\ 0 & \mathbf{Q} & 0 \\ 0 & 0 & \mathbf{TR} \end{bmatrix} \begin{Bmatrix} \overline{\mathbf{u}} \\ \overline{\mathbf{h}} \\ \overline{\mathbf{T}} \end{Bmatrix} = \begin{Bmatrix} \dot{\mathbf{f}} + \mathbf{c} \\ \mathbf{HG} \\ \mathbf{TG} \end{Bmatrix} \qquad (48)$$

in which $\overline{\mathbf{u}}$, $\overline{\mathbf{h}}$ and $\overline{\mathbf{T}}$ are the nodal values of the basic variables: displacements, relative humidity and temperature, **HU** and **TU** account for shrinkage and thermal dilation effects, respectively; $\mathbf{L}^T$ and **TP** are the coupling matrices representing the influence of the mechanical and thermal fields on the hygral one, respectively; **Q** is the diffusivity matrix accounting for sorption-desorption isotherms; **TH** is the coupling matrix between the hygral and thermal fields in terms of capacity ; **TS** is the matrix of heat capacity; **TR** is the matrix of thermal transmission including the convective term; **c** is the matrix accounting for creep; **HG** is the matrix of humidity variation due to drying and **TG** accounts for heat fluxes.

For further explanations of the above terms the reader is referred to [20, 22, 33]. It is to be noticed that the constitutive relationship given by Equation (40) is in agreement with the thermodynamics of the previously exposed phenomenon, if we consider the term $d\varepsilon_0$ as the autogenous contribution to the total strain tensor rate $d\varepsilon$ and the terms: $d\varepsilon_c$ and $d\varepsilon_{lits}$ as accounting for the relaxation effects at law to medium and high temperatures, respectively. Finally the term: $d\varepsilon_{sh}$, responsible for shrinkage, is supposed to arise entirely from changes to moisture content.

In the following Section the considerations leading to the definition of the new variable, the radiation damage parameter D_r, are outlined.

3 The radiation damage parameter

Nuclear radiation may influence structural and mechanical properties of materials significantly; Hilsdorf *et al.* [11] collecting published experimental data on the effect of nuclear radiation on the properties of concrete, indicate that up to neutron fluences of the order of 10^{19} n/cm^2 the effects of the irradiation are relatively small, while higher fluences may have detrimental effects on concrete strength and the modulus of elasticity. On the other hand, thermal coefficients of expansion, thermal conductivity and shielding properties are proved to be little affected by radiation.

Radiation damage is mainly explained in literature by lattice defects in the aggregates which cause a volume increase in aggregates and concrete, thus leading to fractures and there is evidence that a prolonged exposure to nuclear radiation significantly increases the reactivity of silica-rich aggregates to alkali [12, 27].

The mechanism is explained by the reaction of OH$^-$ ions in the alkaline solution in the micropores of concrete with SiO_2 in aggregates; first the Si–O bonding breaks, next follows the expansion of the aggregates by hydration of SiO_2. The consumption of OH$^-$ ions leads to the dissolution of Ca^{2+} ions into the solution. The Ca^{2+} ions then react with hydrated SiO_2 gels to generate calcium silicate. Rigid calcium silicate shells are therefore formed on the surfaces of the aggregates by successive reactions with OH$^-$ and Ca^{2+} ions and the alkaline solution is possible to penetrate into the aggregates through the calcium silicate shells and to dissolve the SiO_2. Since the rigid shells prevent the deformation of the aggregates, the expansion pressure generated by the penetration of the solution is accumulated in the aggregates, thus leading to cracks in the cement paste and expansion of the concrete.

In the following the most significant effects of radiation on the properties of concrete are summarized, particularly focusing on the decay of the Young modulus, which has been exploited to define radiation damage within NEWCON3D in a phenomenological way, in agreement with the effective stress theory, as a reduced elastic modulus.

3.1 Phenomenological aspects of concrete

Figures 1-3 report the experimental data referred to as results from slow or fast neutron exposure: filled symbols are used for slow neutrons, empty symbols for fast and half-filled symbols where no separation between fast and slow neutrons was possible. Though it is generally expected that fast neutron radiation would lead to more pronounced radiation damage than slow neutrons, this is not always confirmed on the basis of the data.

In Figure 1a the compressive strength of concrete samples f_{cu} from various test series is reported as a fraction of the compressive strength of companion specimens f_{cuo}, which are neither irradiated nor temperature exposed, whereas in Figure 1b the same strength values are presented, but here concrete compressive strength is related to the strength of companion specimens, f_{cuT}, which are not irradiated but heated, in order to isolate the temperature effect from the one induced by radiation.

From this collection of data it may be concluded that some concretes can resist neutron radiation higher than 5×10^{19} n/cm^2 without decay in strength, whereas others exhibit a decay at a considerably smaller radiation dose. As an average, a neutron fluence higher than 10^{19} n/cm^2 leads to a marked decrease in the concrete compressive strength, although even for a neutron fluence lower than the above mentioned, strength ratios may be lower than unity.

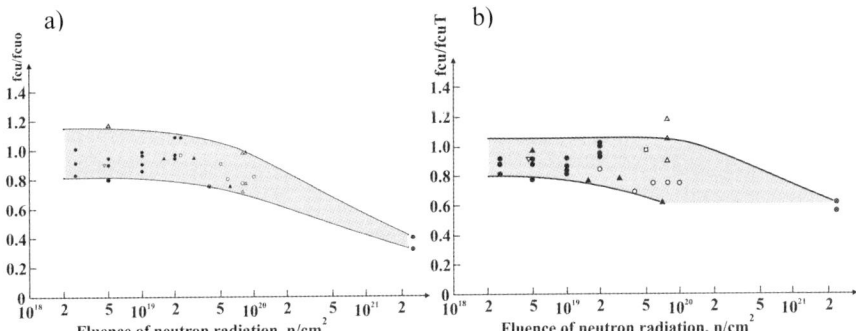

Figure 1: (a) Compressive strength of concrete exposed to neutron radiation f_{cu} related to strength of untreated concrete f_{cuo}; (b) Compressive strength of concrete exposed to neutron radiation f_{cu} related to strength of temperature exposed concrete f_{cuT}

A comparison between the two figures indicates that the observed strength decrease is primarily the result of neutron radiation although there is evidence of some detrimental effect arising from a temperature increase. The experimental data vary over a wide range for a given neutron fluence: for a neutron fluence of 5×10^{19} n/cm² the strength ratios range from 0,72 to 1,05 and from 0,65 to 1,05 for f_{cu}/f_{cuo} and f_{cu}/f_{cuT}, respectively.

Fast neutrons seem to induce the greatest decreases. The effect of neutron radiation on the tensile strength f_{ru} of concrete samples is shown in Figures 2(a) and 2(b). The first one gives the tensile strength of concrete samples after neutron radiation as a fraction of the tensile strength of companion specimens f_{ruo}, which are neither irradiated nor temperature exposed, whereas in Figure 2b the tensile strength is related to the strength of temperature exposed but not irradiated specimens, f_{ruT}. According to Figure 2(a), neutron radiation with a fluence higher than 10^{19} n/cm² may lead to a pronounced decrease in concrete tensile strength (reductions are greater than that for compressive strength, at the same neutron fluence), whereas by comparing the two figures it emphasises that neutron radiation has caused a considerable part of the observed strength decrease.

Even for the tensile strength, the individual strength values vary over a wide range: for a neutron fluence of 5×10^{19} n/cm² the observed strength ratios range between 0,2 and 0,82, 0,33 and 0,98 for f_{ru}/f_{ruo} and f_{ru}/f_{ruT}, respectively.

In both cases the large scatter observed in the reported experimental data can be most likely ascribed to differences in the composition of the samples and concrete making materials, apart from test procedures.

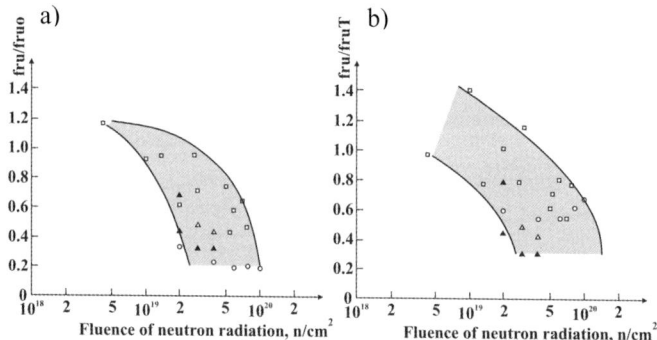

Figure 2: (a) Tensile strength of concrete exposed to neutron radiation f_{ru} related to strength of untreated concrete f_{ruo}; (b) Tensile strength of concrete exposed to neutron radiation f_{ru} related to strength of temperature exposed concrete f_{ruT}

Figure 3 shows the effect of neutron radiation on the concrete elastic modulus; the modulus of irradiated concrete E_c is given as a fraction of the modulus of companion specimens E_{co} neither irradiated nor temperature exposed. The available data are not sufficient to separate the effects of radiation and of heating for the samples, which in many cases they undergo in nuclear vessels conditions, however it seems reasonable that the same considerations made for the previous quantities hold also in the case of the Young's modulus, *i.e.* similarly the strength decay is primarily due to neutron radiation. A neutron fluence lower than 10^{19} n/cm^2 again leads to a slight decrease in the elastic modulus compared to the modulus of untreated companion specimens. With increasing neutron fluences, the modulus decreases.

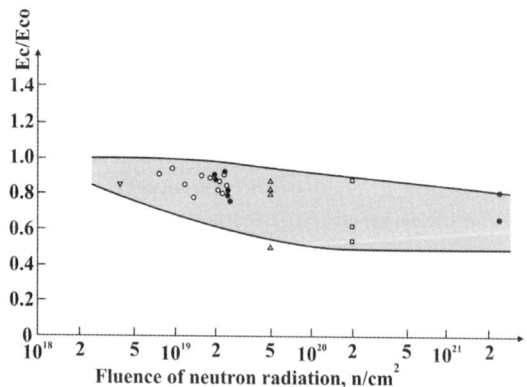

Figure 3: Modulus of elasticity of concrete after neutron radiation E_c related to modulus of elasticity of untreated concrete E_{co}

By differentiating between fast or slow neutrons actions, Gray [6] found that for fast neutron fluences between 7×10^{18} and 3×10^{19} n/cm^2 the modulus of irradiated concrete was between 10 and 20% less than that for not irradiated unheated concrete. Alexander [1] reported similar reductions in E values for slow neutron fluencies of about 2×10^{19} n/cm^2.

An enveloping curve of the collected experimental data has been used to express the decay in the elastic modulus and, at the same time, to define the radiation damage parameter in agreement with Equation (41), as long as $\overline{E}$ is E_c. Particularly, for safety reasons, the lowest curve reported in Figure 3 has been considered to describe the required trend for the ratio $\overline{E}/E$, supposed to arise only from radiation, since the uncoupling with other thermo-chemo-mechanical effects holds in the experimental investigation, as previously stated.

The enveloping curve is defined so that up to a neutron fluence of 10^{18} n/cm^2 the radiation damage is zero, then it increases with the neutron fluence and finally stabilizes at the maximum value of $0{,}5\,\overline{E}/E$. It was not necessary to extrapolate data in the numerical application.

4 Numerical application: Fluka-NEWCON3D, a combined approach

4.1 Geometry of the problem and Fluka results

The study case takes its origin from the SPES Project currently in development at the National Laboratories of Legnaro (Padua, Italy). The facility will be directed to the production of special radioactive heavy ion beams from a primary proton beam impinging on a target made of fissionable material, where fission reactions are expected to take place; the target therefore ideally represents a point-source of neutrons for the specific problem. The target room is expected to be an underground bunker with wall thicknesses of the order of few meters, estimated by INFN with the aid of Monte Carlo simulations in order to keep the environment under 0,25 mSv/h ambient dose equivalent [13], which is the limit dose prescribed by the National Standards on radioprotection.

In Figure 4 a sketch of the geometry is drawn: the red point represents the uranium carbide target; the neutron source is positioned at 1 m from the floor and 2 m from the directly impinged wall which, for safety reasons, has been estimated as 3,7 m thick.

The portion of concrete shielding modeled with NEWCON3D is the prism emphasized in Figure 4, representative of a typical wall portion. The prism has been restrained in order to allow for dilations/contractions along the wall thickness only; self-weight has been neglected. In this way the mechanical damage is expected to be zero, so that the only contributions to damage come from thermo-chemical effects or radiation.

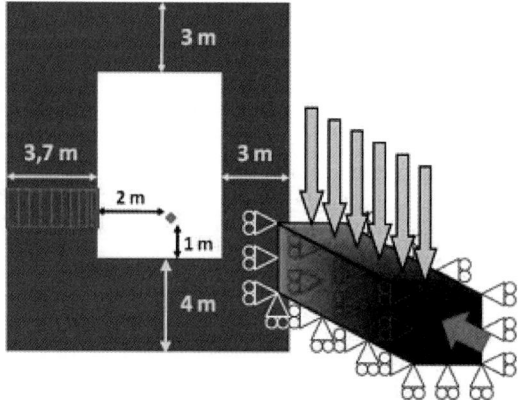

Figure 4: Geometry of the target cave for the SPES Project and static scheme for the model analysed with NEWCON3D

The same geometry has been implemented in Fluka, in order to obtain an estimate of the main physical quantities affecting the problem: neutron fluence and energy deposition. The former, of which the decay in the Young's modulus is a function, is directly responsible for the quantification of the radiation damage; the latter allows determination of the boundary conditions in terms of temperature, at the exposed wall surface. The geometry in Fluka is assigned by means of elemental volumes (planes, spheres, parallelepipeds, regular prisms, *etc.*) and Boolean operations between them (addition, subtraction, union) applied to describe "regions" made of different materials. Each portion of space needs to be assigned to one region only.

The code makes a three-dimensional analysis, though the visualization is two-dimensional. A sketch of the Fluka model is available in Figure 6. The concrete room is supposed to be surrounded by "air" which becomes the vehicle for radiation; in defining the geometry in Fluka the presence of an overall surrounding black sphere, which is assigned by default as the so called material "blackhole", is necessary, in order to "close" the mathematical problem of mass balance of particles. The blackhole material, contained in the library of Fluka, is defined as an extremely highly absorbent material, responsible for entrapping all the scattered particles escaping from the investigated control volume.

The material "air" has a known chemical composition and it is produced by the already implemented library in the code; for concrete the chemical composition of an ordinary concrete in percentage by weight provided in [16] has been adopted (Table 1).

Element	% by weight [%]	Element	% by weight [%]
Hydrogen	0.64	Phosphorus	0.09
Oxygen	45.36	Sulphur	0.09
Carbon	-	Potassium	0.64
Sodium	1.76	Calcium	12.66
Magnesium	3.66	Titanium	0.47
Aluminium	5.88	Iron	0.13
Silicon	20.90	Nickel	7.64
Concrete density [g/cm^3] 2.33			

Table 1: Assumed chemical composition for the ordinary concrete implemented in Fluka.

The target of fissionable material has the dimension of seven disks (uranium and carbon compounds) for a total mass of 30 g, a radius of 2 cm and thickness of about 1,3 mm, which makes it assumable as an ideal point-source for neutrons.
The impinging beam starts at an arbitrary point inside a vacuum pipe; thanks to the vacuum environment, it is therefore supposed to be delivered as not attenuated to the target. The characteristics of the primary proton beam come from the most serious exercise scenario designed by INFN for the SPES facility: a beam of 70 MeV energy, and 300 µA current, so that the proton flux p is the following:

$$p = \frac{300 \cdot 10^{-6}}{1,602 \cdot 10^{-19}} = 1,87 \, 10^{15} \, p/s \qquad (49)$$

In the facility under design at the Legnaro Laboratories, the initial proton driver is a cyclotron with variable energy (15-70 MeV).
Fissions are performed at a rate of 10^{13} fissions per second in the dedicated target cave; the isotopes are then extracted from the target and ionized with an ion source connected to the production target; they are consequently charge breeded and mass separated to fit the required charge state and ion velocity for injection into the system: the pre-accelerator PIAVE and further acceleration is given by the ALPI linac, from which the secondary beam is delivered to the experimental areas.
Fluka results are provided on a regular parallelepiped grid, of the centre of mass for each element, the dimensions of which are determined by the required results resolution. A regular grid of small elements 5 cm wide along each direction has been adopted for our study. A subsequent interface program additionally used to pass results from the centre of the mass for each volume of the results' grid to the FE mesh adopted in NEWCON3D, based on the algorithm of the minimum distance.
The stochastic simulation with Fluka has required three cycles per run, with 10^6 primary particles launched at each run, in order to reach acceptable statistics.
In Figure 5 the map of neutron flux density for the control volume corresponding to the mesh analyzed in NEWCON3D (Figure 4) is evidenced, pointing to the neutron source, which is of interest as being expected to be the most exposed; only the first metre of wall seems to be affected by neutron absorption, and these values have been transferred to the corresponding volume in NEWCON3D.

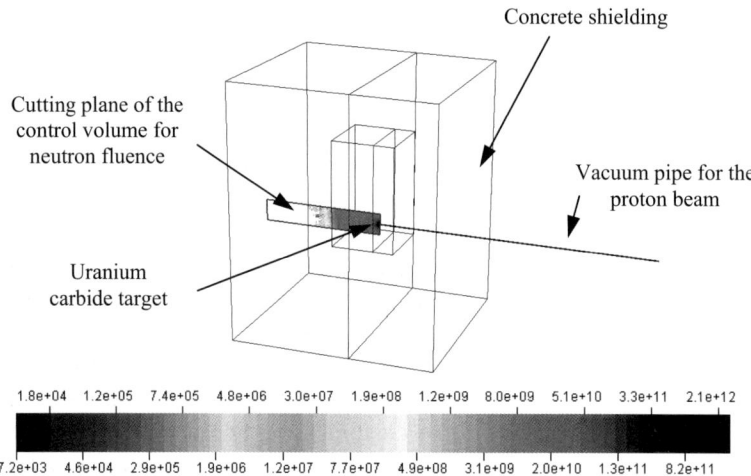

Figure 5: Neutron flux density for the control volume corresponding to the FE mesh implemented in NEWCON3D. Results on the cutting plane just in front of the source [n/(cm² s)]

In Figure 6 the results in terms of deposited energy are shown, for a wider investigated area of the shielding. Results refer to prompt radiation, *i.e.* the constant component of the total amount of deposited energy, and here are intended per primary incident particle (*i.e.* per proton) and need to be normalized by the factor of Equation (49) in order to reach total values of power density.

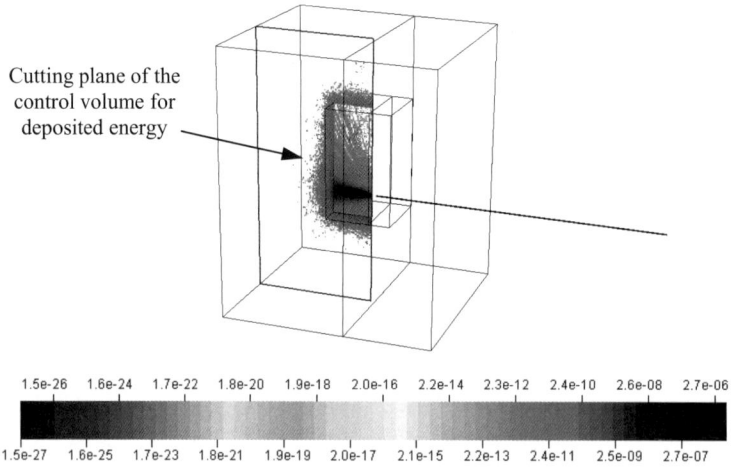

Figure 6: Energy deposition on the directly exposed area of the concrete shielding. Results on the cutting plane just in front of the source [GeV/(cm³ primary)]

For a better visualization, Figure 7 depicts the same values of neutron flux and power density (here normalized), for the same directly exposed investigated prism and the same geometric cutting planes. The order of magnitude of the maximum neutron flux density is proved to be 10^{10} n/(cm^2 s); the maximum power density is 10^{10} GeV/(cm^3 s).

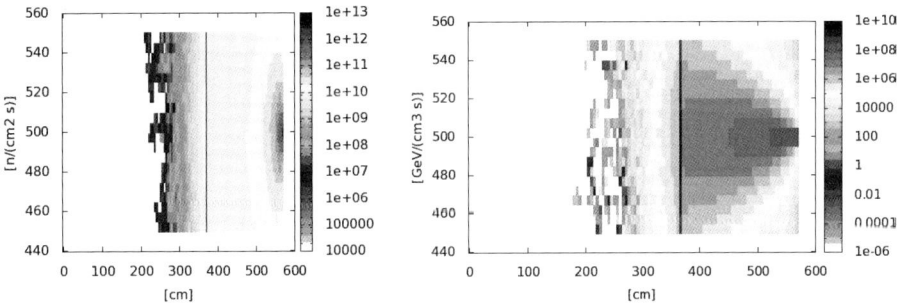

Figure 7: Neutron flux density [n/(cm^2 s)] and power density [GeV/(cm^3 s)] on the cutting plane of the directly exposed control volume

Power density results then have been collected for all the internal bunker surfaces and transformed into heat fluxes, subsequently used to perform a standard transient thermal analysis (see Figure 8) in order to evidence temperature evolution on the main internal walls and particularly at the directly impinged area; in fact, temperature rising due to absorbed radiation possibly affects the hydrogen content in concrete, which is meant to be the main item responsible for neutron attenuation in the medium.

Figure 8 shows that the maximum temperature is encountered at the corner close to the source, where after six months nearly 70°C are reached; in the last picture a time history for the same critical node is reported, which shows a parabolic increase in time. The effect is shown to be not negligible, which is in agreement with the requirements of ANSI/ANS-6.4-1985, according to which radiation heat and subsequent thermal effects are to be taken into account for energy flux densities (power densities) above 10^{10} MeV/(cm^2 s).

It is to be noticed that Figure 7 gives maximum values with the order of magnitude of 10^{10} GeV/(cm^3 s), where cm^3 are to be intended per volume of the small elements defined by the resolution of the result grid (side of the cubes 5 cm), therefore one gets 5×10^{10} GeV/(cm^2 s), which is considerably above the prescribed limit required to neglect thermal effects. In fact, thermal effects are understood to be mainly responsible for the stress state of the shielding, as it will be better illustrated in the results obtained from the thermo-hygro-mechanical analysis. The thermal analysis justifies also the working period of the facility assumed in the study, *i.e.* six months, nearly 4500 hours per year, longer durations being unacceptable for the material. Once the temperature field was obtained, it was necessary to re-launch Fluka in order to have a better estimate of the neutron flux for a changed geometry, where concrete was supposed to change in density and water content according to the experimental curves of Figure 9, provided by Kaplan [16] for the same adopted concrete mixture and reported in Table 1.

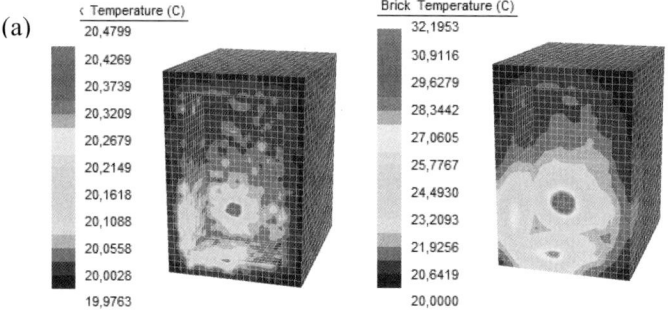

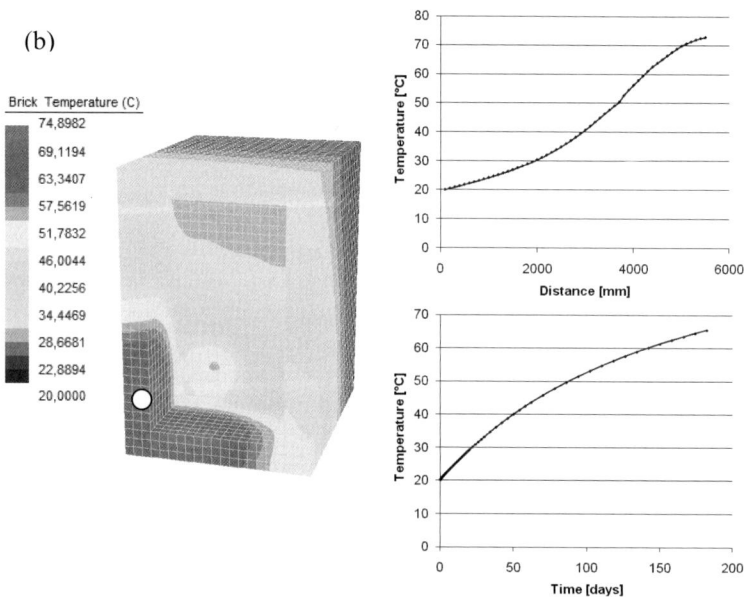

Figure 8: Temperatures on the directly impinged zone after 1 hour, 1 week (a) and 6 months (b) of work for the facility.

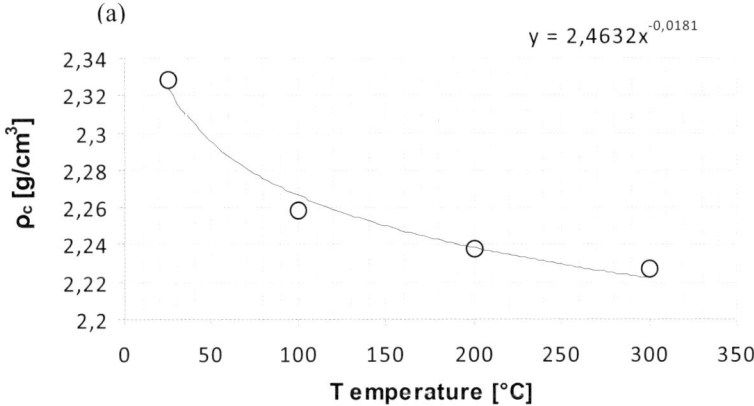

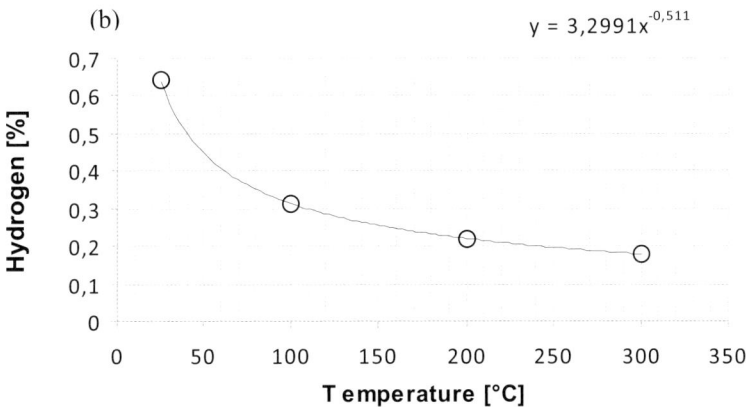

Figure 9: Experimental curves for density loss (a) and hydrogen content (b) in concrete (Table 1) when varying temperature.

A section view of the new Fluka geometry is reported, for several temperature conditions in time, in Figure 10: the shielding is now described as a map of regions still made of concrete, but differentiated by their specific weight and chemical composition. Particularly, three types of concrete have been defined as new materials in the code, in the function of the temperatures reached: they refer to the same chemical composition of the original mixture (see Table 1) except for the hydrogen content and specific weight, as reported in Table 2.

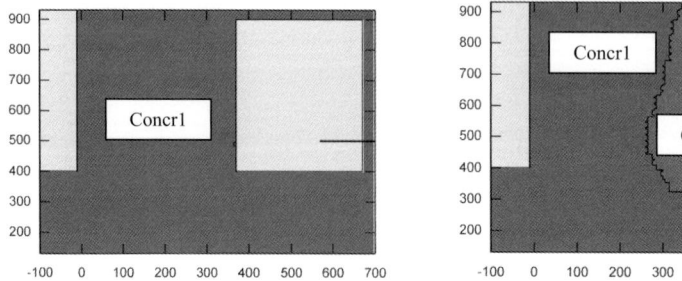

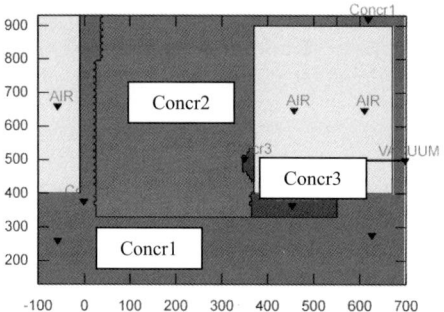

Figure 10: New geometry for the target cave obtained by differentiating concrete properties according to temperature rise: concr1, concr2, concr3

Concrete	Temperature range [°C]	ρ_c [g/cm3]	H [%]
Concr1	<20	2,33	0,64
Concr2	20-50	2,31	0,54
Concr3	50-75	2,29	0,40

Table 2: Definition of modified concretes for updating the radiation field

Slight variations in concrete properties, however, are proved not to bring significant changes in the radiation field, for the loss in hydrogen (the intrinsic moderating substance for neutrons) not justifying a relevant alteration in the shielding properties of the concrete, at least for the investigated period of time. As a check, in Figure 11 the neutron flux after six months and with the new geometry is compared to the same quantity as is the original problem, the critical value maintaining the order of 10^{10} n/(cm^2 s).

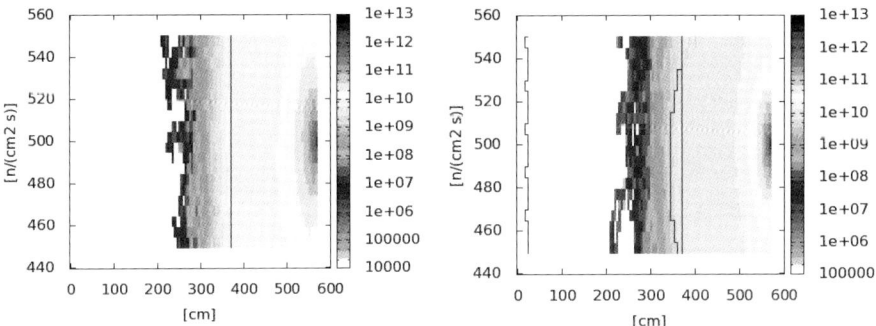

Figure 11: Neutron flux density compared to the one obtained by considering the new geometry, after six months

4.2. Implementation into NEWCON3D

As already shown, the attenuation process for the incident radiation on concrete is a phenomenon that evolves in time: neutron fluence [n/cm^2] is a flux density [n/(cm^2 s)] integrated over the time duration of radiation; instead power density [GeV/(cm^3 s)] varies with the irradiation profile of the facility and in general is due to a constant component, so called "*prompt radiation*", instantaneously provided when the facility works (zero otherwise), and a "*delayed*" component represented by the decay of radioactive particles, which is the time-variable amount. Consequently, the problem is numerically treated as a time dependant process where radiation damage and temperatures at the boundary need to be calculated for several time steps.

The adopted discretization consists of 4037 nodes and 800 elements, 20-node isoparametric brick elements (Figure 4); temperature is applied at the internal free prism surface. Initial conditions are homogeneous and consider an internal relative humidity of 60% and a temperature of 20°C.

The coupling between the hygral, the thermal and the mechanical fields provided by NEWCON3D has allowed capture of the variation in time for several variables (Figures 12-15; only a portion of the model is shown here): relative humidity seems not to be much affected by the six months prolonged radiation, whereas the temperature rise is understood to be of interest, leading to thermal gradients up to 50°C (impinged area); for this reason the irradiation profile for the SPES facility should not exceed 4500 hours/year, *i.e.* approximately six months of continuous service within the investigated exercise scenario.

As regards damage, it is intended to result from thermo-chemical effects only, since the neutron fluence, for the investigated time span, always stays under the critical value of 10^{19} n/cm^2, which is expected to mark the beginning of the first evidence of concrete damage by radiation; therefore the thermal aspect of the problem, once more, represents the most restrictive condition to limit the service period of the facility.

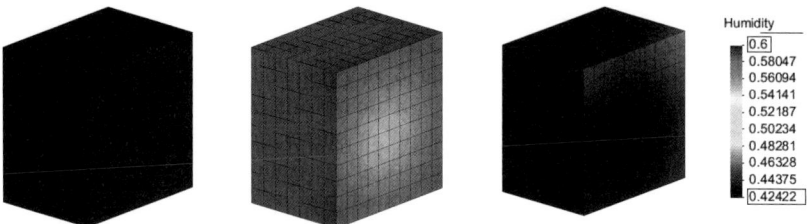

Figure 12: Contour maps of relative humidity after one hour, one week, six months

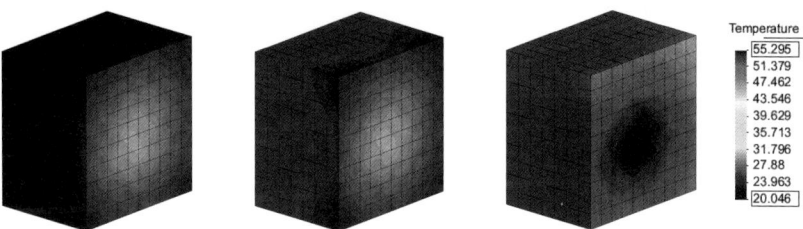

Figure 13: Contour maps of temperature after one hour, one week, six months [°C]

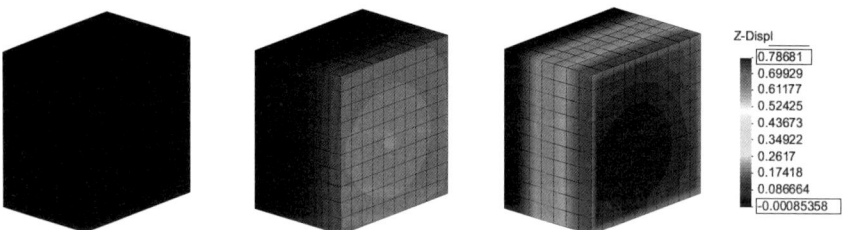

Figure 14: Contour maps of displacements along the prism length after one hour, one week, six months [mm]

5 Conclusions

Nuclear radiation is known to affect concrete in its mechanical behavior over particular threshold quantities of radiation fluence. Even thermal collateral effects, due to heat production by absorbed radiation, can be of interest, thus requiring a fully coupled hygro-thermo-mechanical approach for this type of problem to better understand the response of concrete as a shielding for nuclear facilities. In this work a study case is represented by the SPES Project, an ongoing research project at the INFN Laboratories in Legnaro (Padua, Italy) directed towards the production of a special type of neutron-rich nuclei, with possible future applications in medicine as well.

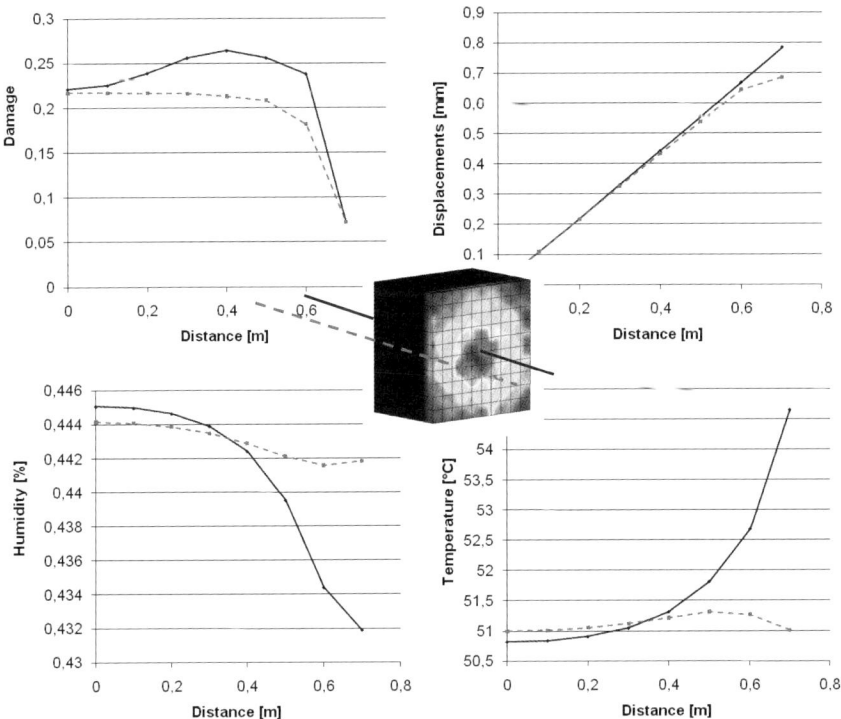

Figure 15: Distribution of the main variables after 6 months along the first 0.7 m wall thickness (reference lines: prism central axis and parallel axis at the border)

A collection of the most relevant experimental results on neutron irradiated concrete has allowed the formulation of radiation damage in the context of the FE research code NEWCON3D assessing the coupled hygro-thermo-mechanical behavior of concrete.

Results from a Monte Carlo code developed by CERN and INFN, Fluka, have been properly interfaced with the numerical code in order to investigate the evolution of humidity, temperature and mechanical quantities for a typical concrete wall portion. Specifically, the numerical analyses have helped identify that 4500 hours per year is the maximum time span for continuous work of the SPES, in order to stay within admissible thermal gradients for the material; in fact thermal effects due to radiation appear to be the most critical ones and undesirable temperatures represent the most strictest constraints in achieving a safe exercise scenario.

Further steps in the research will be devoted to studying the long term behaviour of the concrete construction under cyclic irradiations up to the final decommissioning of the nuclear structure. On the other hand, a meso-scale modeling of concrete is envisaged to be of interest, particularly for a more realistic representation of the building material, as well as for understanding the contribution of aggregates and interfacial transition zone (ITZ) in the characterization and definition of radiation shielding properties.

References

[1] S.C. Alexander, "Effects of irradiation on concrete: final results", Atomic Energy Research Establishment, Harwell, United Kingdom Atomic Energy Authority, 1963.
[2] P. Baggio, C.E. Majorana, B.A. Schrefler, "Thermo-hygro-mechanical analysis of concrete", Int J Num Meth Fluids, 20, 573-595, 1995.
[3] Z.P. Bazant, C. Huet, "Thermodynamic functions for ageing viscoelasticity: integral form without internal variables", Int J Solids Structures, 36, 3993-4016, 1999.
[4] A. Fassò, A. Ferrari, J. Ranft, P.R. Sala, CERN-2005-10, INFN/TC_05/11, SLAC-R-773, "FLUKA: a multi-particle transport code", 2005.
[5] B. Gerard, G. Pijaudier-Cabot, C. Laborderie, "Coupled diffusion-damage modeling and the implications on failure due to strain localization", Int J Solids Structures, 35(31-32), 4107-4120, 1998.
[6] B.S. Gray, "The effect of reactor radiation on cements and concrete", in Conference on Pre-stressed Concrete Pressure Vessels, Commission of European Communities, Luxembourg, 17-39, 1972.
[7] K. Hayakawa, S. Murakami, Y. Liu, "An irreversible thermodynamics theory for elastic-plastic-damage materials", Eur J Mech, A/Solids, 17(1), 13-32, 1998.
[8] J.R. Harrison, "Nuclear reactor shielding", Temple Press, 1958.
[9] S. Hazanov, "New class of creep-relaxation functions", Int J Solids Structures, 32(2), 165-172, 1995.
[10] S. Hazanov, "On separation of energies in viscoelasticity", Mech Res Commun, 24(2), 167-177, 1997.
[11] H.K. Hilsdorf, J. Kropp, H.J. Koch, "The effects of nuclear radiation on the mechanical properties of concrete", in Douglas McHenry International Symposium on Concrete and Concrete Structures, Amer Concr Inst Special Publication SP 55-10, Detroit, Michigan, 223-251, 1978.
[12] T. Ichikawa, H. Koizumi, "Possibility of radiation-induced degradation of concrete by alkali-silica reaction of aggregates", J of Nucl Sci Technol, 39(8), 880-884, 2002.
[13] INFN-LNL-224, "SPES Selective Production of Exotic Species: Executive summary", A. Covello, G. Prete. (Editors), Laboratori Nazionali di Legnaro, 2008.
[14] P. Ireman, A. Klarbring, N. Stromberg, "A model of damage coupled to wear", Int J Sol Str, 40, 2957-2974, 2003.
[15] M.D. Kachanov, "Time of Rupture Process under Creep Conditions", Izvestia Akademii Nauk, 8, 26-31, 1958. (in Russian)
[16] M.F. Kaplan, "Concrete radiation shielding: nuclear physics, concrete properties, design and construction", John Wiley & Sons, New York, 1989.

[17] A.N. Komarovskii, "Shielding materials for nuclear reactors", Pergamon Press, London, 1961.
[18] J. Lemaitre, J.-L. Chaboche, "Mechanics of solid materials", Dunod, 1988. (in French)
[19] C.E. Majorana, A. Saetta, R. Scotta, R. Vitaliani, "Mechanical and durability models for lifespan analysis of bridges", IABSE Sym.: Extending the lifespan of structures, San Francisco, CA, USA, 23-25, 1253-1258, 1995.
[20] C.E. Majorana, V.A. Salomoni, "Parametric analyses of diffusion of activated sources in disposal forms", J of Hazard Mat, A113, 45-56, 2004.
[21] C.E. Majorana, V.A. Salomoni, S. Secchi, "Effects of mass growing on mechanical and hygrothermic response of three-dimensional bodies", J Mat Process Technol, PROO64/1-3, 277-286, 1997.
[22] C.E. Majorana, V.A. Salomoni, B.A. Schrefler, "Hygrothermal and mechanical model of concrete at high temperature", Mat Str, 31, 378-386, 1998.
[23] J. Mazars, "Application de la mecanique de l'endommagement au comportament non lineaire et la rupture du beton de structure", Ph.D. dissertation, L.M.T., Université de Paris, France, 1984.
[24] J. Mazars, "Description of the behaviour of composite concretes under complex loadings through continuum damage mechanics", in J.P. Lamb, (Editor), Proc. 10th U.S. National Congress of Applied Mech, ASME, 1989.
[25] J. Mazars, G. Pijaudier–Cabot, "Continuum Damage Theory – Application to Concrete", J Engrg Mech ASCE, 115, 345-365, 1989.
[26] J. Mazars, G. Pijaudier-Cabot, "From damage to fracture mechanics and conversely: a combined approach", Int J Solids Structures, 33, 3327-42, 1996.
[27] D.J. Naus, "Primer on durability of nuclear power plant reinforced concrete structures: a review of pertinent factors", Oak Ridge National Laboratory, U.S. Nuclear Regulatory Commission Office of Nuclear Regulatory Research Washington, DC, 2007.
[28] W. Nechnech, J.M. Reynouard, F. Meftah, "On modeling of thermo-mechanical concrete for the finite element analysis of structures submitted to elevated temperatures", in De Borst, Mazars, Pijaudier-Cabot, van Mier, (Editors), Proc. Fracture Mech of Concrete Structures, 271-278, 2001.
[29] V.A. Salomoni, C.E. Majorana, G.M. Giannuzzi, A. Miliozzi, "Thermal-fluid flow within innovative heat storage concrete systems for solar power plants", Int J Num Meth for Heat and Fluid Flow (Special Issue), 18(7/8), 969-999, 2008.
[30] V.A. Salomoni, C.E. Majorana, G.A. Khoury, "Stress-Strain Experimental Based Modelling of Concrete under High Temperature Conditions", in B.H.V. Topping, (Editor), "Civil Engineering Computations: Tools and Techniques", Saxe-Coburg Publications, Stirlingshire, UK, Chapter 14, 319-345, 2007. doi:10.4203/csets.16.14
[31] V.A. Salomoni, C.E. Majorana, G. Mazzucco, G. Xotta, G.A. Khoury, "Multiscale modelling of concrete as a fully coupled porous medium", in J.T. Sentowski, (Editor), Concrete Materials: properties, performance and applications, Chapter 3, Nova Science Publishers Inc., 2009.

[32] V.A. Salomoni, G. Mazzucco, C.E. Majorana, "Mechanical and Durability Behavior of Growing Concrete Structures", Engrg Comput, 24(5), 536-561, 2007.
[33] B.A. Schrefler, L. Simoni, C.E. Majorana, "A general model for the mechanics of saturated-unsaturated porous materials", Mat Str, 22, 323-334, 1989.
[34] J.K. Shultis, R.E. Faw, "Radiation shielding", Prentice Hall PTR, 1996.
[35] A.J. Staverman, F. Schwarzl, "Thermodynamics of viscoelastic behaviour", in Proc Acad Sci Amst, The Netherlands, B55, 474-490, 1952.
[36] A.J. Staverman, "Thermodynamics of linear viscoelastic behavior", in Proc Second International Congress on Rheology, Academic Press, inc., New York, 134-138, 1954.
[37] Y. Weitsman, "A continuum diffusion model for viscoelastic materials", J Phys Chem, 94, 961-968, 1990.

Chapter 3

©Saxe-Coburg Publications, 2011.
Civil and Structural Engineering Computational Technology
B.H.V. Topping and Y. Tsompanakis, (Editors)
Saxe-Coburg Publications, Stirlingshire, Scotland, 65-98.

Development of Realistic Three-Dimensional Track Models for Railway Vehicle Dynamic Analyses

J. Pombo and J. Ambrósio
IDMEC/IST
Technical University of Lisbon, Portugal

Abstract

The dynamic analysis of complex, three-dimensional, railway systems involves the construction of three independent models: the track model; the vehicle model; and the wheel-rail contact model. In this chapter, a methodology for the accurate representation of the track geometry is presented. It involve, not only the description of the design track layout, but also the representation of its irregularities. For this purpose, a methodology that includes the track imperfections, measured experimentally by the railroad companies, in the definition of the track model is developed. The objective is to obtain a realistic representation of the track, which is essential to study the dynamic behaviour of the railway vehicles. This approach allows one to tackle some sensible issues in the railway industry that involve the damage to vehicles caused by the track conditions and the infrastructure deterioration arising from the trainsets' operation. A multibody formulation is used here to describe the rigid bodies, the kinematic joints and the suspension elements that constitute the vehicle model. The complex interaction that is developed between the wheels of the rolling stock and the rails is studied using a generic wheel-rail contact model. Such a formulation, during the dynamic analysis, allows determination of the contact points location and the respective normal and tangential forces, even for the most general three-dimensional motion of the wheelset. The methodologies described in this work are applied to study the dynamic behaviour of the railway vehicle ML95, which is operated by the Lisbon metro company. The studies are performed in real operating conditions when travelling between two of the metro stations. The track irregularities are obtained experimentally with a vehicle equipped for the purpose. The accuracy and suitability of the methodologies is demonstrated through the comparison of the dynamic analysis results against those obtained by experimental testing and with the commercial code ADAMS/Rail.

Keywords: railway dynamics, multibody systems, track irregularities, vehicle-track interaction, wheel-rail contact.

1 Introduction

The dynamic analysis of railway [1-6], tramway [7-9], roller coaster or any other type of rail guided vehicles requires an accurate description of the track geometry. The track is composed of two rails, which can be viewed as two 'parallel' lines defined in a plane that sits in the track centerline spatial curve, also called the reference path. The basic ingredient to define the track is, therefore, the geometry of the reference path, which must include the vertical gradients, horizontal curves and cant.

Besides the description of the design track layout, a realistic track model also involves the representation of its imperfections. The track irregularities can be perceived as deviations from the parallel curves, representing the rails. In the literature, there are two distinct approaches to including the track irregularities in the computational models for railway dynamics. The first one considers the track irregularities as absolute measured values, which are obtained with railway vehicles specially equipped to measure the track imperfections. An alternative approach consists of introducing the track irregularities as stochastic data characterized by, e.g., Power Spectral Density (PSD).

When using the first approach, some authors [1] use the absolute irregularities as a function of distance of the course and represent the track imperfections by the alignment and the longitudinal level, defined on the track centerline as the mean value between left and right rails, and by the cant and gauge irregularities, defined as deviations from the nominal (design) values. In the commercial tool ADAMS/Rail a similar method is used, i.e., the track irregularities are expressed as deviations applied on the track centerline [10]. Another procedure is used in the commercial program VAMPIRE [11,12]. Here, the track irregularities are characterized by several parameters: a) cross level (cant) irregularities; b) curvature irregularities; c) lateral irregularities; d) vertical irregularities; and e) gauge variation. Within SIMPACK software there are three types of tracks that can be used [13]: standard, cartographic and measured. The standard track enables the user to define standard manoeuvres, e.g. curve entries or crossovers, using only a few parameters. More complex tracks of arbitrary structure can be modelled using cartographic tracks. In addition, measured tracks allow tracks to be generated from measured track data.

The second approach used to include the track irregularities in the computational models consists of using stochastic data. In this case, the track irregularities are not considered as absolute values defined as functions of the distance run. Instead, they are represented as functions of wavelength or frequency. PSD are widely used for this purpose [1,14,15]. It can be said that a PSD is the square of the Fourier transform of a signal. A PSD can be used either directly as an input for power spectral analysis or they can be retransformed into track irregularities as a function of distance. However, in this last case, the original signal will not be obtained rather a signal that is statistically equivalent.

In this work an alternative method to include track irregularities as absolute measured values in railway vehicle dynamic analyses is presented. The objective is to create fully three-dimensional track models where the track irregularities are defined separately on each rail, instead of including them on the track centerline as the mean value between left and right rails. This methodology starts by expressing the track centerline in a parametric form [16-18]. Then, a moving reference frame is defined in the track centerline with its axis in the intersections of the normal, osculating and rectifying planes [19]. The introduction of the cant angle and its variation along the track is also implemented.

As the purpose here is to obtain realistic track models, the experimentally acquired track irregularities must be included in the model. The strategy adopted for the parameterization of the track centerline serves as the basis to describe the left and right rail geometries. This allows us to define separately the spatial curves representing the geometry of the left and right rails and to consider the track irregularities in such a definition. Each rail curve is then parameterized as a function of its respective arc length using a piecewise cubic interpolation scheme.

The track flexibility, or the deformability of its foundations, is another feature that can be included in the track model. Despite this phenomenon not being addressed in this work, there are authors devoting their attention to this issue [20-23].

Besides the track model, the study of a complete railway system involves the construction of the vehicle model and of the wheel-rail contact model. In this work, a multibody formulation [24-30] is used to represent the railway vehicle, including the masses and inertias of the structural elements, and the characteristics of the suspensions. Moreover, the vehicle-track interaction is studied through an appropriate wheel-rail contact formulation [31-33], which is used to compute the normal and tangential forces that develop in the contact area.

The methodology presented here for the accurate representation of three-dimensional track models is applied to study the dynamic behaviour of the railway vehicle ML95 in realistic operating conditions. The influence of the measured track irregularities on vehicle performance is analysed and the results obtained are compared against those available from experimental testing and with the commercial package ADAMS/Rail.

2 Parameterisation of a three-dimensional track centerline

Depending on the specific application, the reference path of the track geometry can be described by a number of types of parametric curves. For railway and light track vehicles the description of the nominal geometry of the track is generally done by putting together straight and circular track segments, interconnected by transition segments that ensure the continuity of the first and second derivatives in the transition points [3]. Moreover, these transition elements are responsible for the smooth variation of the lateral accelerations on vehicles, when they move from a straight track to a circular track or between two track segments of the same type with different radius or orientations. For other applications, parametric curves that interpolate a given number of control points can be used to define the track geometry. In any case, the complete characterization of the tracks also requires the definition of the cant angle variation along the reference path. For flat tracks, the cant angle φ in a given point of the reference path is measured in the plane perpendicular to the reference path, between a line that seats on both rails and the horizontal plane, as shown in Figure 1(a). For tracks with a full spatial geometry, a new definition of this angle is introduced. It is proposed that the osculating plane of the reference track plays the role of the horizontal plane of the flat track in measuring the cant angle [16-18], as depicted in Figure 1(b).

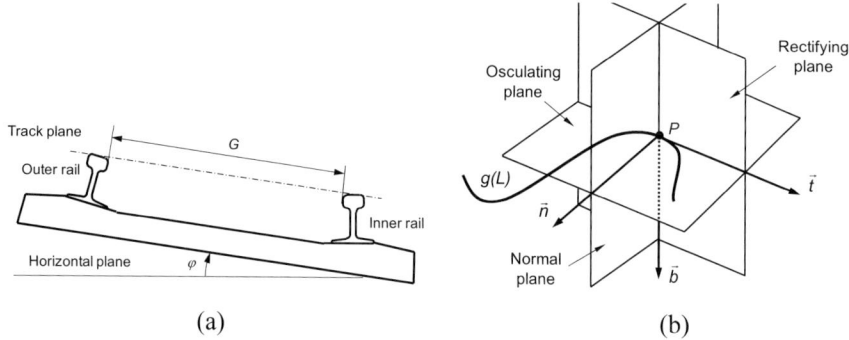

Figure 1: (a) Cant angle and gauge; (b) Moving Frenet frame of a parametric curve

2.1 Parameterization approach

As the objective here is to create fully three-dimensional track models, it is necessary to develop a methodology that allows accurate definition the spatial geometry of the track centerline. It is proposed to parameterize the geometry of the centerline using a piecewise cubic interpolation scheme. For this purpose, the user only has to provide a set of control points that are representative of the centerline geometry and the correspondent cant angle in each point. This information is used to parameterize the spatial curve as a function of the track length using shape preserving splines [34,35]. Then, the track cant angle is also parameterized as a function of the track length and the principal unit vectors (**t,n,b**), which define the Frenet frames associated with the track centerline after the cant angle rotation, are calculated. A schematic representation of the methodology used to parameterize the track centerline is presented in Figure 2.

The use of shape preserving splines instead of other common piecewise cubic interpolation schemes avoids several problems. The application of cubic splines [36,37] to interpolate a set of control points describing the track geometry can lead to undesired oscillations in the track model [16,38]. With Akima splines [37,39] no unwanted oscillations are introduced by the interpolation algorithm in the geometric description of the spatial curve. Nevertheless, this interpolation scheme only ensures C^1 continuity between spline segments, *i.e.*, it only guarantees that the parameterized curve is continuous and has a continuous first derivative. Therefore, the continuity of the centerline curvature is not ensured, which conflicts with the properties required for track geometries. The parameterization with shape preserving splines guarantees the consistency with the concavity of the data, which is rather useful when it is important to preserve the convex and concave regions implied by the control points. Furthermore, this interpolation scheme also ensures C^2 continuity between spline segments, *i.e.*, it guarantees that the parameterized curve has a continuous curvature.

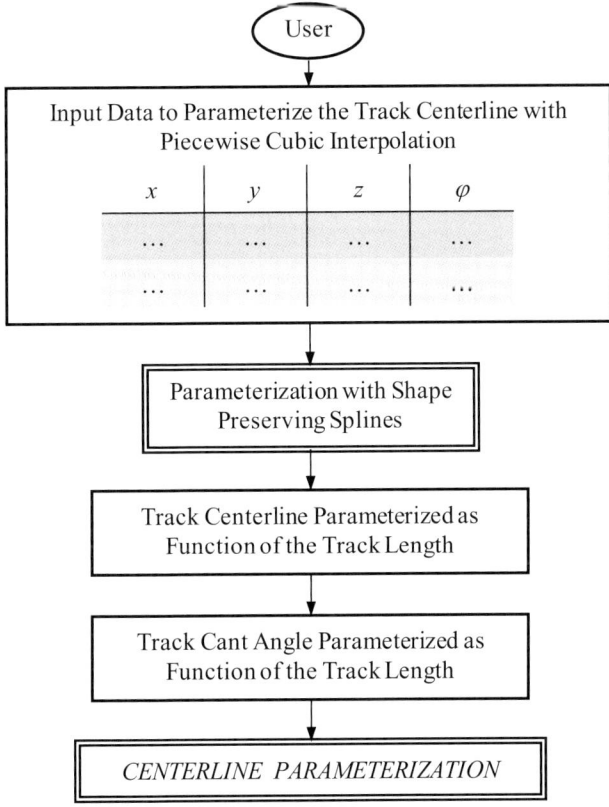

Figure 2: Parameterization of the three-dimensional track centerline

The track parameterization with analytical segments, which use straight, circular and transition curves, does not introduce unwanted oscillations on the track geometry. However, this description is rather complex to model railways with large slopes or with vertical curves. Some of the commercial codes that adopt this description impose the fact that the tracks are basically horizontal in order to avoid difficulties [10,14]. Therefore, the track description with analytical segments cannot be used to parameterize fully spatial geometries like the ones used on roller coasters.

2.2 Cant angle contribution

When parameterizing the track centerline, the cant variation along the track has to be taken into account. For tracks with a full spatial geometry it is not clear what the reference plane is relative to which cant angle should be defined. According to

Figure 1(b), it is noticeable that the osculating plane [19], associated to the track centerline, can be the reference plane used to define the cant angle.

The cant angle φ is measured in the normal plane as the angle between the principal unit normal vector **n**, which lies in the osculating plane as represented in Figure 1(b), and the unit normal vector $\mathbf{n}_c$, which is associated with the new curve moving frame $(\mathbf{t}_c, \mathbf{n}_c, \mathbf{b}_c)$ after the cant angle rotation, as shown in Figure 3. According to this methodology, the new components of the principal unit vectors, after the cant angle rotation, are expressed as [16-18]:

$$\begin{cases} \mathbf{t}_c = \mathbf{t} \\ \mathbf{n}_c = \mathbf{n}\ cos(\varphi) - \mathbf{b}\ sin(\varphi) \\ \mathbf{b}_c = \mathbf{n}\ sin(\varphi) + \mathbf{b}\ cos(\varphi) \end{cases} \quad (1)$$

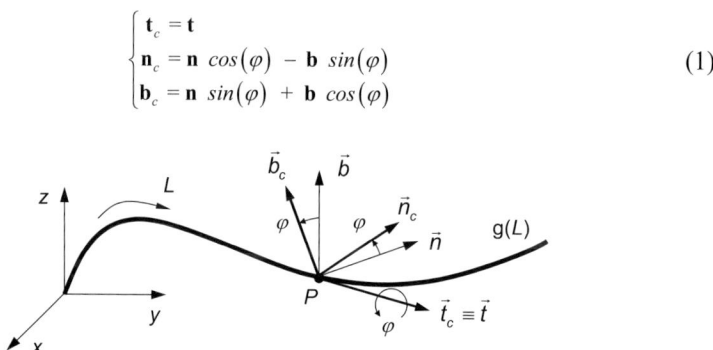

Figure 3: Cant angle contribution to the track model

3 Rail space curves

The dynamic analysis of any type of rail-guided vehicle requires an accurate and suitable description of the track geometry. After the parameterization of the track centerline, it is necessary to implement an appropriate methodology for the description of each rail, in the general case of a fully three-dimensional track. It is proposed here to use a pre-processor to build the track model by defining the nominal geometry of both left and right rails based on the interpolation of a discrete number of points, which are representative of their space curves. The two rails are considered as separate geometric entities in order to account for the track irregularities on their definition.

To achieve computational efficiency, the pre-processor generates, in a tabular manner and as a function of the left and right rails arc lengths, all the track position data and other general quantities required for the dynamic analysis of rail-guided vehicles. At every time step, during the dynamic simulations, the program interpolates both rail databases in order to obtain all the necessary information to analyse the complex interaction between the wheels and rails.

3.1 Track pre-processor

The purpose of the track pre-processor is to define the track model in the form of two parameterized curves that represent the nominal geometry of the left and right rails space curves. In the pre-processor a parametric track description, based on shape preserving splines, is implemented. The information is organized in two databases where all quantities, necessary to define both left and right rails curves, are obtained as functions of the arc length of each rail. The methodology can be summarized as follows:

1) The geometry of the track centerline is parameterized as a function of the track length using shape preserving splines, as described in Section 2. The user input consists of a set of control points that are representative of the track geometry and the cant angle at each nodal point.
2) The track cant angle is also parameterized as a function of the track length and the principal unit vectors ($\mathbf{t}_c, \mathbf{n}_c, \mathbf{b}_c$), which define the reference frame associated with the track centerline after the cant angle rotation, are calculated.
3) The track irregularities, measured experimentally, can also be included in the track model. For this purpose, the user only has to supply, as input data for the pre-processor, all irregularities parameters organized in a tabular manner. Then, the track irregularities are parameterized as continuous functions of the track length using the piecewise cubic interpolation scheme.
4) The coordinates of a set of control points that are representative of the left and right rails space curves are then calculated.
5) The rails space curves are parameterized, as a function of their respective arc lengths, using shape preserving splines. A reference frame associated with each rail space curve is defined taking into account the track cant angle and the rail inclination.
6) An output database for each rail is created where all parameters necessary to define the geometric characteristics of the spatial curves of both rails are stored. Each database has a track length step as small as desired by the user. During dynamic analyses the rails databases are used by the multibody code to find the necessary geometric information at each time step by performing interpolations on the data contained in the tables.

A schematic representation of the methodology used in the track pre-processor is presented in Figure 4. The formulations inherent to each one of the procedures reported in the flowchart are presented in the following.

3.2 Track irregularities

The track irregularities represent the deviations of the track from its design geometry and result from construction imperfections, usage operations and change in the foundations. These irregularities are characterized by the lateral and vertical displacement of each rail, as shown in Figure 5. Nevertheless, in the railway industry, the track irregularities are not, in general, represented by these degrees-of-freedom. Instead, they are commonly characterized by the alignment, longitudinal level, cant variation and gauge variation. Such parameters are measured experimentally by the railway industry using special track recording vehicles [1,3].

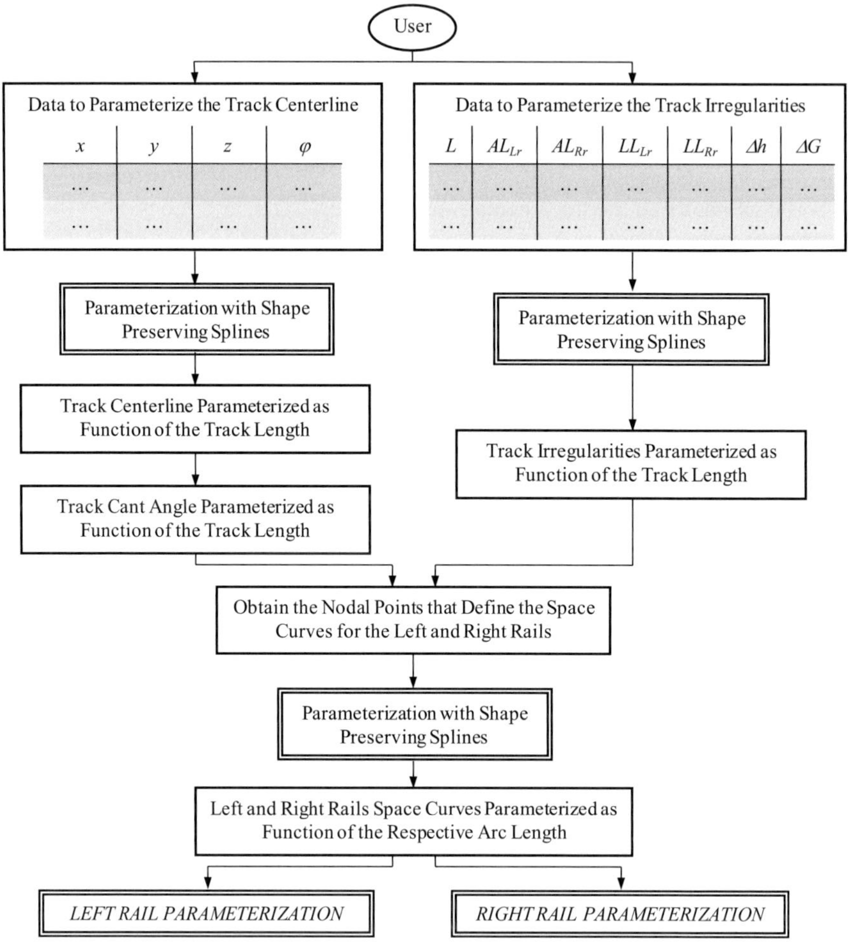

```
                                    ┌─────────┐
                                    │  User   │
                                    └────┬────┘
              ┌──────────────────────────┴──────────────────────────┐
              ▼                                                     ▼
```

Data to Parameterize the Track Centerline			
x	y	z	φ
...	...	...	...
...	...	...	...

Data to Parameterize the Track Irregularities					
L	AL_{Lr}	AL_{Rr}	LL_{Lr}	LL_{Rr}	Δh	ΔG
...	...	...	...	...	...	...
...	...	...	...	...	...	...

Parameterization with Shape Preserving Splines

Parameterization with Shape Preserving Splines

Track Centerline Parameterized as Function of the Track Length

Track Irregularities Parameterized as Function of the Track Length

Track Cant Angle Parameterized as Function of the Track Length

Obtain the Nodal Points that Define the Space Curves for the Left and Right Rails

Parameterization with Shape Preserving Splines

Left and Right Rails Space Curves Parameterized as Function of the Respective Arc Length

LEFT RAIL PARAMETERIZATION **RIGHT RAIL PARAMETERIZATION**

Figure 4: Parameterization of the left and right rails

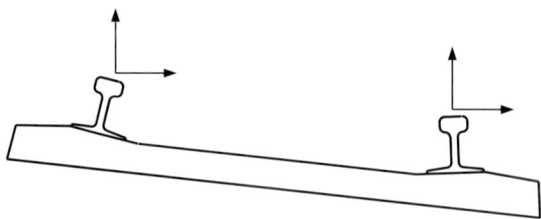

Figure 5: Degrees-of-freedom associated to the left and right rails

3.2.1 Alignment

The alignment of the left and right rails, respectively AL_{Lr} and AL_{Rr}, represents the lateral displacement of each rail along the track with respect to the design configuration, as depicted in Figure 6. The alignment is one of the major causes of lateral vibration in railway vehicles. These irregularities usually result from initial crookedness of rails, track construction, maintenance procedures and from accumulated lateral track movements under traffic loads [4].

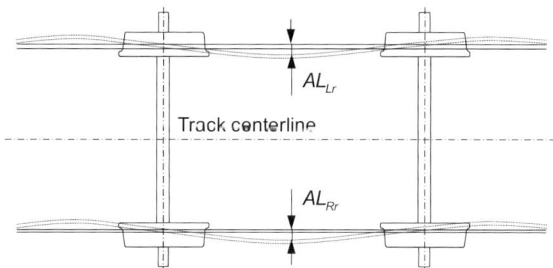

Figure 6: Track alignment

3.2.2 Longitudinal level

The longitudinal level for the left and right rails, respectively LL_{Lr} and LL_{Rr}, represents the vertical displacement of each rail along the track with respect to its design configuration, as shown in Figure 7. The longitudinal level has little influence on lateral vehicle dynamics. It results mainly from the existence of low joints between the rails, rails assembling, track flexibility due to the weight of the vehicles and from the thermal loads imposed on the track [4].

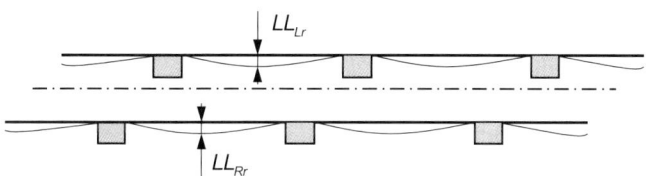

Figure 7: Longitudinal level

3.2.3 Cant variation

The cant variation Δh_t represents the difference between the design and the real cant measured on the track. In the same way, it is also possible to define the cant angle variation $\Delta \varphi$. As to the alignment, the cant variation is one of the major causes of lateral vibration in railway vehicles. These irregularities usually result from rails assembling, track flexibility due to the weight of the vehicles and from the thermal loads imposed on the track [4].

3.2.4 Gauge variation

The gauge variation ΔG represents the difference between the nominal and the real gauge measured on the track. It plays an important role in the lateral stability of railway vehicles. The gauge variation results primarily from track construction, maintenance procedures and from relative lateral movements of rails under traffic loads [4]. The gauge irregularities are always associated with track alignment imperfections.

3.2.5 Parameterization of the track irregularities

The track irregularities must be considered when defining realistic track models. In general, these irregularities are obtained experimentally by a measuring vehicle and provided by the railway companies in a table where all irregularities parameters are stored in columns as a function of the track length L.

A typical form of presenting the irregularities data as it is provided by the track measuring device is shown in Table 1. This information is defined in a discrete manner and it is used as input data for the track pre-processor described here. Then, the irregularities parameters are parameterized as continuous functions of the track length L by using a piecewise cubic interpolation scheme. In this text, the quantities with subscript $(.)_{Lr}$ refer to the left rail and the quantities with subscript $(.)_{Rr}$ refer to the right rail.

L	AL_{Lr}	AL_{Rr}	LL_{Lr}	LL_{Rr}	Δh	ΔG
0.0	......	......	......	......	......	......
0.1	......	......	......	......	......	......
0.2	......	......	......	......	......	......
.....	......	......	......	......	......	......

Table 1: Structure of the track irregularities data

3.3 Parameterisation of the rail space curves

As previously described, the track centerline and the track irregularities are parameterized as functions of the track length L. Moreover, as shown in Figure 3, a moving frame $(\mathbf{t_c}, \mathbf{n_c}, \mathbf{b_c})_L$, which accounts for the cant angle rotation, is defined in order to provide an appropriate track referential at every point. With reference to Figure 8, the Cartesian coordinates of the control points Q and R that characterize the geometry of the left and right rails space curves, respectively, are written as:

$$\mathbf{r}_{Lr}^{Q} = \mathbf{r}_c + \mathbf{A}_c \, \mathbf{s}_{Lr}^{\prime Q} \quad ; \quad \mathbf{r}_{Rr}^{R} = \mathbf{r}_c + \mathbf{A}_c \, \mathbf{s}_{Rr}^{\prime R} \tag{2}$$

where $\mathbf{r}_c$ is the global position vector that defines the location of the origin of the centerline reference frame $(\mathbf{t_c}, \mathbf{n_c}, \mathbf{b_c})_L$ and $\mathbf{A}_c$ is the transformation matrix that defines the orientation of the track centerline referential with respect to the inertial frame (x,y,z), given by [25]:

$$\mathbf{A}_c = \begin{bmatrix} \mathbf{t}_c & \mathbf{n}_c & \mathbf{b}_c \end{bmatrix} \quad (3)$$

The quantities $\mathbf{s}_{Lr}^{\prime Q}$ and $\mathbf{s}_{Rr}^{\prime R}$ are, respectively, the position vectors that define the location of the control points Q and R of the rail space curves with respect to the centerline Frenet frame, as shown in Figure 8. Considering the track irregularities parameters described previously, the local coordinates of the position vectors are defined as:

$$\mathbf{s}_{Lr}^{\prime Q} = \begin{Bmatrix} 0 \\ \dfrac{D}{2} + AL_{Lr} + \dfrac{\Delta G}{2} \\ LL_{Lr} \end{Bmatrix} \ ; \ \mathbf{s}_{Rr}^{\prime R} = \begin{Bmatrix} 0 \\ -\dfrac{D}{2} + AL_{Rr} - \dfrac{\Delta G}{2} \\ LL_{Rr} \end{Bmatrix} \quad (4)$$

where D represents the distance between the nominal wheel-rail contact points on the left and right side.

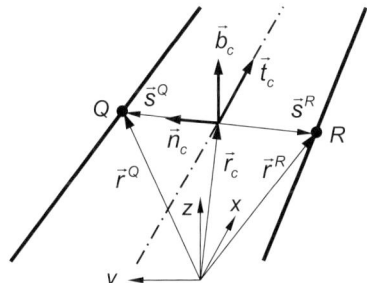

Figure 8: Nodal points that define the rail space curves

Once the control points that characterize the geometry of the left and right rails space curves are obtained, a methodology similar to the one proposed for the parameterization of the track centerline can be applied. Each rail space curve is parameterized as a function of its arc length using shape preserving splines.

3.4 Rail space curve databases

The direct online use of the formulation necessary for the parametric description of the rails space curves is neither practical nor efficient from the computational point of view. In fact, as the track model is to be used in the framework of a dynamic analysis program, the evaluation of the complete methodology presented in Figure 4 at every time step, with all parameterizations and interpolations required, would be a heavy burden on the code. An alternative approach is the construction of two lookup tables where all quantities necessary for the definition of the left and right rails space curves are tabulated as functions of their arc lengths. In this way, before starting the dynamic simulation the track pre-processor is used in order to make all the necessary calculations required to define the simple and compact databases that fully characterize the curves and referentials for the track model. As shown in Figure 9, the track model is characterized by the Cartesian coordinates of the space curves associated with the left and right rails and by the reference frames associated with these curves.

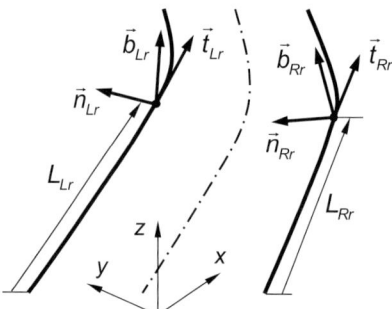

Figure 9: Representation of the left and right rail space curves

A length parameter step ΔL, adopted for the construction of both left and right rails databases, has to be chosen. Then, the pre-processor program automatically generates the tables, which are functions of the left and right rails arc lengths, with all the track position data and other geometric parameters necessary to define the track model as required by the multibody code. At every time step during the dynamic simulations, the program interpolates the left and right rails databases to obtain all the necessary information to analyze the interaction between the vehicle and the track. Therefore, the length step ΔL used in the table construction must be small enough to ensure that the interpolation still guarantees the quality of the results, but not so small that it deteriorates the numerical efficiency of the program.

L	x	y	z	t_x	t_y	t_z	n_x	n_y	n_z	b_x	b_y	b_z
0.00	...	...	...	...	...	...	...	...	...	...	...	...
0.05	...	...	...	...	...	...	...	...	...	...	...	...
0.10	...	...	...	...	...	...	...	...	...	...	...	...
....	...	...	...	...	...	...	...	...	...	...	...	...

Table 2: Database for a rail space curve

The structure of a rail database obtained with the pre-processor program described here is presented in Table 2 for a rail arc length step size of $\Delta L = 0.05$ m. Each rail database consists of a table with 13 columns associated with geometric parameters. The first column of the database corresponds to the rail length L, with a step size ΔL, being the correspondent Cartesian coordinates (x,y,z) of the curve stored in the three columns that follow. Columns 5 through 7 contain the information about the Cartesian components of the unit tangent vector **t**. The next three columns of the rail database contain the Cartesian components of the principal unit normal vector **n**, and the three final columns store the components of the binormal vector **b**. These three unit vectors are used to define the moving reference frame associated with each rail space curve, as shown in Figure 9.

4 Wheel-rail contact formulation

The complex contact forces that develop in the wheel-rail interface strongly influence the dynamic behaviour of railway vehicles. Despite the complexity of the physical phenomena involved, the demands of increasing speeds, better comfort and greater load capacity do not stop increasing. Such requirements bring new problems to control the wheel-rail wear and to maintain the vehicle stability and reliability in the different operating conditions.

The wheel-rail contact model involves the determination of the location of the contact points on the profiled surfaces of the bodies, the evaluation of the creepages, *i.e.* the normalized relative velocities at the point of contact, and the computation of the contact forces. In this work, the geometry of the wheel and rail surfaces is defined in a parametric form. A new formulation for the accurate prediction of the contact points location on the wheel and rail surfaces, even for the most general three dimensional motion of the wheelset, is developed to support the contact model [31-33].

4.1 Wheel and rail surfaces

In this work the wheel and rail surfaces are considered as sweep surfaces, obtained by dragging plane curves on spatial curves. As a result, the problem of describing the surfaces reduces to the problem of defining plane curves, which represent the cross sections of the wheel and rail. Four independent surface parameters s_r, u_r, s_w and u_w are used to define the geometry of the wheel and rail surfaces, as shown in Figure 10. The parameter s_r represents the arc length of the rail space curve, *i.e.*, it positions the rail cross-section on which the contact point lies, while u_r defines the lateral position of the contact point in the rail profile coordinate system (ξ_r, η_r, ζ_r). The parameter s_w represents the rotation of the wheel profile coordinate system (ξ_w, η_w, ζ_w) with respect to the wheelset coordinate system $(\xi_{ws}, \eta_{ws}, \zeta_{ws})$, *i.e.*, it defines the rotation of the contact point, while u_w defines the lateral position of the contact point in the wheel profile coordinate system [32,33,40,41]. Note that the origin of the wheel profile and wheel coordinate systems is the same and that $\eta_w = \eta_{ws}$ are always coincident. Furthermore, ζ_w is always in the plane of the rail cross-section in which the contact point is located and pointing in the opposite direction from such contact point. In the text, subscripts $(.)_r$ and $(.)_w$ refer to the rail and wheel, respectively, whereas the subscript $(.)_{ws}$ refer to the wheelset.

The problem of finding the position of each contact point consists of finding the four parameters s_r, u_r, s_w and u_w associated with such point. For the sake of computational efficiency, the position of two potential points of contact on each wheel, one on the flange and other on the tread, are always monitored. By potential contact points the authors mean that the pair of points, on the wheel and on the rail, are the ones that are closer to each other if no contact occurs. If contact effectively takes place then the appropriate set of forces is applied.

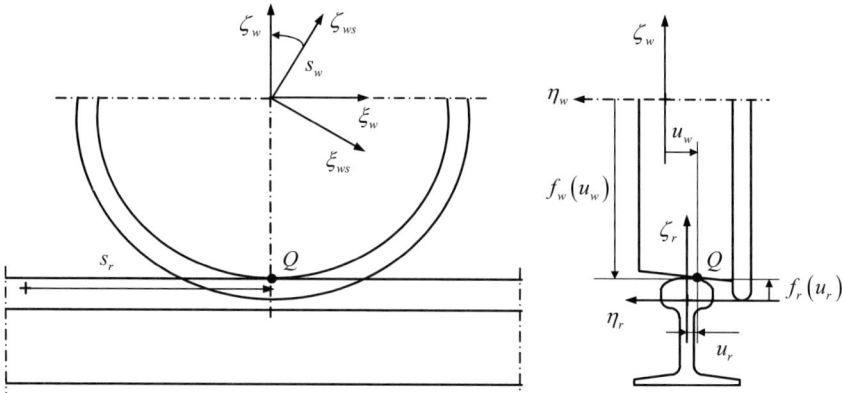

Figure 10: Wheel and rail surface parameters

4.1.1 Rail surface

The rail surface is generated by the two-dimensional curve that defines the rail profile, when it is moved along the rail space curve. The location of the origin and the orientation of the rail profile coordinate system, defined respectively by the vector $\mathbf{r}_r$ and the transformation matrix $\mathbf{A}_r$, are uniquely determined using the surface parameter s_r [42]. Using this description, the global position vector of an arbitrary point Q on the rail surface is written as:

$$\mathbf{r}_r^Q = \mathbf{r}_r + \mathbf{A}_r\, \mathbf{s}_r'^Q \qquad (5)$$

where $\mathbf{s}_r'^Q$ is the local position vector that defines the location of the contact point Q on the rail surface with respect to the profile coordinate system. Note that arising from the above-mentioned description of the rail geometry, the following relations hold:

$$\mathbf{r}_r = \mathbf{r}_r(s_r); \quad \mathbf{A}_r = \mathbf{A}_r(s_r); \quad \mathbf{s}_r'^Q = \{0 \quad u_r \quad f_r(u_r)\}^T \qquad (6)$$

where f_r is the function that defines the rail profile. The transformation matrix $\mathbf{A}_r$ can be expressed in terms of the set of three orthogonal vectors $(\mathbf{t}_r, \mathbf{n}_r, \mathbf{b}_r)$ that define the moving reference frame associated with the rail space curve, written as [25]:

$$\mathbf{A}_r = \mathbf{A}_r(s_r) = [\mathbf{t}_r(s_r) \quad \mathbf{n}_r(s_r) \quad \mathbf{b}_r(s_r)] \qquad (7)$$

The unit vectors are expressed uniquely in terms of the rail arc length, i.e., as a function of the surface parameter s_r. The Cartesian components of these vectors are obtained from the respective rail database previously described.

Generally, the function f_r, which defines the rail profile at each cross section, is not given by simple analytical functions. Here the rail profile is parameterized as a function of the surface parameter u_r using a piecewise cubic interpolation scheme. Hence, to obtain $f_r(u_r)$ the user only has to define a set of control points that are representative of the rail profile geometry, as shown in Figure 11.

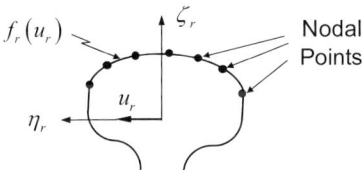

Figure 11: Rail profile parameterization using piecewise cubic interpolation schemes

This methodology is general since it allows not only the use of any type of rail profile, obtained from direct measurements or by design requirements, but also allows a change in the rail profile online during the analysis, if needed. For instance, using this formulation, the geometry of switches or crossings [43] can be modelled by a profile that changes as a function of the track length parameter. Also the change of the rail profile as a function of time or of the location along the track, arising from a wear law for instance, can be performed without any particular modification to the formulation presented here.

4.1.2 Wheel surface

The wheel is a surface of revolution that is obtained by a complete rotation, about the wheel axis, of the two-dimensional curve that defines the wheel profile [44]. The location of the origin and the orientation of the wheelset reference frame are defined, respectively, by the vector $\mathbf{r}_{ws}$ and the transformation matrix $\mathbf{A}_{ws}$. The global position vector of an arbitrary point Q on the wheel surface is written as:

$$\mathbf{r}_w^Q = \mathbf{r}_{ws} + \mathbf{A}_{ws}\left(\mathbf{h}_w + \mathbf{A}_w\,\mathbf{s}_w^{\prime Q}\right) \tag{8}$$

where $\mathbf{h}_w = \{0 \quad \tfrac{1}{2}H \quad 0\}^T$ is the local position vector that defines the location of wheel profile coordinate systems with respect to the wheelset reference frame and H is the lateral distance between wheels profiles origin. The transformation matrix that defines the orientation of the wheel profile coordinate system with respect to the wheelset frame is given by [19]:

$$\mathbf{A}_w = \begin{bmatrix} \cos s_w & 0 & \sin s_w \\ 0 & 1 & 0 \\ -\sin s_w & 0 & \cos s_w \end{bmatrix} \tag{9}$$

The quantity $\mathbf{s}'^Q_w$ is the local position vector that defines the location of the contact point Q on the wheel profile coordinate system, written as:

$$\mathbf{s}'^Q_w = \{ 0 \quad u_w \quad f_w(u_w) \}^T \tag{10}$$

where f_w is the function that defines the wheel profile. Since in general f_w is not given by analytical functions, it is proposed to parameterize the wheel profile using a piecewise cubic interpolation scheme, as described for the rail surface. Hence, the user has to define a set of control points that are representative of the wheel profile geometry, as shown in Figure 12. Note that the wheel profile is represented by two functions f^t_w and f^f_w that parameterize the wheel tread and flange, respectively. This methodology is general since it allows not only the use of any type of wheel profile, obtained from direct measurements or by design requirements, but also allows the user to change the wheel profile online during the analysis, if needed. For instance, using this formulation, the polygonization of the wheel [45] can be modelled by a profile that changes as a function of the wheel rotation. Also the change of the wheel profile as a function of time, due to some wear law for instance, can be performed without any particular modification to the formulation presented here.

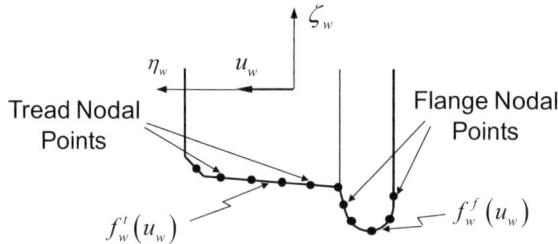

Figure 12: Wheel profile parameterization with piecewise cubic interpolation

4.2 Wheel-rail contact model

The accurate prediction of the location of the contact points between wheel and rail surfaces is fundamental for the calculation of the contact forces. Since the wheel and the rail have profiled surfaces, the online prediction of the contact points location is not simple, especially when the most general three-dimensional motion of the wheelset with respect to the rails is considered. Furthermore, the large number of parameters needed, which include the shape of the surfaces in contact, relative contact velocities and physical properties of the materials, unavoidably lead to complex contact theories. In this work, a multibody methodology that leads to an accurate description of the wheel-rail contact problem and that is computationally efficient is used [31-33].

The wheel-rail contact model proposed involves the determination of the location of the contact points on the profiled surfaces of the bodies, the evaluation of the creepages, *i.e.* the normalized relative velocities at the point of contact, and the computation of the contact forces. The generic contact detection formulation that is used here to determine the coordinates of the wheel-rail contact points adopts a parametric description for the geometry of the wheel and rail surfaces [16,31-33]. This allows the user to perform the dynamic analyses of railway vehicles using profiles obtained from direct measurements or by design requirements. The numerical implementation of this wheel-rail contact model leads to a fast and efficient algorithm where the coordinates of the contact points are obtained online, during dynamic analysis, not requiring the use of pre-calculated lookup tables.

An elastic force model to calculate the normal contact force in the wheel-rail interface is used to account for the dissipation of energy during contact [46,47]. The tangential creep forces and moments that develop in the wheel-rail interface can be evaluated using, alternatively, the Kalker linear theory [48,49], the Heuristic non-linear model [50] or the Polach formulation [51].

5 Multibody model of the railway vehicle

In the following, a general description of the vehicle model is presented emphasizing the mechanical elements that are relevant for the studies carried out here. The railway vehicle considered is the ML95 trainset, shown in Figure 13, that is used by the Lisbon metro company for passengers' service. It is an electrical three-car unit composed of two powered end vehicles and one intermediate trailer unit.

Figure 13: ML95 railway vehicle

As a result of their high structural stiffness, the carbody, the bogies frames, the wheelsets and the axleboxes are considered here as rigid bodies. These bodies are connected by a set of kinematic joints, responsible for controlling their relative motion, and by a group of rigid or flexible links that compose the vehicle suspensions.

5.1 Rigid bodies

The three-dimensional model of the ML95 trailer vehicle, represented in Figure 14, is described using eleven rigid bodies, identified with their respective numbers. The initial position of each body is given by the location of its center of mass with respect to the global reference frame (x,y,z). The mass, the inertia properties with respect to the three principal axes and the initial position of each rigid body are presented in Table 3.

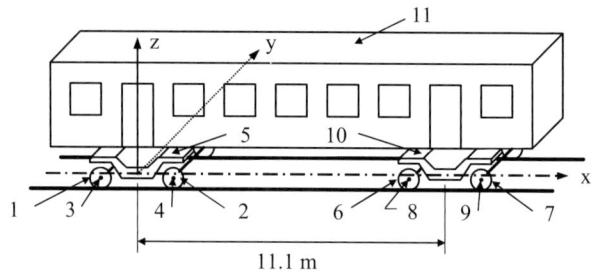

Figure 14: Model of the ML95 trailer vehicle

ID	Rigid Bodies	Mass (Kg)	Inertia Properties (Kg.m^2) $I_{\xi\xi}/I_{\eta\eta}/I_{\zeta\zeta}$	Initial Position (m) $x_0/y_0/z_0$
1	Rear wheelset Rear Bogie	933	461.4/61.6/461.4	-1.05/0.0/0.430
2	Front wheelset Rear Bogie	933	461.4/61.6/461.4	1.05/0.0/0.430
3	Rear axleboxes Rear Bogie	176	144.5/2.2/144.5	-1.05/0.0/0.430
4	Front axleboxes Rear Bogie	176	144.5/2.2/144.5	1.05/0.0/0.430
5	Rear Bogie Frame	1982	1398.5/2667.0/2667.0	0.00/0.0/0.460
6	Rear wheelset Front Bogie	933	461.4/61.6/461.4	10.05/0.0/0.430
7	Front wheelset Front Bogie	933	461.4/61.6/461.4	12.15/0.0/0.430
8	Rear axleboxes Front Bogie	176	144.5/2.2/144.5	10.05/0.0/0.430
9	Front axleboxes Front Bogie	176	144.5/2.2/144.5	12.15/0.0/0.430
10	Front Bogie Frame	1982	1398.5/2667.0/2667.0	11.10/0.0/0.460
11	Carbody	11160	14953/225365/224995	5.55/0.0/1.849

Table 3: Mass, inertia properties and initial positions of each rigid body

5.2 Kinematic joints

The kinematic joints connect the bodies of the multibody system in order to restrain some of their relative motions [25]. In the multibody model of the ML95 trailer vehicle, four revolute joints are used to connect each pair of axleboxes to a wheelset, only allowing the relative rotation between the connected bodies, representing in this form the roller bearings that exist at the extremities of each wheelset. All other relative motions between the system components are limited by force elements.

5.3 Primary suspension

The primary suspension of a railway vehicle consists of a group of flexible links that ensure the transmission of forces between the bogie frame and the axleboxes assembled at the extremities of the wheelsets. In the ML95 trailer vehicle, the primary suspension is modelled with a three-dimensional spring-damper element, which has stiffness and damping properties defined for the vertical, longitudinal and lateral directions, as shown in Figure 15. The characteristics of these spring and damper elements, the numbers of the bodies that they connect and the local coordinates of the attachment points are presented in Table 4. This table only presents the characteristics of the suspension elements for the rear bogie. The elements of the front bogie have exactly the same characteristics.

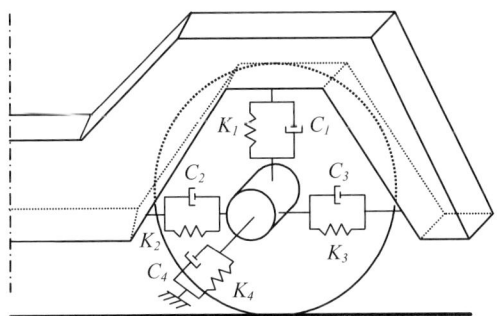

Figure 15: Primary suspension model of the ML95 trailer vehicle

Suspension Element Direction	Spring Stiffness 10^6 N/m	Damping Coef. 10^3 N.s/m	Bogie Location	Bodies		Attachment Pts Local Coords (m)	
						Body i	Body j
				i	j	$\xi_i/\eta_i/\zeta_i$	$\xi_j/\eta_j/\zeta_j$
Vertical	K_1 1.28	C_1 55.41	Rear Right	3	5	0.0/-0.9/0.0	-1.05/-0.9/0.22
			Rear Left	3	5	0.0/ 0.9/0.0	-1.05/ 0.9/0.22
			Front Right	4	5	0.0/-0.9/0.0	1.05/-0.9/0.22
			Front Left	4	5	0.0/ 0.9/0.0	1.05/ 0.9/0.22
Longitud.	K_2 6.21	C_2 86.29	Rear Right	3	5	0.0/-0.9/0.0	-0.775/-0.9/-0.03
			Rear Left	3	5	0.0/ 0.9/0.0	-0.775/ 0.9/-0.03
			Front Right	4	5	0.0/-0.9/0.0	0.775/-0.9/-0.03
			Front Left	4	5	0.0/ 0.9/0.0	0.775/ 0.9/-0.03
Longitud.	K_3 6.21	C_3 86.29	Rear Right	3	5	0.0/-0.9/0.0	-1.325/-0.9/-0.03
			Rear Left	3	5	0.0/ 0.9/0.0	-1.325/ 0.9/-0.03
			Front Right	4	5	0.0/-0.9/0.0	1.325/-0.9/-0.03
			Front Left	4	5	0.0/ 0.9/0.0	1.325/ 0.9/-0.03
Lateral	K_4 2.06×10^6	C_4 70.29×10^3	Rear Right	3	5	0.0/-0.9/0.0	-1.05/-1.1/-0.03
			Rear Left	3	5	0.0/ 0.9/0.0	-1.05/ 1.1/-0.03
			Front Right	4	5	0.0/-0.9/0.0	1.05/-1.1/-0.03
			Front Left	4	5	0.0/ 0.9/0.0	1.05/ 1.1/-0.03

Table 4: Primary suspension elements used in the model of the ML95 trailer vehicle

5.4 Secondary suspension

The secondary suspension of a railway vehicle consists of a set of springs and dampers that transmit the forces between the carbody and the bogies providing, in the process, the isolation of the passenger compartment from the vibrations induced by the track. In the ML95 vehicle model, the carbody is supported by the two bogies through four airsprings. These components are modelled here with three pairs of spring-damper elements acting in the vertical, longitudinal and lateral directions, as shown in Figure 16. In parallel with the airsprings, four vertical dampers are used for the stabilization of the carbody. The characteristics of the spring and damper elements that compose the secondary suspension model, the bodies that they connect and the local coordinates of the attachment points are presented in Table 5.

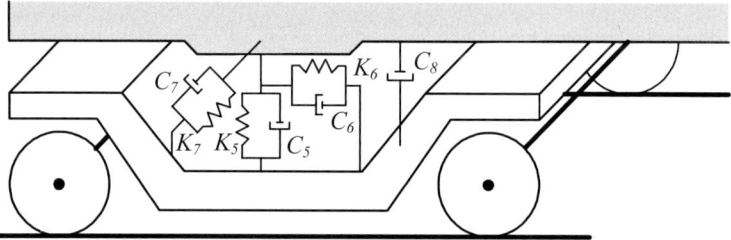

Figure 16: Secondary suspension model of the ML95 trailer vehicle

Suspension Element	Spring Stiffness 10^3N/m	Damping Coeffic. 10^3N.s/m	Bodies Connected		Attachment Points Local Coords (m)	
			i	j	Body i $\xi_i/\eta_i/\zeta_i$	Body j $\xi_j/\eta_j/\zeta_j$
Airspring Vertical	K_5 250	C_5 47.54	5	11	0.0/-0.935/0.09	-5.55/-0.935/-1.009
			5	11	0.0/ 0.935/0.09	-5.55/ 0.935/-1.009
			10	11	0.0/-0.935/0.09	5.55/-0.935/-1.009
			10	11	0.0/ 0.935/0.09	5.55/ 0.935/-1.009
Airspring Longitudinal	K_6 75	C_6 26.04	5	11	0.2/-0.935/0.38	-5.55/-0.935/-1.009
			5	11	0.2/ 0.935/0.38	-5.55/ 0.935/-1.009
			10	11	-0.2/-0.935/0.38	5.55/-0.935/-1.009
			10	11	-0.2/ 0.935/0.38	5.55/ 0.935/-1.009
Airspring Lateral	K_7 75	C_7 26.04	5	11	0.0/-1.135/0.38	-5.55/-0.935/-1.009
			5	11	0.0/ 1.135/0.38	-5.55/ 0.935/-1.009
			10	11	0.0/-1.135/0.38	5.55/-0.935/-1.009
			10	11	0.0/ 1.135/0.38	5.55/ 0.935/-1.009
Vertical Damper		C_8 21.00	5	11	0.48/-1.130/-0.07	-5.07/-1.130/-1.009
			5	11	-0.48/ 1.130/-0.07	-6.03/ 1.130/-1.009
			10	11	0.48/-1.130/-0.07	6.03/-1.130/-1.009
			10	11	-0.48/ 1.130/-0.07	5.07/ 1.130/-1.009

Table 5: Secondary suspension elements in the model of the ML95 trailer vehicle

5.5 Bogie-carbody connection

In railway vehicles, the bogie-carbody connection is enforced by a group of flexible links that have steering functions and that do not transmit vertical loads. In the ML95 trailer vehicle, the transmission of traction and braking efforts between each bogie and the carbody is done by two traction rods. These traction rods are modelled using spring elements, as depicted in Figure 17. Each spring connects the bogie frame to the center plate acting in the longitudinal direction. The lateral stability of the carbody is obtained by the action of two transversal dampers assembled between the center plate and each bogie frame. The characteristics of the spring and damper elements, which compose the bogie-carbody connection model, the numbering of the bodies connected by them and the local coordinates of the attachment points are presented in Table 6.

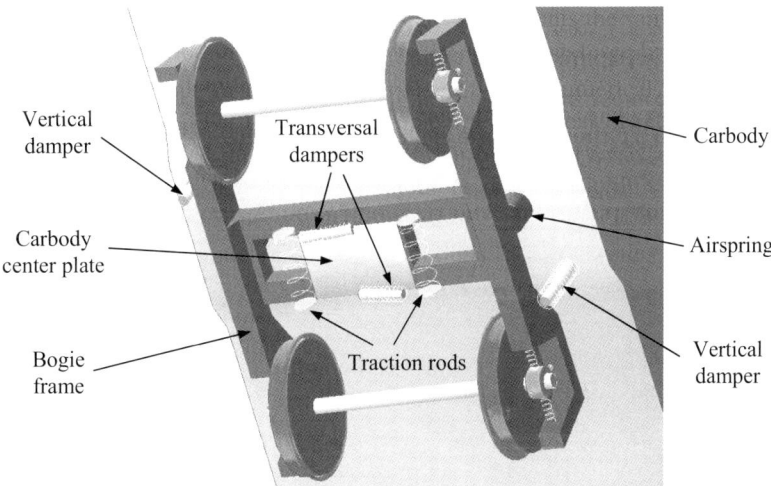

Figure 17: Bogie-carbody connection model for the ML95 trailer vehicle

Connection Element	Spring Stiffness 10^6 N/m	Damping Coeffic. 10^3 N.s/m	Bodies Connected		Attachment Points Local Coords (m)	
					Body i	Body j
			i	j	$\xi_i/\eta_i/\zeta_i$	$\xi_j/\eta_j/\zeta_j$
Traction Rod	K_9 1.905		5	11	0.245/-0.44/-0.21	-5.795/-0.44/-1.599
			5	11	-0.245/ 0.44/-0.21	-5.305/ 0.44/-1.599
			10	11	0.245/-0.44/-0.21	5.305/-0.44/-1.599
			10	11	-0.245/ 0.44/-0.21	5.795/ 0.44/-1.599
Transversal Damper		C_{10} 12.0	5	11	0.225/-0.285/-0.27	-5.325/ 0.04/-1.599
			5	11	-0.225/ 0.285/-0.27	-5.775/-0.04/-1.599
			10	11	0.225/-0.285/-0.27	5.775/ 0.04/-1.599
			10	11	-0.225/ 0.285/-0.27	5.325/-0.04/-1.599

Table 6: Bogie-carbody connection elements in the model of ML95 trailer vehicle

6 Dynamic analysis in real operation conditions

All methodologies proposed in this work are implemented in the general multibody code DAP-3D that is used for the dynamic analysis of railway and other types of rail-guided vehicles [16].

6.1 Simulation conditions

The ML95 trailer vehicle is studied here in real operating conditions when travelling between the Lisbon metro stations of Anjos and Arroios. The track between such stations has a straight geometry using a length of 666 m. The track irregularities are obtained experimentally with a vehicle of the Lisbon metro company equipped for the purpose. The data collected consists of a table where all irregularities parameters are stored in columns as a function of the distance travelled along the track. The information about the track geometry and its irregularities is used as input for the railway pre-processor in order to build a realistic track model, using the methodology described in this work. During the track model assemblage, the distance step adopted for the left and right rails databases construction is $0.05m$.

When the vehicle model is assembled on the track, as depicted in Figure 18, the system should be in static equilibrium in order to avoid discontinuities in the results when the dynamic analysis starts. For the purpose of obtaining this equilibrium, no track irregularities are considered in the first $50m$ of the track model.

Figure 18: ML95 trailer vehicle running on a straight track

In order to evaluate the influence of the track irregularities on the vehicle performance, two dynamic analyses are carried out. In the first one, an ideal track model is considered, having a perfect straight geometry without irregularities. In the second analysis, the vehicle is studied on the realistic track model that includes the measured track irregularities. In both cases the vehicle is assembled on the track with an initial misalignment of $2mm$ with respect to the track centerline. An initial forward velocity of $16m/s$ ($58Km/h$) is assigned to the vehicle model, corresponding to the service conditions. The creep forces are computed here using the Polach formulation.

Track Models for Railway Vehicle Dynamic Analyses 87

6.2 Influence of track irregularities on vehicle performance

6.2.1 Wheelset results

The lateral displacement of the leading wheelset, with respect to the track centerline, when running on the straight track with and without irregularities is presented in Figure 19. On the perfect track it can be observed that, after the initial misalignment of *2mm*, the wheelset presents a damped oscillating motion and returns to the centre of the track after about 5 seconds. Despite running at a velocity much lower than the vehicle critical speed, no decaying oscillations are observed in the wheelset motion when running on the realistic track. This is explained by the fact that the track irregularities induce disturbances in the wheelset which means that its motion is a dynamic response to such excitations. Notice that the coincidence of the results at the beginning of the analysis arises from the fact that no track irregularities are considered in the first *50m* of the realistic track model.

The lateral and vertical contact forces on the left and right wheels of the front wheelset of the vehicle are shown in Figure 20 and Figure 21, respectively. The comparative results obtained when the vehicle is running on the straight track with and without irregularities are presented for comparison. The high oscillations observed in the contact forces obtained with the realistic track, are a direct consequence of the existence of track irregularities. In the perfect track it is observed that the contact forces are almost constant.

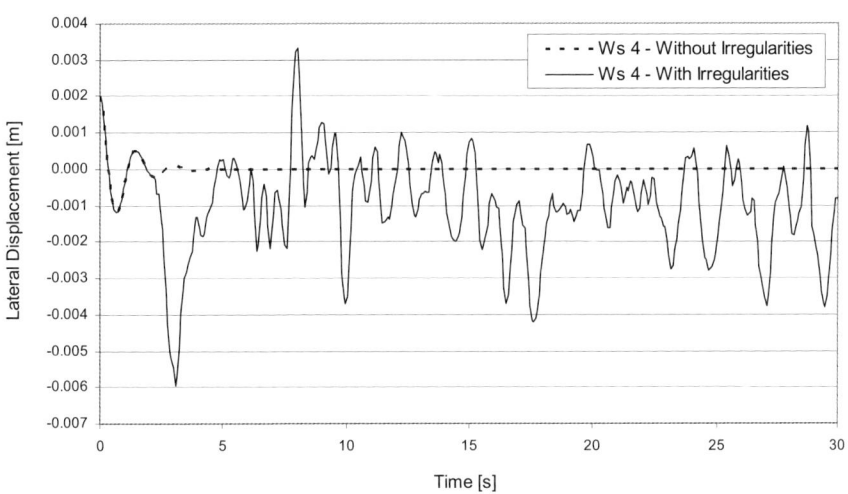

Figure 19: Lateral displacement of the leading wheelset of the vehicle

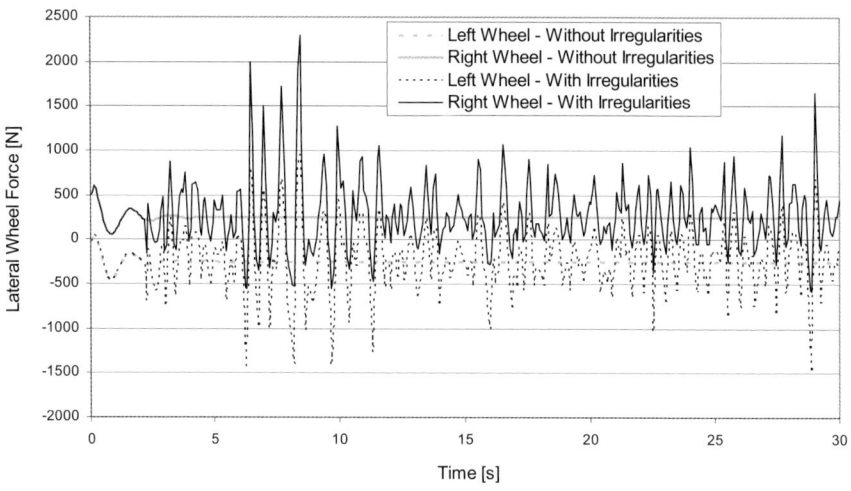

Figure 20: Lateral contact forces on the leading wheelset of the vehicle

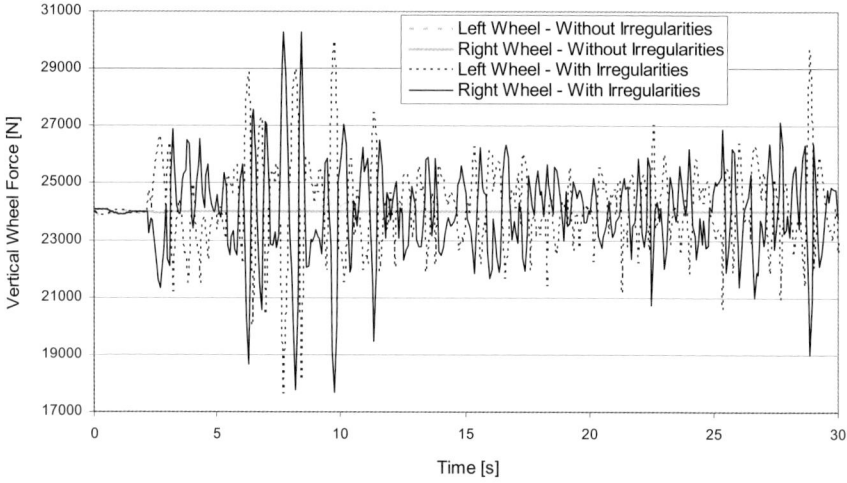

Figure 21: Vertical contact forces on the leading wheelset of the vehicle

6.2.2 Carbody results

The objective here is to assess how the track irregularities influence the comfort in the passengers' compartment. For this purpose, the lateral and vertical accelerations on the carbody are presented in Figure 22 and Figure 23, respectively. In the perfect track, the passengers travelling in the vehicle are subjected to almost no accelerations. However, when the imperfections of the track are present, some noticeable accelerations are observed in the carbody. Therefore, the structural elements below the carbody and the two levels of suspension do not isolate the passengers' compartment from all the vibrations arising from the vehicle interaction with the realistic track.

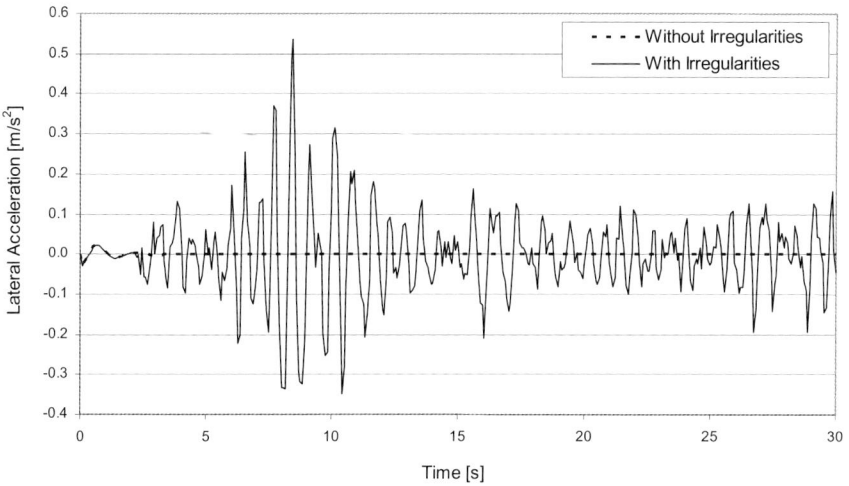

Figure 22: Lateral accelerations on the carbody

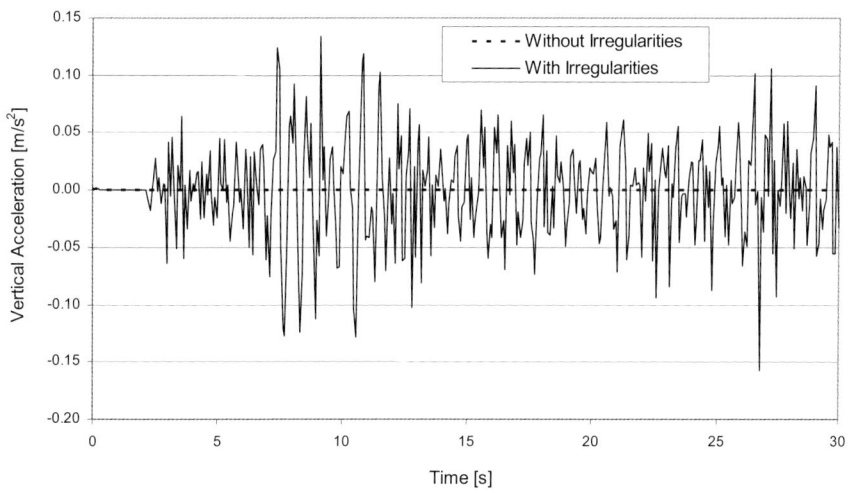

Figure 23: Vertical accelerations on the carbody

6.3 Validation of the methodology

6.3.1 Comparison with the commercial tool ADAMS/Rail

In order to validate the methodology proposed here and implemented in the multibody program DAP-3D, the results are compared with the ones obtained from the commercial program ADAMS/Rail [14,15] for the same scenario. The strategy adopted for the construction of the ML95 vehicle model in ADAMS/Rail follows the same multibody approach as is described here. The only difference is related to the fact that in ADAMS/Rail airspring models are available, whereas in DAP-3D these elements are modelled with the system of springs and dampers depicted in Figure 16.

The track model in ADAMS/Rail has a straight geometry and the measured track irregularities are considered. The wheel-rail contact model used in the ADAMS/Rail simulation is the Table Book method [14,15], whereas in DAP-3D the creep forces are computed by the Polach formulation.

The lateral contact forces on the right wheel of the leading wheelset of the ML95 vehicle are shown in Figure 24. There is a reasonable agreement between the results, which present identical medium values and similar variation ranges, with the ADAMS/Rail contact forces being slightly higher. Also the peaks in the results, arising from the track irregularities, are larger in the commercial code. The differences between the results can be explained by two main reasons. The first is related to the fact that the creep force models used are not the same. The second reason arises from the different approaches used by the two programs when including the irregularities in the track model. This suggests that the isolated higher peaks in the forces are a result of the track parameterization process rather than being a physical characteristic of the system.

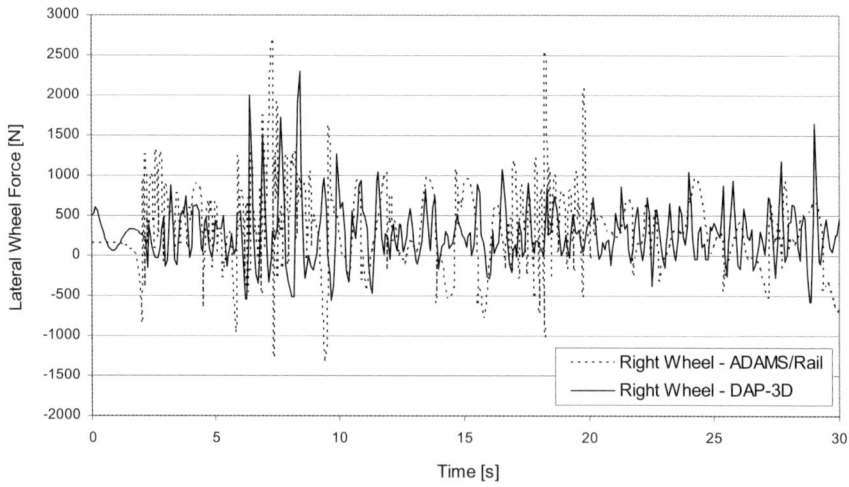

Figure 24: Lateral contact forces on the right wheel of the leading wheelset

Track Models for Railway Vehicle Dynamic Analyses   91

In Figure 25, the normal contact forces on the right wheel of the leading wheelset are presented. The results present identical medium values but the range variation of the normal forces computed by DAP-3D is higher than the one obtained with the commercial code. Besides the two main reasons previously described, another important factor contributes to this difference. It is related to the different formulations used to calculate the normal contact forces that develop at the wheel-rail interface. Notice that the results are almost coincident at the beginning of the analysis, where no track irregularities are considered.

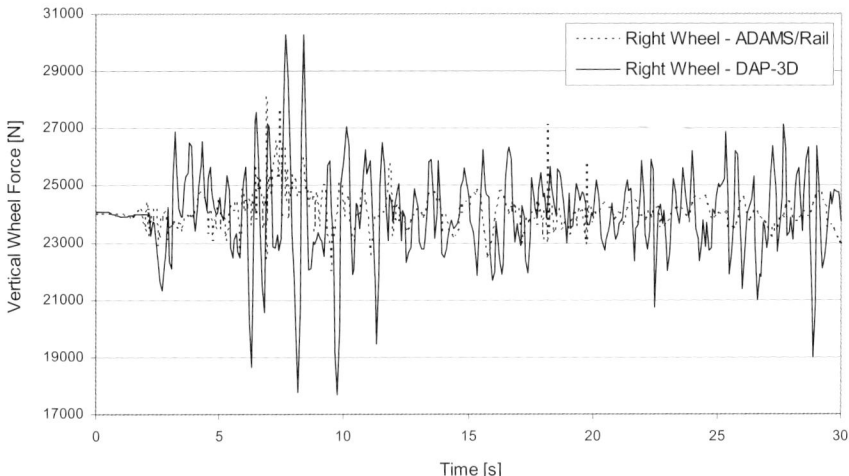

Figure 25: Normal contact forces on the right wheel of the leading wheelset

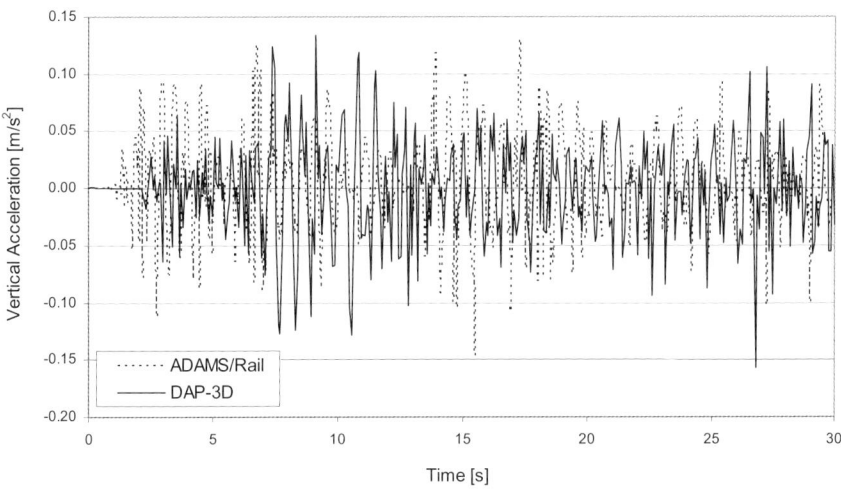

Figure 26: Vertical accelerations on the carbody - Comparison with ADAMS/rail

The vertical accelerations on the carbody of the ML95 vehicle are presented in Figure 26 where a good agreement between the results is observed.

6.3.2 Comparison with experimental data

The results obtained with DAP-3D are now compared with the experimental results in terms of the carbody accelerations. The lateral and vertical accelerations on the carbody of the ML95 vehicle are presented in Figure 27 and Figure 28, respectively. The figures show a reasonable agreement between the results, which present identical medium values and similar variation ranges. The exceptions are the kinks in the results arising from the track irregularities. It is observed that, especially in the vertical accelerations, such discontinuities are significantly higher in the results obtained experimentally.

A Fast Fourier Transform (FFT) is applied to the results presented in Figure 27 and Figure 28, in order to study the frequency contents of these signals. The spectra for the lateral and vertical accelerations, obtained experimentally and with DAP-3D, are shown in Figure 29 and Figure 30, respectively. The lateral acceleration spectra show a good agreement. In fact, the most meaningful frequency of the experimental results is $0.9Hz$ whereas, for the numerical simulation results, it corresponds to $1.1Hz$. For the vertical accelerations the spectra exhibit a reasonable agreement. The most meaningful frequencies for the experimental and numerical results are about 1.2 and $1.7Hz$, respectively. Notice that the most meaningful frequencies exhibit higher values in the results obtained with DAP-3D. This suggests that the suspensions of the vehicle model are stiffer than the real ones.

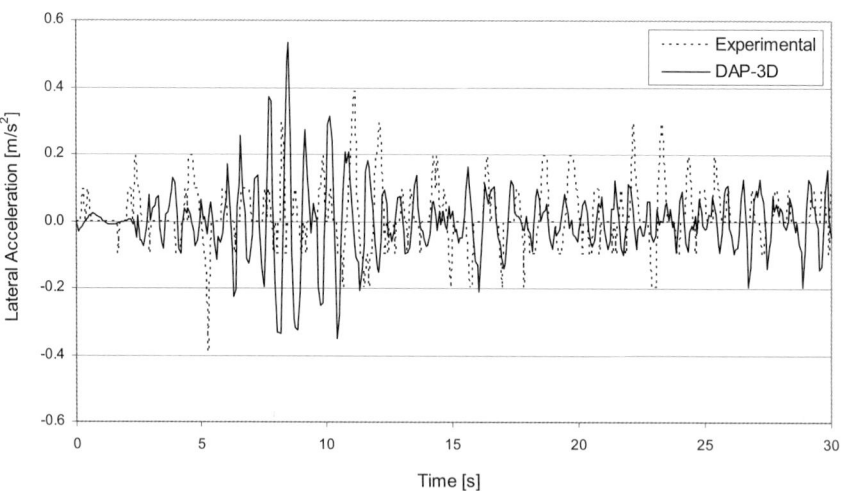

Figure 27: Lateral acceleration on carbody - Comparison with experimental data

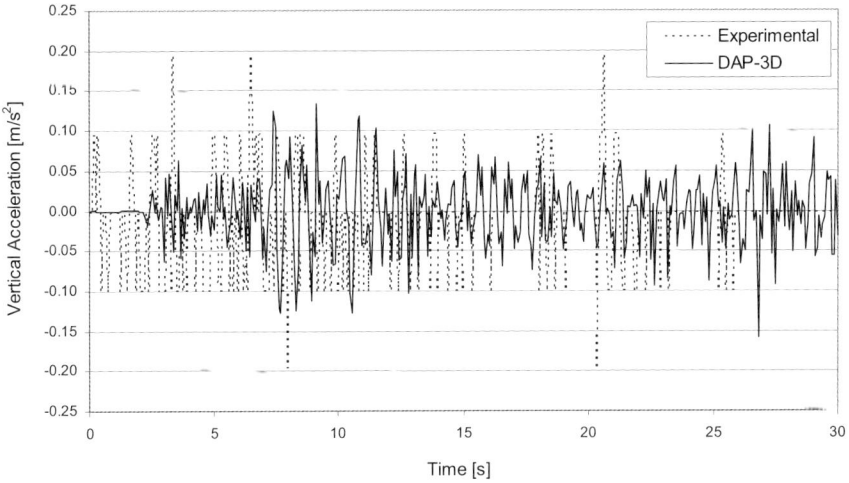

Figure 28: Vertical acceleration on carbody - Comparison with experimental data

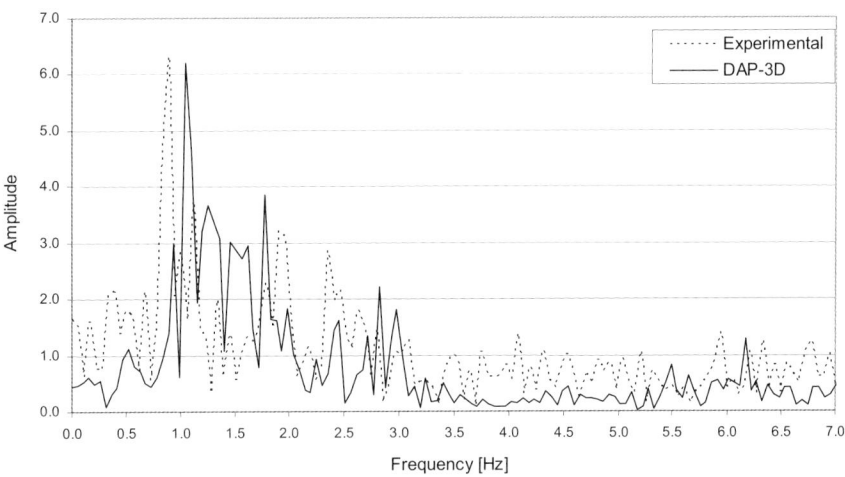

Figure 29: Spectra of the lateral accelerations on the carbody

7 Conclusions

The objective of this chapter is to describe the spatial geometric representation of tracks and to present their computational implementation in a form suitable to the multibody formulation adopted to study railway systems. Hence, a methodology for the realistic representation of the three-dimensional track geometry is proposed. It involves not only the description of the design track layout but also the representation of its irregularities.

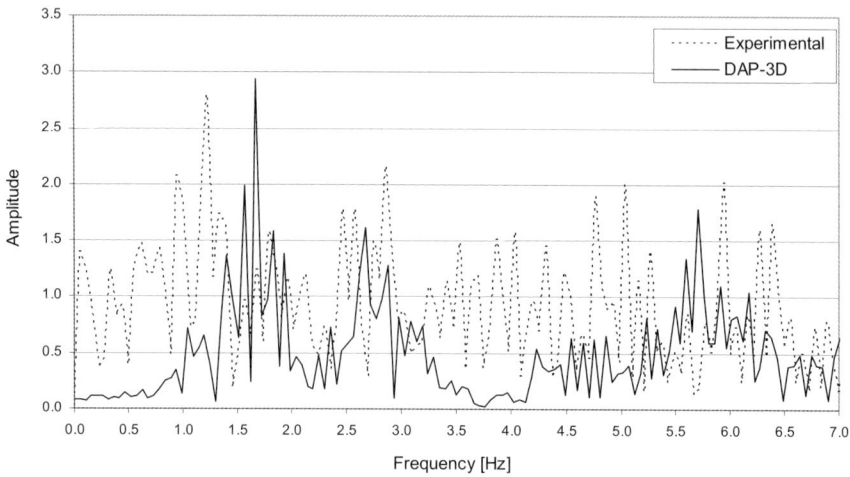

Figure 30: Spectra of the vertical accelerations on the carbody

Depending on the specific application envisaged, the track model can be defined by different types of parametric curves. In this work, the track geometric description adopted uses Frenet frames that provide the appropriate track referential at every point. Furthermore, shape preserving splines are used to interpolate a set of control points that are representative of the track geometry. It is also proposed that, for tracks with a full spatial geometry, the cant angle is measured with respect to the osculating plane. Then, a pre-processing strategy is used to obtain the accurate description of two parameterized curves that represent the geometry of the left and right rails. These space curves are modelled by adding, to the rails perfect geometry, small perturbations resulting from the track irregularities, which are measured experimentally.

The complex contact forces that develop in the wheel-rail interface and that strongly influence the dynamic behaviour of railway vehicles are computed here through a suitable wheel-rail contact formulation. Also the characteristics of the railway vehicle as suspensions, masses and inertias of the structural elements are considered in the analyses performed here.

It is observed that the track irregularities increase the wheel-rail contact forces. If these forces reach sufficiently high values, they will contribute to an increase in the track imperfections, which originate higher contact forces. High wheel-rail contact forces can also cause rail fracture, due to the structural fatigue of the material, damage to the track foundations and deterioration of the wheel and rail profiles. Hence, the reduction in the track irregularities allows the user to limit the loads applied on the rolling stock, reducing the structural fatigue in the mechanical elements that constitute the vehicles and reducing the wear on wheels and rails.

The results also show that the track imperfections significantly reduce the comfort levels in the carbody. In this sense, the strategies that seek to increase the passengers' comfort in a given railway network, must concern not only the rolling stock characteristics, but also must address the progressive improvement of the tracks, minimizing their imperfections.

When comparing the experimental data, and the results from the commercial program, with the dynamic analysis results, it is observed that the models and methodologies developed here are not only qualitatively but also quantitatively correct. It is believed that the differences that are observed result from uncertainties associated with the experimental tests and to the modelling methodologies.

The case tests in this work showed that the proposed methodology provides a reliable and accurate track model. This methodology is appropriate to the needs of the railway operators since the required input data is the information that is standard in the railway industry. It is also shown that the formulation integrates, in an efficient way, the main requirements associated with the dynamic analysis of railway vehicles. Therefore, the models allow extraction of results that would be either impossible to measure experimentally or for which very costly experimental procedures would be required.

Another advantage of the proposed track model is that the time required for the dynamic simulation of the rail-guided vehicles is completely independent of the track complexity. Any descriptive form of parametric curves is dealt with within the track pre-processor, while the dynamic analysis program only has to proceed with linear interpolations of the rail databases. By ensuring that the arc-length step is small enough, the linear interpolation procedure does not introduce any significant error in the geometric description of the rails.

Acknowledgements

The support of Fundação para a Ciência e Tecnologia (FCT) through the Post-Doc grant with the reference BPD/19066/2004 made this work possible and it is gratefully acknowledged. The authors also want to acknowledge the valuable discussions with Prof. José Escalona, University of Seville, and with Prof. Ahmed Shabana, University of Illinois at Chicago.

References

[1] E. Andersson, M. Berg, S. Stichel, "Rail Vehicle Dynamics, Fundamentals and Guidelines", Royal Institute of Technology (KTH), Stockholm, Sweden, 1998.
[2] British-Rail, Research, "Introduction to Railway Vehicle Dynamics", AEA Technology plc., London, United Kingdom, 1997.
[3] R.V. Dukkipati, J.R. Amyot, "Computer-Aided Simulation in Railway Dynamics", M. Dekker Inc., New York, New York, 1988.
[4] V.K. Garg, R.V. Dukkipati, "Dynamics of Railway Vehicle Systems", Academic Press, New York, New York, 1984.
[5] S. Iwnicki, "Handbook of Railway Vehicle Dynamics", Taylor & Francis, London, UK, 2006.

[6] A. Shabana, K. Zaazaa, H. Sugiyama, "Railroad Vehicle Dynamics: A Computational Approach", Taylor & Francis, London, UK, 2008.
[7] S. Lee, T. Park, K. Moon, K. Chung, K. Kim, "A study on the manoeuvrability analysis of the bi-modal tram using the all-wheel steering electronic control unit", Vehicle System Dynamics, 48(1), 113-131, 2010.
[8] C. Schindler, M. Schwickert, A. Simonis, "Structural safety of trams in case of misguidance in a switch", Vehicle System Dynamics, 48(8), 967-981, 2010.
[9] G. Shen, J. Zhou, L. Ren, "Enhancing the resistance to derailment and side-wear for a tramway vehicle with independently rotating wheels", Vehicle System Dynamics, 44(S1), 641-651, 2006.
[10] MDI, Mechanical and Dynamics, "ADAMS/Rail 9.1.1 ADtranz Milestone I - Release Notes", Ann Arbor, Michigan, 1999.
[11] AEA Technology plc, "VAMPIRE User Manual - V 4.32", Derby, UK, 2004.
[12] DeltaRail Group Ltd, "VAMPIRE Pro User Manual - V 5.02", Derby, UK, 2006.
[13] G. Hippmann, M. Duke, "Advances in Track Modelling", SIMPACK News, 2003.
[14] MDI, Mechanical and Dynamics, "ADAMS/Rail 9.1 - Technical Manual", Ann Arbor, Michigan, 1995.
[15] MDI, Mechanical and Dynamics, "Using ADAMS/Rail 9.1", Ann Arbor, Michigan, 1998.
[16] J. Pombo, "A Multibody Methodology for Railway Dynamics Applications", PhD Dissertation, Instituto Superior Técnico, Lisbon, Portugal, 2004.
[17] J. Pombo, J. Ambrósio, "General Spatial Curve Joint for Rail Guided Vehicles: Kinematics and Dynamics", Multibody Systems Dynamics, 9(3), 237-264, 2003.
[18] J. Pombo, J. Ambrósio, "Modelling Tracks for Roller Coaster Dynamics", Int. J. Vehicle Design, 45(4), 470-500, 2007.
[19] M.E. Mortenson, "Geometric Modeling", Wiley, New York, New York, 1985.
[20] R. Chamorro, J. Escalona, M. González, "An Approach for Modeling Long Flexible Bodies with Application to Railroad Dynamics", Multibody Systems Dynamics, 2011. doi:10.1007/s11044-011-9255-x
[21] A.A. Shabana, R. Chamorro, C. Rathod, "A Multibody System Approach for Finite Element Modeling of Rail Flexibility in Railroad Vehicle Applications", Institution of Mechanical Engineers, Part K: J. Multi-body Dynamics, 222, 1-15, 2008.
[22] A.A. Shabana, G. Sanborn, "An Alternative Simple Multibody System Approach for Modelling Rail Flexibility in Railroad Vehicle Dynamics", Institution of Mechanical Engineers, Part K: J. Multi-body Dynamics, 223, 107-120, 2009.
[23] A.A. Shabana, J.R. Sany, "A Survey of Rail Vehicle Track Simulations and Flexible Multibody Dynamics", Nonlinear Dynamics, 26, 179-210, 2001.
[24] E. Haug, "Computer Aided Kinematics and Dynamics of Mechanical Systems", Allyn and Bacon, Boston, Massachussetts, 1989.
[25] P.E. Nikravesh, "Computer-Aided Analysis of Mechanical Systems", Prentice-Hall, Englewood Cliffs, New Jersey, 1988.

[26] M. Pereira, J. Ambrósio, "Computational Dynamics in Multibody Systems", Kluwer Academic Publishers, Dordrecht, The Netherlands, 1995.
[27] J. Pombo, J. Ambrósio, "Dynamic Analysis of a Railway Vehicle in Real Operation Conditions Using a New Wheel-Rail Contact Detection Model", Int. J. Vehicle Systems Modelling and Testing, 1(1/2/3), 79-105, 2005.
[28] R.E. Roberson, R. Schwertassek, "Dynamics of Multibody Systems", Springer-Verlag, Berlin, Germany, 1988.
[29] W. Schiehlen, "Advanced Multibody System Dynamics - Simulation and Software Tools", Kluwer Academic Publishers, Dordrecht, The Netherlands, 1993.
[30] A.A. Shabana, "Dynamics of Multibody Systems", 2nd Edition, Cambridge University Press, Cambridge, United Kingdom, 1998.
[31] J. Pombo, J. Ambrósio, "A Computational Efficient General Wheel-Rail Contact Detection Method", Journal of Mechanical Science and Technology, The Separate Volume of KSME International Journal, 19(1), Special Edition, 411-421, 2005.
[32] J. Pombo, J. Ambrósio, "Application of a Wheel-Rail Contact Model to Railway Dynamics in Small Radius Curved Tracks", Multibody Systems Dynamics, 19(1-2), 91-114, 2008.
[33] J. Pombo, J. Ambrósio, M. Silva, "A New Wheel-Rail Contact Model for Railway Dynamics", Vehicle System Dynamics, 45(2), 165-189, 2007.
[34] L.D. Irvine, S.P. Marin, P.W. Smith, "Constrained Interpolation and Smoothing", Constructive Approximation, 2, 129-151, 1986.
[35] C.A. Micchelli, P.W. Smith, J. Swetits, J.D. Ward, "Constrained Lp Approximation", Constructive Approximation, 1, 93-102, 1985.
[36] V.B. Anand, "Computer Graphics and Geometric Modeling for Engineering", J. Wiley, New York, 1994.
[37] C. De Boor, "A Practical Guide to Splines", Springer-Verlag, New York, 1978.
[38] J. Ambrósio, "Trainset Kinematic: A Planar Analysis Program for the Study of the Gangway Insertion Points", Technical Report IDMEC/CPM/97/005, Institute of Mechanical Engineering, Instituto Superior Técnico, Lisbon, Portugal, 1997.
[39] H. Akima, "A New Method of Interpolation and Smooth Curve Fitting Based on Local Procedures", Association for Computing Machinery, 17(4), 589-602, 1970.
[40] A.A. Shabana, K.E. Zaazaa, J.L. Escalona, J.R. Sany, "Modeling Two-Point Wheel/Rail Contacts Using Constraint and Elastic-Force Approaches", Proceedings of the IMECE'02: 2002 ASME International Mechanical Engineering Congress and Exposition, New Orleans, Louisiana, November 17-22, 2002.
[41] A.A. Shabana, K.E. Zaazaa, J.L. Escalona, J.R. Sany, "Dynamics of the Wheel/Rail Contact Using a New Elastic Force Model", Technical Report #MBS02-3-UIC, Department of Mechanical Engineering, University of Illinois, Chicago, 2002.

[42] M. Berzeri, J.R. Sany, A.A. Shabana, "Curved Track Modeling Using the Absolute Nodal Coordinate Formulation", Technical Report #MBS00-4-UIC, Department of Mechanical Engineering, University of Illinois, Chicago, 2000.
[43] E. Kassa, C. Andersson, J. Nielsen, "Simulation of dynamic interaction between train and railway turnout", Vehicle System Dynamics, 44(3), 247-258, 2006.
[44] A.A. Shabana, M. Berzeri, J.R. Sany, "Numerical Procedure for the Simulation of Wheel/Rail Contact Dynamics", Journal of Dynamic Systems Measurement and Control-Transactions of the ASME, 123(2), 168-178, 2001.
[45] A. Johansson, C. Andersson, "Out-of-round railway wheels - A study of wheel poligonization through simulation of three-dimensional wheel-rail interaction and wear", Vehicle System Dynamics, 43(8), 539-559, 2005.
[46] H.M. Lankarani, P.E. Nikravesh, "A Contact Force Model with Hysteresis Damping for Impact Analysis of Multibody Systems", AMSE Journal of Mechanical Design, 112, 369-376, 1990.
[47] H.M. Lankarani, P.E. Nikravesh, "Continuous Contact Force Models for Impact Analysis in Multibody Systems", Nonlinear Dynamics, 5, 193-207, 1994.
[48] J.J. Kalker, "Survey of Wheel-Rail Rolling Contact Theory", Vehicle System Dynamics, 8(4), 317-358, 1979.
[49] J.J. Kalker, "Three-Dimensional Elastic Bodies in Rolling Contact", Kluwer Academic Publishers, Dordrecht, The Netherlands, 1990.
[50] Z.Y. Shen, J.K. Hedrick, J.A. Elkins, "A Comparison of Alternative Creep Force Models for Rail Vehicle Dynamic Analysis", in J.K. Hedrick, (Editor), 8th IAVSD Symposium on Dynamics of Vehicles on Road and Tracks, Swets and Zeitlinger, Cambridge, Massachussetts, 591-605, 1983.
[51] O. Polach, "A Fast Wheel-Rail Forces Calculation Computer Code", Vehicle System Dynamics, Supplement, 33, 728-739, 1999.

Chapter 4

Dynamic Analysis of Beam Structures under Moving Loads: A Review of the Modal Expansion Method

Z. Dimitrovová
UNIC, Department of Civil Engineering
Universidade Nova de Lisboa, Portugal

Abstract

In this chapter, the dynamic analysis of beam structures under moving loads is presented. Hamilton's principle is implemented as an efficient tool for obtaining the governing equations for transversal vibrations in beam structures induced by moving loads. The determination of elastic constants representing the effect of an elastic half space is included. The influence of the axial force is also considered. The concept of the dynamic stiffness matrix is posted as a general principle for finite, semi-infinite and infinite beams. Its implementation in beam structures composed of several sub-domains is developed. Advantages and disadvantages of modal expansion in series governed by damped and undamped modes are stated.

Keywords: dynamic stiffness matrix, state-space formulation, moving load, moving mass, modal expansion, localized disturbances.

1 Introduction

Dynamic analyses of beam structures under moving loads have attracted the engineering and scientific community from the middle of the 19th century, when railway construction began. Several concerns related to railway lines can be quickly solved on simple models concerning just a beam on a viscoelastic foundation traversed by uniformly moving loads. Since a considerable amount of studies have been published on this subject, only a few pioneering works are mentioned. Dynamic stresses in the beam structure were firstly solved by Krylov [1] and later by Timoshenko [2]. Transversal vibrations in a simply supported beam traversed by a constant force moving at a constant velocity were presented by Inglis [3], Lowan [4] and later on, other solutions have been given by Koloušek [5] and Frýba [6]. These works employ mainly modal expansion methods. The first solutions for infinite beams were presented by Timoshenko [7]. In [8] the effect of viscous damping in the foundation is discussed. The case of a load variable in time is presented in [9].

The first high speed line, called Shinkansen, was inaugurated in Japan in 1964. The train circulated at a cruise speed of 210km/h between Tokyo and Osaka. Since then, a systematic growth of the high-speed network has begun. The continuing development of vehicles and infrastructures has resulted in increased speed. The world speed record is attributed to the magnetic-levitation Maglev Train (581km/h in 2003 in Japan). This technology avoids the wheel/rail contact, which provides higher speeds while consuming less energy, but the very high cost makes them prohibitive to implement in existing rail networks. The speed record for the conventional wheeled train in test operations is held by TGV France (574.8 km/h in 2007). With increasing speed dynamic effects become more important. Some of the consequences of excessive vibrations on the track and foundation soils are: aggravation of the wear on the vehicle's wheels and of the rail route itself; instability of railway vehicles; passenger discomfort; propagation of said vibrations to neighbouring buildings, where they can cause discomfort to habitants and/or affect precision devices in industrial buildings. These issues motivate increasing effort by the scientific community on the moving load problems. Closed form analytical solutions, if obtainable, are always desirable because of their numerical efficiency and physical insight into the problem.

The chapter is organized in the following way: three important aspects that one should be aware of before starting analyses in this field are described in Section 2; governing equations are developed in a very general form in Section 3; detailed analysis of vibration modes and modal co-ordinates determination is presented in Section 4; additional issues are discussed in Section5 and the chapter is concluded in Section 6.

2 Three introductory issues

2.1 Standard finite elements

It has been proven in [10] that standard finite elements induce an error in natural frequencies that is a consequence of polynomial shape functions. This error is invariant if normalized by the total number of degrees of freedom increased by one and aggravates with the polynomial degree. Our concern is on transversal vibrations, but for the sake of simplicity, the following explanation pertains to longitudinal vibrations.

Let longitudinal vibrations of an elastic rod fixed on both ends be assumed. The stiffness and consistent mass matrices for an element of unitary axial stiffness ($EA=1$) and unitary mass per length ($\mu=1$), for linear and quadratic shape functions, can be obtained by exact integration:

$$\mathbf{K}_{p=1} = \frac{1}{h}\begin{bmatrix} 1 & -1 \\ -1 & 1 \end{bmatrix}, \quad \mathbf{K}_{p=2} = \frac{1}{3h}\begin{bmatrix} 7 & -8 & 1 \\ -8 & 16 & -8 \\ 1 & -8 & 7 \end{bmatrix} \quad (1)$$

Dynamic Analysis of Beam Structures under Moving Loads

$$\mathbf{M}_{p=1} = \frac{h}{6}\begin{bmatrix} 2 & 1 \\ 1 & 2 \end{bmatrix}, \quad \mathbf{M}_{p=2} = \frac{h}{30}\begin{bmatrix} 4 & 2 & -1 \\ 2 & 16 & 2 \\ -1 & 2 & 4 \end{bmatrix} \quad (2)$$

where h designates the element length and p the polynomial degree. A typical equation for an interior node of a unit length rod ($L=1$) for $p=1$ reads as:

$$\frac{1}{h}(\phi_{n-1} - 2\phi_n + \phi_{n+1}) + \omega_h^2 \frac{h}{6}(\phi_{n-1} + 4\phi_n + \phi_{n+1}) = 0, \quad n=1,...,N \quad (3)$$

where N is the total number of interior nodes, which is equal to the number of modes. The element length is therefore $h=1/(N+1)$. ϕ_n stands for a discrete value of the vibration mode and ω_h is the numerical value of the natural frequency. The boundary conditions dictate that $\phi_0 = \phi_{N+1} = 0$. This set of equations allow for analytical resolution in the form of:

$$\omega_{h,j} = \frac{1}{h}\sqrt{\frac{6(1-\cos(\omega_j h))}{2+\cos(\omega_j h)}}, \quad j=1,...,N \quad (4)$$

where $\omega_j = j\pi$ is the analytical frequency value. A plot of points $\left[\frac{j}{N+1}, \frac{\omega_{h,j}}{\omega_j}\right]$, $j=1...N$, is invariant with respect to N.

For $p=2$ the typical equation for an interior node (after condensation of mid-node degrees of freedom) is:

$$\frac{1}{3h(10-\omega_h^2 h^2)}\left((30+2\omega_h^2 h^2)(\phi_{n-1}+\phi_{n+1}) + (-60+16\omega_h^2 h^2)\phi_n\right)$$
$$+ \omega_h^2 \frac{h}{120(10-\omega_h^2 h^2)}\left(5\omega_h^2 h^2(\phi_{n-1}+\phi_{n+1}) + (400-30\omega_h^2 h^2)\phi_n\right) = 0, \quad n=1,...,\tilde{N} \quad (5)$$

where $\tilde{N}$ is the number of interior nodes (except for mid-nodes). In this case, the total number of natural frequencies corresponds to the total number of nodes, which is $N = 2\tilde{N}+1$. The element length is $h=1/(\tilde{N}+1)$ and boundary conditions are the same as before ($\phi_0 = \phi_{N+1} = 0$). This set of equations has an analytical solution. Two values are obtained, corresponding to the acoustical and optical branches:

$$\omega_{h,j}^{acs} = \frac{2}{h}\sqrt{\frac{13+2\cos(\omega_j h) - \sqrt{124+112\cos(\omega_j h) - 11\cos^2(\omega_j h)}}{3-\cos(\omega_j h)}}, \quad j=1,...,N \quad (6)$$

$$\omega_{h,j}^{opt} = \frac{2}{h}\sqrt{\frac{13+2\cos(\omega_j h) + \sqrt{124+112\cos(\omega_j h) - 11\cos^2(\omega_j h)}}{3-\cos(\omega_j h)}}, \quad j=1,...,N \quad (7)$$

These numerical values are symmetrical around the middle value and the singularity point ($\omega_{h,\tilde{N}+1}^{acs} = \sqrt{10}/h = \sqrt{10}/(\tilde{N}+1)$), respectively. A plot of points $\left[\dfrac{j}{N+1}, \dfrac{\omega_{h,j}}{\omega_j}\right]$, $j=1…N$, is again invariant with respect to N. Results are quite different when lumped mass matrix is implemented. Values are summarized in Figure 1.

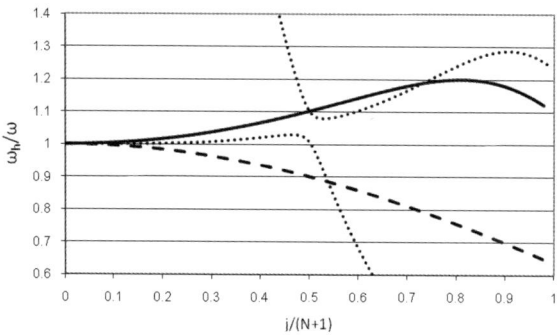

Figure 1: Ratio between the numerical and analytical values of natural frequencies: linear elements, consistent mass (solid black line), linear elements, lumped mass (dashed black lines) and quadratic elements, consistent mass (left and right dotted black lines stand for acoustical and optical branches, respectively)

It is seen in Figure 1 that the error for higher frequencies is quite large and can reach 30%. Nevertheless, if the number of degrees of freedom is at least twice the number of significant frequencies for a particular problem, the error can be kept within a reasonable value. In theory, this error could be avoided by implementation of exact shape functions. It will be seen in the next section that these functions raise several numerical issues.

2.2 Exact shape functions

An implementation of exact shape functions of distributed mass elements in dynamic analyses is problematical because of the presence of hyperbolic functions, which require: (i) evaluations with a high number of digits precision and (ii) capacity to deal with very large/small numbers. One can verify that 709 is the largest integer exponent that can be used in Matlab software [11] for evaluation of e^{709}. On the other hand Maple [12] and similar software can work easily with very large and small numbers, and adapt the number of digits precision to an arbitrary value, regardless of hardware and operating system's specifications. The limitation to double precision can induce unexpected errors, which are very hard to discover. The following example should clarify the problem. Let a cantilever clamped at the left end be assumed. In the Euler-Bernoulli (E-B) formulation, its characteristic equation is given by $1 + \cos\lambda \cosh\lambda = 0$ ([10]), where λ/L is the wave number and L is the cantilever length. The natural undamped j-th mode shape w_j reads as:

Dynamic Analysis of Beam Structures under Moving Loads 103

$$w_j(x) = \sin\left(\frac{\lambda_j}{L}x\right) - \frac{\sin(\lambda_j) + \sinh(\lambda_j)}{\cos(\lambda_j) + \cosh(\lambda_j)}\cos\left(\frac{\lambda_j}{L}x\right)$$
$$-\sinh\left(\frac{\lambda_j}{L}x\right) + \frac{\sin(\lambda_j) + \sinh(\lambda_j)}{\cos(\lambda_j) + \cosh(\lambda_j)}\cosh\left(\frac{\lambda_j}{L}x\right)$$
(8)

from which the displacement value at the free (right) end is

$$w_j(L) = \frac{2\sin\lambda_j \sinh\lambda_j}{\sinh\lambda_j - \sin\lambda_j}$$
(9)

It is possible to write Equation (8) in the following form:

$$w_j(x) = \frac{1}{\sin\lambda_j - \sinh\lambda_j}\left(-\sin(\lambda_j)\sinh\left(\frac{\lambda_j}{L}x\right) - \sinh(\lambda_j)\sin\left(\frac{\lambda_j}{L}x\right)\right.$$
$$\left. + \cos\left(\frac{\lambda_j}{L}x - \lambda_j\right) - \cosh\left(\frac{\lambda_j}{L}x - \lambda_j\right) - \cos(\lambda_j)\cosh\left(\frac{\lambda_j}{L}x\right) + \cosh(\lambda_j)\cos\left(\frac{\lambda_j}{L}x\right)\right)$$
(10)

	λ	Eq.(8)100	Eq.(10)100	Eq.(9)100	Eq.(8)20	Eq.(10)20	Eq.(9)20
m1	1.87510	2.72444	2.72444	2.72444	2.72444	2.72444	2.72444
m2	4.69409	-1.96374	-1.96374	-1.96374	-1.96374	-1.96374	-1.96374
m3	7.85476	2.00155	2.00155	2.00155	2.00155	2.00155	2.00155
m4	10.99554	-1.99993	-1.99993	-1.99993	-1.99993	-1.99993	-1.99993
m5	14.13717	2.00000	2.00000	2.00000	2.00000	2.00000	2.00000
m6	17.27876	-2.00000	-2.00000	-2.00000	-2.00000	-2.00000	-2.00000
m7	20.42035	2.00000	2.00000	2.00000	2.00000	2.00000	2.00000
m8	23.56194	-2.00000	-2.00000	-2.00000	-2.00000	-2.00000	-2.00000
m9	26.70354	2.00000	2.00000	2.00000	2.00000	2.00000	2.00000
m10	29.84513	-2.00000	-2.00000	-2.00000	-2.00000	-2.00000	-2.00000
m11	32.98672	2.00000	2.00000	2.00000	2.00000	2.00000	2.00000
m12	36.12832	-2.00000	-2.00000	-2.00000	-2.00000	-2.00000	-2.00000
m13	39.26991	2.00000	2.00000	2.00000	1.99800	2.00000	2.00000
m14	42.41150	-2.00000	-2.00000	-2.00000	-2.00000	-2.00000	-2.00000
m15	45.55309	2.00000	2.00000	2.00000	1.00000	2.00000	2.00000
m16	48.69469	-2.00000	-2.00000	-2.00000	0.00000	-2.00000	-2.00000
m17	51.83628	2.00000	2.00000	2.00000	0.00000	2.00000	2.00000
m18	54.97787	-2.00000	-2.00000	-2.00000	0.00000	-2.00000	-2.00000
m19	58.11946	2.00000	2.00000	2.00000	0.00000	2.00000	2.00000
m20	61.26106	-2.00000	-2.00000	-2.00000	0.00000	-2.00000	-2.00000

Table 1: Deflection values at the free cantilever end

Table 1 summarizes deflection values of the first 20 modes at the free end of the cantilever calculated by the three formulas and different digits precision assigned as 100 and 20 in columns 3-5 and 6-8, respectively.

Cells with significant error are highlighted. It is seen that the original formula (8) leads to a large unexpected error in double precision software. An alternative reformulation (10), avoiding subtraction of very large/small numbers or using a higher number of digits precision must be implemented. Equation (10) solves this particular problem. An alternative mode shape for a generic beam is given in [13].

Natural frequencies search has no simple adaptation avoiding very large/small numbers. One may object that beam structures usually do not require many mode shapes to achieve acceptable results. However, beams on elastic foundation exhibit deflection shapes that are more accentuated, requiring the superposition of many modes to represent them correctly. This is shown in the following example. Two standard European rails UIC60 of $L=100$m model the cantilever (properties are summarized in Table 2). An axle force of $P=200$kN travels at a speed of $v=50$m/s from the clamped to the free end. The elastic foundation is characterized by the Winkler constant $k=1$MN/m^2. Deflection shapes at $t=0.5$s (when the force is distant 25m from the clamped end) are shown in Figure 2. The results are for the superposition of the first n modes (in Figure 2a, $n=1,2,...,8$; in figure 2b, $n=49,50$). It is seen that 50 mode shapes are necessary for sufficiently accurate results.

Property	Beam (2 UIC60)
Young's modulus E (GPa)	210
Poisson's ratio v	0.3
Density ρ (kg/m^3)	7800
Transversal cross-section area A (m^2)	$153.68 \cdot 10^{-4}$
Geometrical moment of inertia I (m^4)	$6110 \cdot 10^{-8}$
Bending stiffness EI (MNm^2)	12.831
Mass per unit length μ (kg/m)	119.8704

Table 2: Characteristics of 2UIC60 rails

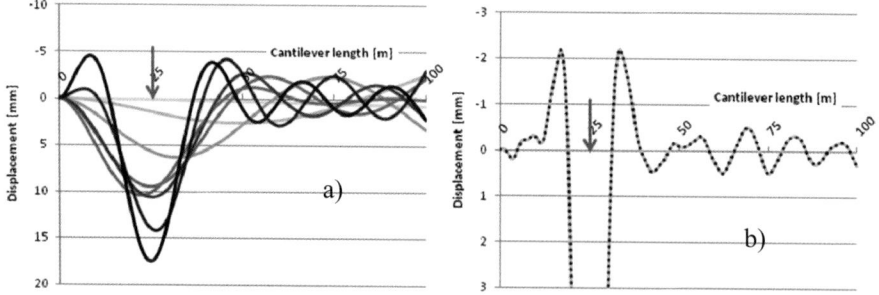

Figure 2: Cantilever on elastic foundation: a) superposition for the fist n modes ($n=1,2,...,8$ from lighter to darker grey solid lines); b) for the first 49 (grey solid line) and 50 modes (black dotted line)

The same test is presented in Figure 3, but the cantilever is considered without the elastic foundation and the moving force is reduced to $P=1$kN in order to keep the displacement values within the range of validity of the linear theory.

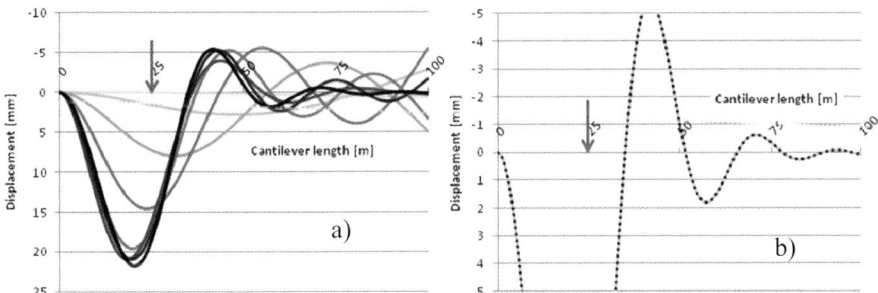

Figure 3: Cantilever without elastic foundation: a) superposition for the fist n modes (n=1,2,...,8 from lighter to darker grey solid lines); b) for the first 8 (grey solid line) and 9 modes (black dotted line)

2.3 Results obtained using a standard finite element software

Let the error reported in Section 2.1 be exemplified on transversal vibrations. It is necessary to point out that transversal vibrations are fundamentally different from longitudinal ones, because in higher frequencies, the effect of the shear deformation and the rotary inertia is significant even on thin beams. Commercial software is protected against inappropriate usage: for instance, in ANSYS, the effect of the rotary inertia cannot be deactivated, while the shear deformation is optional. Thus it is thus not correct to evaluate the error of higher frequency values obtained by the E-B formulation. The numerical error will be analysed on a homogeneous simply supported beam because, in this case, the wave number is $\lambda_j/L=j\pi/L$ and the natural frequencies can be expressed analytically for: the E-B, the Timoshenko (T), the Timoshenko-Rayleigh (T-R) and the Rayleigh (R) theory in Equations (11-14).

$$\omega_j^2 = \left(\frac{j\pi}{L}\right)^4 \frac{EI}{\mu} + \frac{k}{\mu} \qquad (11)$$

$$\omega_j^2 = \left(\left(\frac{j\pi}{L}\right)^4 \frac{EI}{\mu} + \left(\frac{j\pi}{L}\right)^2 \frac{EI}{\mu G\tilde{A}} + \frac{k}{\mu}\right)\left(1 + \left(\frac{j\pi}{L}\right)^2 \frac{EI}{G\tilde{A}}\right)^{-1} \qquad (12)$$

$$\omega_j^2 = \frac{G\tilde{A}}{2\mu r^2}\left\{1 + \frac{kr^2}{G\tilde{A}} + \left(\frac{j\pi}{L}\right)^2\left(\frac{EI}{G\tilde{A}} + r^2\right) - \left(\left(1 - \frac{kr^2}{G\tilde{A}}\right)^2\right.\right.$$
$$\left.\left. -2\frac{kr^2}{G\tilde{A}}\left(\frac{j\pi}{L}\right)^2\left(\frac{EI}{G\tilde{A}} - r^2\right) + 2\left(\frac{j\pi}{L}\right)^2\left(\frac{EI}{G\tilde{A}} + r^2\right) + \left(\frac{j\pi}{L}\right)^4\left(\frac{EI}{G\tilde{A}} - r^2\right)^2\right)^{1/2}\right\} \qquad (13)$$

$$\omega_j^2 = \left(\left(\frac{j\pi}{L}\right)^4 \frac{EI}{\mu} + \frac{k}{\mu}\right)\left(1 + \left(\frac{j\pi}{L}\right)^2 r^2\right)^{-1} \qquad (14)$$

where G is the shear modulus, $\widetilde{A}$ is the reduced transversal cross-section area by the Timoshenko shear coefficient and r is the radius of gyration. In the numerical test, two standard European rails UIC60 with L=100m model the simply supported beam. Their properties are summarized in Tables 2 and 3.

Property	Beam (2 UIC60)
Coefficient of area reduction	0.41
Radius of gyration r (m)	0.063
Shear stiffness $G\widetilde{A}$ (MN)	508.92

Table 3: Additional characteristics of 2UIC60 rails

The beam is discretized into 250, 500, 1000 and 5000 elements and the elastic foundation is omitted. Numerical values of the first 250 and 500 natural frequencies are extracted using the ANSYS software with the Block Lanczos extraction method and the consistent mass matrix, and by the preconditioned conjugate gradient (PCG) Lanczos extraction method and the lumped mass matrix. Results are summarized in Figure 4. Only the results for the first two discretizations cover the full horizontal axis. It is seen that the invariant property is maintained in Figure 4a), but separation into acoustical and optical branches is not verified. The error is much lower than predicted in [10]. There is, however, a very large error in fundamental and higher frequencies when the PCG Lanczos extraction method and lumped mass matrix are used. It reaches 17% in the fundamental value and 40% in the 500-th value for a discretization into 5000 elements.

For an E-B beam on an elastic foundation, Equation (11) indicates that the square of the fundamental frequency is limited by $\omega_{min}^2 = k/\mu$. This is, however, not true in others theories. For example, in the R beam, if the condition:

$$\left(\frac{j\pi}{L}\right)^4 EIr^2 + 2\left(\frac{j\pi}{L}\right)^2 EI - kr^2 \geq 0 \tag{15}$$

is not verified, then the lowest frequency is less than the square root of k/μ and it is not achieved for $j=1$. Considering Equation (14) as a function of j, the extreme values of j and ω^2 are:

$$j_{ex} = \frac{L}{\pi r}\sqrt{-1+\sqrt{1+\frac{kr^4}{EI}}}, \quad \omega_{min}^2 = \frac{2EI}{\mu r^4}\left(\sqrt{1+\frac{kr^4}{EI}}-1\right) \tag{16}$$

j_{ex} equals the real root of Equation (15). The closest integer to j_{ex} designates the mode with the lowest frequency. In the case study considered, using a foundation stiffness of k=100MN/m^2, the order of vibration modes with respect to j is 4, 3, 5, 2, 1, 6, ..., as shown in Figure 5. Therefore, the modes appear in a different order than one would expect. ω_{min}^2=834209.61s^{-2}, k/μ=834234s^{-2} and frequencies of modes 4, 3, 5, 2, and 1 lie between these two values: ω_4^2=834208.62s^{-2} and ω_1^2=834231s^{-2}.

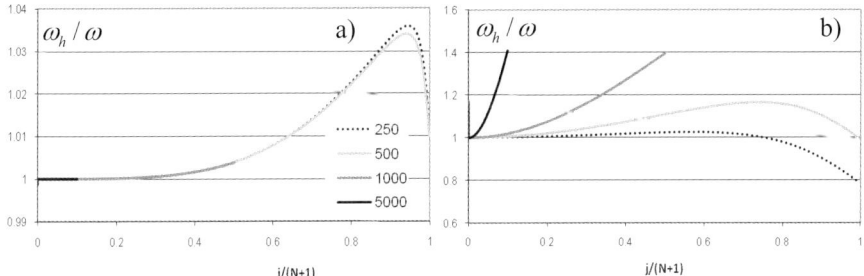

Figure 4: Error in natural frequencies: a) the Block Lanczos extraction method and the consistent mass matrix; b) the PCG Lanczos extraction method and the lumped mass matrix (the number in the legend expresses the number of elements)

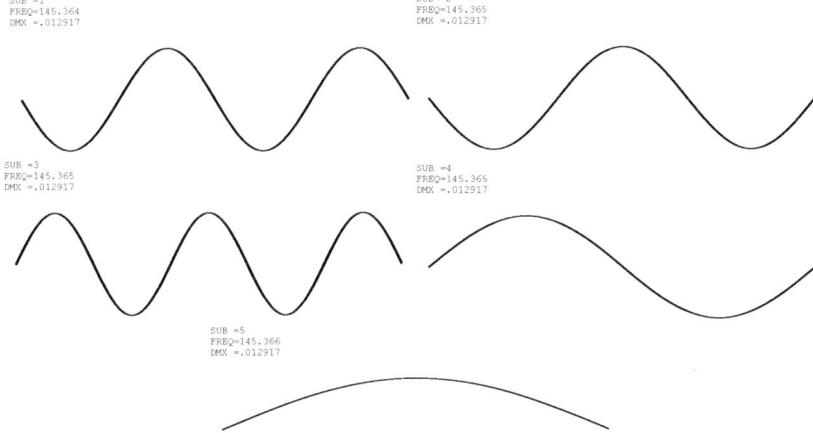

Figure 5: Modes disorder in R beam when Equation (15) is not verified

3 Governing equations

3.1 Soil parameters

When simplified models of railways tracks are under consideration, it is necessary to substitute the effect of the underlying soil. The concern is not on the wave propagation inside the soil, but merely on the deformation properties on the surface. Deformation properties of the soil at the contact with the beam structure can be described by two frequency dependent parameters: $k(\omega)$ (Winkler model) and $k_p(\omega)$ (Filonenko–Borodich, Pasternak or Hetenyi models), that are capable of handling geometric damping. Often static values of these parameters are used in calculation models, removing the frequency dependence and thus the geometric damping.

If a harmonic motion inducing only transversal displacements is assumed, then the deflection w varies inside the soil according to a function $f(z)$ and $w(x,y,z,t)=w(x,y,t)f(z)$, where $w(x,y,t)$ equal to the deflection of the beam/soil contact point. It holds $f(0)=1$ and $f(H)=0$, where H is the so-called active depth, i.e. the depth of the deformable soil. In the static case the function $f(z)$ is usually linear, but in the dynamic case, its shape is frequency dependent and must be determined from the dynamic equilibrium in the vertical direction:

$$\frac{\partial \sigma_z}{\partial z} + \frac{\partial \tau_{xz}}{\partial x} + \frac{\partial \tau_{yz}}{\partial y} = \rho \frac{\partial^2 w}{\partial t^2} \tag{17}$$

where σ and τ stand for normal and tangential stress components, respectively. The components of the deformation tensor are given by:

$$\varepsilon_z = w(x,y,t)\frac{df(z)}{dz}, \quad \gamma_{xz} = \frac{\partial w(x,y,t)}{\partial x}f(z), \quad \gamma_{yz} = \frac{\partial w(x,y,t)}{\partial y}f(z) \tag{18}$$

where ε and γ stand for the extension and engineering distortion, respectively. Therefore, the stress components can be expressed as:

$$\sigma_z = E^{oed} w(x,y,t)\frac{df(z)}{dz}, \quad \tau_{xz} = G\frac{\partial w(x,y,t)}{\partial x}f(z), \quad \tau_{yz} = G\frac{\partial w(x,y,t)}{\partial y}f(z) \tag{19}$$

where the oedometer modulus E^{oed} is given by

$$E^{oed} = E(1-v)/((1+v)(1-2v)) \tag{20}$$

Assuming harmonic vibrations and neglecting the shear stress derivatives, the differential equation for the function $f(z)$ reads as:

$$\frac{d^2}{dz^2}f(z) + \lambda^2 f(z) = 0 \tag{21}$$

where the wave number λ is given by:

$$\lambda = \sqrt{\frac{\omega}{v_p}} = \sqrt[4]{\frac{\omega^2 \rho}{E^{oed}}} \tag{22}$$

and v_p is the velocity of the pressure waves. The solution of Equation (21) is:

$$f(z) = \cos \lambda z - \cotg \lambda H \sin \lambda z \tag{23}$$

The total energy (both potential and kinetic) of the soil can be expressed as:

$$U = \frac{1}{2}\int_\Omega \left\{ \int_0^H \left[E^{oed}\left(\frac{df}{dz}\right)^2 w^2 + Gf^2\left(\left(\frac{\partial w}{\partial x}\right)^2 + \left(\frac{\partial w}{\partial y}\right)^2\right) - \omega^2 \rho f^2 w^2 \right] dz \right\} d\Omega$$

$$= \frac{1}{2}\int_\Omega \left(kw^2 + k_p\left(\left(\frac{\partial w}{\partial x}\right)^2 + \left(\frac{\partial w}{\partial y}\right)^2\right)\right) d\Omega \tag{24}$$

If a sufficiently extensive area Ω is selected, the energy beyond this region can be neglected. In this formulation, the energy attributed to the Pasternak modulus in fact corresponds to the energy of distributed rotational springs. It follows:

$$k(\omega) = \int_0^H E^{oed} \left(\left(\frac{df}{dz}\right)^2 - (\lambda f)^2 \right) dz = \frac{E^{oed}}{H} \lambda H \frac{\cos \lambda H}{\sin \lambda H} \quad (25)$$

$$k_p(\omega) = \int_0^H Gf^2 dz = \frac{1}{2} GH \left(\frac{\lambda H - \sin \lambda H \cos \lambda H}{\lambda H \sin^2 \lambda H} \right) \quad (26)$$

and the vertical stress (the reaction pressure of the soil) at the contact is given by:

$$p_s = kw - k_p \left(\frac{\partial^2 w}{\partial x^2} + \frac{\partial^2 w}{\partial y^2} \right) \quad (27)$$

If λH tends to zero, static values of the Winkler and Pasternak parameters are obtained: $k = E^{oed}/H$ and $k_p = GH/3$. It is necessary to realize that, by adopting static values, the frequency dependence, the geometrical damping and the soil mass will be disregarded. Nevertheless, this is a common approach and will be adopted here.

Regarding the soil damping, the most correct option would be to assume hysteretic damping by elastic soil constants in the form of complex numbers. To simplify matters, viscous damping is usually implemented. In summary, the effect of the viscoelastic foundation can be represented by the soil pressure, which for beam structures takes the following form:

$$p_s = kw - k_p \frac{\partial^2 w}{\partial x^2} + c \frac{\partial w}{\partial t} - c_p \frac{\partial^3 w}{\partial x^2 \partial t} \quad (28)$$

Here c and c_p are distributed damping coefficients of the Winkler and Pasternak-like foundation. The Pasternak contribution is usually omitted, but there is no review work summarizing the importance of this term in railway applications, so far.

3.2 Governing equations

Let a uniform motion of a time dependent load along a finite horizontal beam on a viscoelastic foundation be assumed. The foundation includes both the Winkler and Pasternak viscoelastic contributions, as described in the previous section. The beam is composed of b sub-domains. Within each sub-domain the beam is homogeneous, with a uniform cross section made of a linear elastic material and damping proportional to the velocity of vibration. There are $b-1$ points common to two sub-domains and two boundary nodes. Their coordinates are $x_1=0$, $x_2,..,$ $x_{b+1}=L$. In all of them, localized springs, dampers and masses can be placed. The length of each sub-domain is given by $L_n = x_{n+1} - x_n$. To keep the following deductions relatively simple, prescribed displacements and/or rotations, applied concentrated forces and/or moments and internal hinges/transversal slidings are not taken into account. The load inertia is neglected at this point. Its effect is discussed in Section 5.

In Figure 6, a finite simply supported beam composed of two sub-domains with concentrated mass at the common point is shown. $P(t)$ stands for the moving force, v is its constant velocity, x and w are the spatial coordinate and vertical deflection. The deflection is assumed positive when oriented downward and is measured from the equilibrium position, when the beam is only loaded with its own weight. At zero time ($t=0$), the load is located at the origin of the spatial coordinate x.

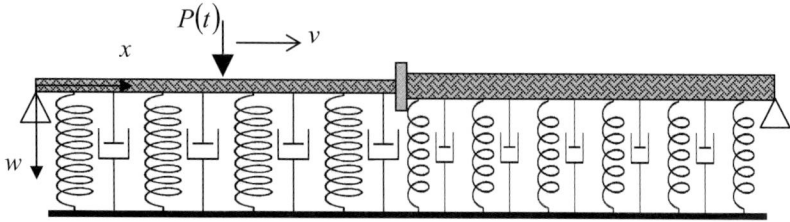

Figure 6: Simply supported beam on foundation composed of two sub-domains

The governing equations for transversal vibrations induced by the moving force can be conveniently derived by the Hamilton principle, see *e.g.* [15] for similar development. The formulation presented is for conservative forces, so the damping influence must be included into the equations derived by analogy. For the sake of simplicity, the functional dependency (x,t) is written only when a particular node coordinate x_n is used. The potential energy U of the beam sub-domains on an elastic foundation with shear contribution governed by Pasternak coefficient k_p, and U^c of the concentrated springs, read as:

$$U = \sum_{n=1}^{b} \frac{1}{2} \int_{x=x_n}^{x_n+L_n} \left(EI_n \left(\frac{\partial \psi_n}{\partial x}\right)^2 + G\tilde{A}_n \left(\frac{\partial w_n}{\partial x} - \psi_n\right)^2 + N_n \left(\frac{\partial w_n}{\partial x}\right)^2 \right. $$
$$\left. + k_n w_n^2 + k_{p,n} \left(\frac{\partial w_n}{\partial x}\right)^2 - 2 p w_n \right) dx \qquad (29)$$

$$U^c = \sum_{n=1}^{b+1} \frac{1}{2} \left(k_{L,n} w^2(x_n) + k_{R,n} \psi^2(x_n) \right) \qquad (30)$$

where $k_{L,n}$, $k_{R,n}$, are stiffness and damping coefficients of the concentrated linear and rotational springs and dampers. N_n is the axial force, assumed positive as traction, $\psi_n(x,t)$ is the bending rotation and $p(x,t)$ is a general distributed transversal external load. The kinetic energy T of the beam sub-domains and T^c of the concentrated masses are:

$$T = \sum_{n=1}^{b} \frac{1}{2} \int_{x=x_n}^{x_n+L_n} \left(\rho_n I_n \left(\frac{\partial \psi_n}{\partial t}\right)^2 + \rho_n A_n \left(\frac{\partial w_n}{\partial t}\right)^2 \right) dx \qquad (31)$$

$$T^c = \sum_{n=1}^{b+1} \frac{1}{2} \left(J_n \left(\frac{\partial \psi(x_n)}{\partial t}\right)^2 + m_n \left(\frac{\partial w(x_n)}{\partial t}\right)^2 \right) \qquad (32)$$

where m_n and J_n are the mass and mass moment of inertia of the concentrated masses. Within sub-domains it is convenient to substitute $\rho_n I_n$ by $\mu_n r_n^2$.

It holds, for the dynamic equilibrium:

$$\delta \int_{t=t_1}^{t_2} \left(U + U^c - T - T^c \right) dt = 0 \tag{33}$$

where δ designates variation. By variation of the terms in the potential energy, one obtains:

$$\delta U = \sum_{n=1}^{b} \int_{x=x_n}^{x_n+L_n} \left(EI_n \frac{\partial \psi_n}{\partial x} \delta\left(\frac{\partial \psi_n}{\partial x}\right) + G\tilde{A}_n \left(\frac{\partial w_n}{\partial x} - \psi_n\right) \delta\left(\frac{\partial w_n}{\partial x} - \psi_n\right) \right.$$
$$\left. + N_n \frac{\partial w_n}{\partial x} \delta\left(\frac{\partial w_n}{\partial x}\right) + k_n w_n \delta w_n + k_{p,n} \frac{\partial w_n}{\partial x} \delta\left(\frac{\partial w_n}{\partial x}\right) - 2 p \delta w_n \right) dx \tag{34}$$

$$\delta U^c = \sum_{n=1}^{b+1} \left(k_{L,n} w(x_n) \delta w(x_n) + k_{R,n} \psi(x_n) \delta \psi(x_n) \right) \tag{35}$$

Exchanging the order of the variation and the derivate and carrying out the integration by parts (the beam characteristics are constant within each sub-domain):

$$\delta U = \sum_{n=1}^{b} \left[EI_n \frac{\partial \psi_n}{\partial x} \delta \psi_n + \left(k_{p,n} \frac{\partial w_n}{\partial x} + N_n \frac{\partial w_n}{\partial x} + G\tilde{A}_n \left(\frac{\partial w_n}{\partial x} - \psi_n\right) \right) \delta w_n \right]_{x=x_n}^{x_n+L_n}$$
$$+ \sum_{n=1}^{b} \int_{x=x_n}^{x_n+L_n} \left(\left(-EI_n \frac{\partial^2 \psi_n}{\partial x^2} - G\tilde{A}_n \left(\frac{\partial w_n}{\partial x} - \psi_n\right) \right) \delta \psi_n + \left(-N_n \frac{\partial^2 w_n}{\partial x^2} \right. \right. \tag{36}$$
$$\left. \left. - k_{p,n} \frac{\partial^2 w_n}{\partial x^2} - G\tilde{A}_n \left(\frac{\partial^2 w_n}{\partial x^2} - \frac{\partial \psi_n}{\partial x}\right) + k_n w_n - p \right) \delta w_n \right) dx$$

Similarly, for the kinetic energy:

$$\delta T = \sum_{n=1}^{b} \int_{x=x_n}^{x_n+L_n} \left(\rho_n I_n \frac{\partial \psi_n}{\partial t} \delta\left(\frac{\partial \psi_n}{\partial t}\right) + \rho_n A_n \frac{\partial w_n}{\partial t} \delta\left(\frac{\partial w_n}{\partial t}\right) \right) dx \tag{37}$$

$$\delta T^c = \sum_{n=1}^{b+1} \left(J_n \frac{\partial \psi(x_n)}{\partial t} \delta\left(\frac{\partial \psi(x_n)}{\partial t}\right) + m_n \frac{\partial w(x_n)}{\partial t} \delta\left(\frac{\partial w(x_n)}{\partial t}\right) \right) \tag{38}$$

The time integration must be added to carry out the integration by parts:

$$\int_{t=t_1}^{t_2} \delta(T + T^c) dt = -\int_{t=t_1}^{t_2} \left(\int_{x=x_n}^{x_n+L_n} \left(\mu_n r_n^2 \frac{\partial^2 \psi_n}{\partial t^2} \delta \psi_n + \mu_n \frac{\partial^2 w_n}{\partial t^2} \delta w_n \right) dx \right) dt$$
$$- \int_{t=t_1}^{t_2} \left(\sum_{n=1}^{b+1} \left(J_n \frac{\partial^2 \psi(x_n)}{\partial t^2} \delta \psi(x_n) + m_n \frac{\partial^2 w(x_n)}{\partial t^2} \delta w(x_n) \right) \right) dt \tag{39}$$

Employing Equation (33) and grouping the corresponding terms, two coupled governing equations (so-called equations of motion) for unknown displacement and rotation fields are obtained in each sub-domain:

$$\mu_n r_n^2 \frac{\partial^2 \psi_n}{\partial t^2} - EI_n \frac{\partial^2 \psi_n}{\partial x^2} - G\tilde{A}_n \left(\frac{\partial w_n}{\partial x} - \psi_n \right) = 0 \quad n = 1,...,b \tag{40}$$

$$\mu_n \frac{\partial^2 w_n}{\partial t^2} - G\tilde{A}_n \left(\frac{\partial^2 w_n}{\partial x^2} - \frac{\partial \psi_n}{\partial x} \right) - N_n \frac{\partial^2 w_n}{\partial x^2} - k_{p,n} \frac{\partial^2 w_n}{\partial x^2} + k_n w_n$$

$$- c_{p,n} \frac{\partial^3 w_n}{\partial t \partial x^2} + c_n \frac{\partial w_n}{\partial t} = p, \quad n = 1,...,b \tag{41}$$

where the foundation damping terms were added by analogy with the stiffness terms. Continuity conditions in the common sub-domain nodes are the following:

$$w_{n-1}(x_n) = w_n(x_n), \quad n = 2,...,b \tag{42a}$$

$$\psi_{n-1}(x_n) = \psi_n(x_n), \quad n = 2,...,b \tag{42b}$$

$$\left(k_{p,n-1} \frac{\partial w_{n-1}}{\partial x} + c_{p,n-1} \frac{\partial^2 w_{n-1}}{\partial t \partial x} + N_{n-1} \frac{\partial w_{n-1}}{\partial x} + G\tilde{A}_{n-1} \left(\frac{\partial w_{n-1}}{\partial x} - \psi_{n-1} \right) \right)\bigg|_{x=x_n}$$

$$- \left(k_{p,n} \frac{\partial w_n}{\partial x} + c_{p,n} \frac{\partial^2 w_n}{\partial t \partial x} + N_n \frac{\partial w_n}{\partial x} + G\tilde{A}_n \left(\frac{\partial w_n}{\partial x} - \psi_n \right) \right)\bigg|_{x=x_n} \tag{42c}$$

$$+ k_{L,n} w(x_n) + c_{L,n} \frac{\partial w(x_n)}{\partial t} + m_n \frac{\partial^2 w(x_n)}{\partial t^2} = 0, \quad n = 2,...,b$$

$$\left(EI_{n-1} \frac{\partial \psi_{n-1}}{\partial x} \right)\bigg|_{x=x_n} - \left(EI_n \frac{\partial \psi_n}{\partial x} \right)\bigg|_{x=x_n}$$

$$+ k_{R,n} \psi(x_n) + c_{R,n} \frac{\partial \psi(x_n)}{\partial t} + J_n \frac{\partial^2 \psi(x_n)}{\partial t^2} = 0, \quad n = 2,...,b \tag{42d}$$

where $c_{L,n}$, $c_{R,n}$, are the damping coefficients of the concentrated linear and rotational dampers, added by analogy with localized springs. The boundary conditions are:

$$w_1(x_1) = 0 \quad \text{or} \tag{43a}$$

$$- \left(k_{p,1} \frac{\partial w_1}{\partial x} + c_{p,1} \frac{\partial^2 w_1}{\partial t \partial x} + N_1 \frac{\partial w_1}{\partial x} + G\tilde{A}_1 \left(\frac{\partial w_1}{\partial x} - \psi_1 \right) \right)\bigg|_{x=x_1}$$

$$+ k_{L,1} w(x_1) + c_{L,1} \frac{\partial w(x_1)}{\partial t} + m_1 \frac{\partial^2 w(x_1)}{\partial t^2} = 0 \tag{43b}$$

$$w_b(x_b) = 0 \quad \text{or} \tag{43c}$$

$$\left(k_{p,b}\frac{\partial w_b}{\partial x}+c_{p,b}\frac{\partial^2 w_b}{\partial t \partial x}+N_b\frac{\partial w_b}{\partial x}+G\tilde{A}_b\left(\frac{\partial w_b}{\partial x}-\psi_b\right)\right)\bigg|_{x=x_b}$$

$$+k_{L,b}w(x_b)+c_{L,b}\frac{\partial w(x_b)}{\partial t}+m_b\frac{\partial^2 w(x_b)}{\partial t^2}=0 \quad (43d)$$

$$\psi_1(x_1)=0 \text{ or} \quad (43e)$$

$$-\left(EI_1\frac{\partial \psi_1}{\partial x}\right)\bigg|_{x=x_1}+k_{R,1}\psi(x_1)+c_{R,1}\frac{\partial \psi(x_1)}{\partial t}+J_1\frac{\partial^2 \psi(x_1)}{\partial t^2}=0 \quad (43f)$$

$$\psi_b(x_b)=0 \text{ or} \quad (43g)$$

$$\left(EI_b\frac{\partial \psi_b}{\partial x}\right)\bigg|_{x=x_b}+k_{R,b}\psi(x_b)+c_{R,b}\frac{\partial \psi(x_b)}{\partial t}+J_b\frac{\partial^2 \psi(x_b)}{\partial t^2}=0 \quad (43h)$$

Equations (40-41) can be uncoupled, so either of the following can be used

$$EI_n\frac{\partial^4 w_n}{\partial x^4}-\mu_n r_n^2\frac{\partial^4 w_n}{\partial t^2 \partial x^2}+\left(1+\frac{\mu_n r_n^2}{G\tilde{A}_n}\frac{\partial^2}{\partial t^2}-\frac{EI_n}{G\tilde{A}_n}\frac{\partial^2}{\partial x^2}\right)$$

$$\left(\mu_n\frac{\partial^2 w_n}{\partial t^2}-N_n\frac{\partial^2 w_n}{\partial x^2}+k_n w_n-k_{p,n}\frac{\partial^2 w_n}{\partial x^2}+c_n\frac{\partial w_n}{\partial t}-c_{p,n}\frac{\partial^3 w_n}{\partial t \partial x^2}\right) \quad (44)$$

$$=p+\frac{\mu_n r_n^2}{G\tilde{A}_n}\frac{\partial^2 p}{\partial t^2}-\frac{EI_n}{G\tilde{A}_n}\frac{\partial^2 p}{\partial x^2}, \quad n=1,\ldots,b$$

$$EI_n\frac{\partial^4 \psi_n}{\partial x^4}-\mu_n r_n^2\frac{\partial^4 \psi_n}{\partial t^2 \partial x^2}+\left(1+\frac{\mu_n r_n^2}{G\tilde{A}_n}\frac{\partial^2}{\partial t^2}-\frac{EI_n}{G\tilde{A}_n}\frac{\partial^2}{\partial x^2}\right)$$

$$\left(\mu_n\frac{\partial^2 \psi_n}{\partial t^2}-N_n\frac{\partial^2 \psi_n}{\partial x^2}+k_n \psi_n-k_{p,n}\frac{\partial^2 \psi_n}{\partial x^2}+c_n\frac{\partial \psi_n}{\partial t}-c_{p,n}\frac{\partial^3 \psi_n}{\partial t \partial x^2}\right) \quad (45)$$

$$=\frac{\partial p}{\partial x}, \quad n=1,\ldots,b$$

It is seen that, except for the right hand side, Equations (44-45) are exactly the same, as expected. Similar uncoupling can be done in continuity and boundary conditions. For the complete problem definition, initial conditions must also be stated. The beam bending moment and the shear and transversal forces are:

$$M_n=-EI_n\frac{\partial \psi_n}{\partial x} \quad (46)$$

$$V_n=G\tilde{A}_n\left(\frac{\partial w_n}{\partial x}-\psi_n\right) \quad (47)$$

$$Q_n=G\tilde{A}_n\left(\frac{\partial w_n}{\partial x}-\psi_n\right)+\left(N_n+k_{p,n}\right)\frac{\partial w_n}{\partial x}+c_{p,n}\frac{\partial^2 w_n}{\partial t \partial x} \quad (48)$$

The transversal force Q_n enters the nodal equilibrium in the transversal direction (Equation (42c)) and therefore it is used in the dynamic stiffness matrix assembly, therefore its values are only important in a sub-domain end nodes.

4 Solution

4.1 Solution methods

The transient behaviour of general one-dimensional distributed dynamic systems, like vibrations on beam structures induced by moving loads, are often studied by implementing the Fourier method of variable separation and assuming the existence of free harmonic vibrations, which for the R-T formulation presented above means that:

$$w(x,t) = w(x)e^{i\omega t} \text{ and } \psi(x,t) = \psi(x)e^{i\omega t} \tag{49}$$

where $i = \sqrt{-1}$. The frequency ω of these vibrations is named as the natural frequency and it is determined from the eigenvalue problem obtained from the homogeneous governing equation (both Equations (44) or (45) can be used) by substitution of Equation (49). Natural vibration modes are independent in the sense that an excitation of one mode will never cause motion of a different mode. Then the transient response in the time domain is expressed as an infinite series of these modes, where each vibration mode (function of the spatial coordinate x) is multiplied by a generalized displacement (modal coordinate, amplitude function) that is a function of time.

This solution method is called the eigenvalues expansion, the modal expansion, the modal superposition, the modal analysis, the normal mode analysis, and so on. If the governing equations can be written in a form of a self-adjoint system, then the modal analysis is facilitated by the orthogonality relations among system eigenfunctions. As a consequence, an eigenfunction series representation of system response decouples the original equations of motion into a set of independent equations governing the unknown time-dependent generalized displacements. The convergence of the series solution is guaranteed by the completeness of system eigenfunctions. The analysis steps should follow this order strictly: find eigensolutions, establish orthogonality conditions, define normalization coefficients (modal mass) and determine modal coordinates, which can be accomplished by integral transformations (*e.g.* Laplace transformation), by implementation of the Duhamel integral or by other standard solution methods of ordinary differential equations.

Conventional modal expansion is not directly applicable to non-self-adjoint systems, because the eigenfunctions are not orthogonal. For certain non-self-adjoint distributed systems, the closed form transient response can be expressed by a series of bi-orthogonal eigenfunctions in a state space formulation. In this generalized modal analysis, the completeness of the state space eigenfunctions is often assumed

Dynamic Analysis of Beam Structures under Moving Loads   115

without justification [20]. Augmented spatial operations are introduced in [21] to account for general external, initial and boundary disturbances.

Nevertheless, even in self-adjoint systems, modal expansion is commonly governed by undamped vibration modes, because this allows their determination within the real domain and completeness of the eigenspace is guaranteed. In that case, uncoupling the equations of motion is only possible if the system possesses so-called Rayleigh's damping. It is important to realize that, in such a case, expansion over damped eigenmodes brings no improvement.

4.2 Natural frequencies and mode shapes

Let damping terms be disregarded for now. By substitution of Equation (49) into (44), it holds within each sub-domain:

$$EI_n \frac{d^4 w_n}{dx^4} + \mu_n r_n^2 \omega^2 \frac{d^2 w_n}{dx^2} + \left(1 - \frac{\mu_n r_n^2 \omega^2}{G\tilde{A}_n} - \frac{EI_n}{G\tilde{A}_n}\frac{d^2}{dx^2}\right) \\ \left(-\mu_n \omega^2 w_n - N_n \frac{d^2 w_n}{dx^2} + k_n w_n - k_{p,n}\frac{d^2 w}{dx^2}\right) = 0, \quad n = 1,\ldots,b \tag{50}$$

Equation (50) is verified by $e^{s_n x}$, thus:

$$s_n^4 + B_n s_n^2 + D_n = 0, \quad n = 1,\ldots,b \tag{51}$$

where

$$B_n = \frac{G\tilde{A}_n \mu_n r_n^2 \omega^2 + (N_n + k_{p,n})(\mu_n r_n^2 \omega^2 - G\tilde{A}_n) + EI_n(\mu_n \omega^2 - k_n)}{EI_n(G\tilde{A}_n + N_n + k_{p,n})} \tag{52}$$

$$D_n = \frac{(\mu_n \omega^2 - k_n)(\mu_n r_n^2 \omega^2 - G\tilde{A}_n)}{EI_n(G\tilde{A}_n + N_n + k_{p,n})} \tag{53}$$

s_n is named the wave number (compare with the previously used λ/L in Sections 2.2 and 2.3). In most cases, the four roots of Equation (51) are either real or imaginary and can be expressed analytically as:

$$s_{n,1} = i\sqrt{\frac{B_n}{2} + \sqrt{\left(\frac{B_n}{2}\right)^2 - D_n}}, \quad s_{n,2} = -i\sqrt{\frac{B_n}{2} + \sqrt{\left(\frac{B_n}{2}\right)^2 - D_n}} \tag{54a}$$

$$s_{n,3} = \sqrt{-\frac{B_n}{2} + \sqrt{\left(\frac{B_n}{2}\right)^2 - D_n}}, \quad s_{n,4} = -\sqrt{-\frac{B_n}{2} + \sqrt{\left(\frac{B_n}{2}\right)^2 - D_n}} \tag{54b}$$

The general form of the mode shape within each sub-domain can be written as:

$$w_n(x) = \sum_{l=1}^{4} C_{n,l} e^{s_{n,l} x} \tag{55a}$$

According to particular numerical values, mode representation in purely real domain is possible in the following forms:

$$D_n < \left(\frac{B_n}{2}\right)^2 \ \& \ \sqrt{\left(\frac{B_n}{2}\right)^2 - D_n} > \left|\frac{B_n}{2}\right| \Rightarrow \qquad (55b)$$

$$w_n(x) = C_{n,1}\cos(\bar{s}_{n,1}x) + C_{n,2}\sin(\bar{s}_{n,1}x) + C_{n,3}\cosh(s_{n,3}x) + C_{n,4}\sinh(s_{n,3}x)$$

$$D_n \leq \left(\frac{B_n}{2}\right)^2 \ \& \ \sqrt{\left(\frac{B_n}{2}\right)^2 - D_n} < \frac{B_n}{2} \Rightarrow \qquad (55c)$$

$$w_n(x) = C_{n,1}\cos(\bar{s}_{n,1}x) + C_{n,2}\sin(\bar{s}_{n,1}x) + C_{n,3}\cos(\bar{s}_{n,3}x) + C_{n,4}\sin(\bar{s}_{n,3}x)$$

$$D_n > \left(\frac{B_n}{2}\right)^2 \Rightarrow \qquad (55d)$$

$$w_n(x) = e^{\mathrm{Re}(s_{n,5})x}\left(C_{n,1}\cos(\mathrm{Im}(s_{n,5})x) + C_{n,2}\sin(\mathrm{Im}(s_{n,5})x)\right)$$

The validity of Equations (55b-d) is dependent on the value of the still unknown natural frequency. Equation (55d) defines a mode shape using alternative values to those specified in Equations (54), because in this case all roots will have a real and an imaginary part. The $s_{n,5}$ value can be determined from the original roots by methods of complex analysis. It corresponds to a root with a negative real part. All four roots are located on a circle with the radius equal to $\sqrt[4]{D_n}$. Two roots with the positive real part form a non-physical solution, therefore $C_{n,3}=C_{n,4}=0$. Such a mode resembles the deflection shape of an infinite beam and, in our case, will affect only a small part of the corresponding sub-domain. The corresponding frequency will be below the square root of k_n/μ_n, but the reasoning is different than in the case study of Section 2.3.

Since the homogeneous part of Equations (44-45) is the same, the general relation for the bending rotation can be written using the same roots and constants as:

$$\psi_n(x) = \sum_{l=1}^{4} C_{n,l}\hat{s}_{n,l} e^{s_{n,l}x} \qquad (56)$$

The relation between $s_{n,l}$ and $\hat{s}_{n,l}$ can be obtained by substitution into original coupled Equations (40) or (41):

$$\hat{s}_{n,l} = s_{n,l} + \frac{1}{G\widetilde{A}_n s_{n,l}}\left((N_n + k_{p,n})s_{n,l}^2 + \mu_n\omega^2 - k_n\right) = \frac{G\widetilde{A}_n s_{n,l}}{G\widetilde{A}_n - EI_n s_{n,l}^2 - \mu_n r_n^2 \omega^2} \qquad (57)$$

It is seen that it is closely related to the right-hand side of Equation (44), as expected. The bending moment and the transversal force are:

$$M_n = -EI_n \sum_{l=1}^{4} C_{n,l}\hat{s}_{n,l} s_{n,l} e^{s_{n,l}x} \qquad (58)$$

$$Q_n = G\tilde{A}_n \left(\frac{dw_n}{dx} - \psi_n \right) + \left(N_n + k_{p,n} \right) \frac{dw_n}{dx}$$
$$= G\tilde{A}_n \left(\sum_{l=1}^{4} C_{n,l} \left(s_{n,l} - \hat{s}_{n,l} \right) e^{s_{n,l}x} \right) + \left(N_n + k_{p,n} \right) \left(\sum_{l=1}^{4} C_{n,l} s_{n,l} e^{s_{n,l}x} \right) \quad (59)$$

which can be rewritten to:

$$Q_n = EI_n \left(\hat{s}_{n,1} \hat{s}_{n,3} \left(C_{n,1} s_{n,3} e^{s_{n,1}x} + C_{n,3} s_{n,1} e^{s_{n,3}x} \right) \right.$$
$$\left. \hat{s}_{n,2} \hat{s}_{n,4} \left(C_{n,2} s_{n,4} e^{s_{n,2}x} + C_{n,4} s_{n,2} e^{s_{n,4}x} \right) \right) \quad (60)$$

Mode shape descriptions in different sub-domains are connected by the continuity Equations (42), and by the fact that the frequency of the harmonic movement must be the same. The boundary Equations (43) must also be verified. In order to reduce the number of equations to be solved, a local dynamic stiffness matrix can be derived for a general sub-domain. The global dynamic stiffness matrix is then assembled by the direct stiffness method and localized disturbances from Equations (42-43) simply can be added. Alternatively, local disturbances can be added to modal coordinates determination, as explained in Section 4.3. Following this method, a set of homogeneous equations is obtained for unknown displacements and rotations of the sub-domains common points. The ω values ensuring the nullity of the determinant, i.e. the existence of a non-trivial solution, are named as the natural frequencies, ω_j. Substituting $\omega=\omega_j$ back into the global matrix, unknown nodal displacements and rotations can be determined.

A quite general expression of the local dynamic stiffness matrix can be found in [16]. Here, for the sake of simplicity, the following analysis is reduced to an E-B beam on a Winkler foundation without the effect of the normal force. In such a case mode shape in most cases can be written by Equation (55b) and:

$$\bar{s}_{n,1} = s_{n,3} = \frac{\lambda_n}{L_n} = \sqrt[4]{\frac{\mu_n \omega^2 - k_n}{EI_n}} \quad (61)$$

that immediately allow concluding that for $\mu_n \omega^2 - k_n < 0$ Equation (55d) must be used and

$$s_{n,5} = (-1+i)\sqrt[4]{\frac{k_n - \mu_n \omega^2}{4EI_n}} \quad (62)$$

The frequency connection between sub-domains is written as:

$$\omega = \sqrt{\frac{\lambda_1^4}{L_1^4} \frac{EI_1}{\mu_1} + \frac{k_1}{\mu_1}} = \sqrt{\frac{\lambda_2^4}{L_2^4} \frac{EI_2}{\mu_2} + \frac{k_2}{\mu_2}} = \ldots = \sqrt{\frac{\lambda_b^4}{L_b^4} \frac{EI_b}{\mu_b} + \frac{k_b}{\mu_b}} \quad (63)$$

which implies that the natural frequency values are limited by $\omega_{min}^2 = (k/\mu)_{min}$. It can also be concluded that the first vibration modes excite only the softest part of the structure, while in the stronger sub-domains deflection shape is almost negligible, described by Equation (55d). The full structure is affected by the vibration modes only after the natural frequencies overpass the square root of $(k/\mu)_{max}$. This implies that, if there is a large difference between $(k/\mu)_{min}$ and $(k/\mu)_{max}$, a very high number of modes must be used in modal superposition.

The local dynamic stiffness matrix of the n-th sub-domain can be calculated in the following way. The degrees of freedom are represented in Figure 7a). Excitation with unit amplitude and given circular frequency ω is assumed in the direction of one of the degrees of freedom, while the other degrees of freedom are kept fixed. Figure 7b) exemplifies implementation of the first degree of freedom and orientation of the corresponding terms of the stiffness matrix.

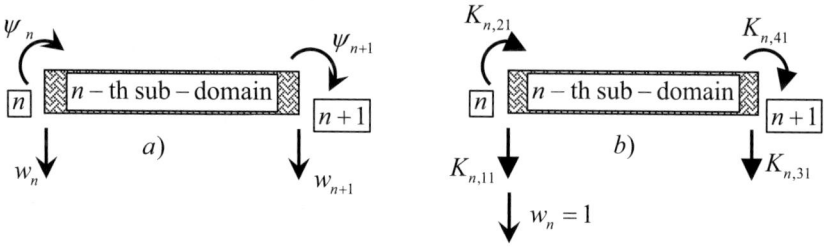

Figure 7: a) Degrees of freedom, b) construction of the local dynamic stiffness matrix of the n-th sub-domain

For such an excitation, member-end generalized harmonic forces in the steady-state regime are calculated exploiting Equations (58) and (60). The procedure is repeated for the other degrees of freedom. Obviously:

$$K_{n,1m} = -Q_{n,m}(x_n), \quad K_{n,2m} = M_{n,m}(x_n) \tag{64a}$$

$$K_{n,3m} = Q_{n,m}(x_{n+1}), \quad K_{n,4m} = -M_{n,m}(x_{n+1}) \tag{64b}$$

where m designates the order of the degree of freedom that was used for excitation. The dynamic stiffness matrix is a symmetric 4x4 matrix composed by harmonic functions with amplitudes shown below in Equation (65).

$$\begin{Bmatrix} -Q_n(x_n) \\ M_n(x_n) \\ Q_n(x_{n+1}) \\ -M_n(x_{n+1}) \end{Bmatrix} = EI_n \begin{bmatrix} F_{n,6}/L_n^3 & -F_{n,4}/L_n^2 & F_{n,5}/L_n^3 & F_{n,3}/L_n^2 \\ & F_{n,2}/L_n & -F_{n,3}/L_n^2 & F_{n,1}/L_n \\ & & F_{n,6}/L_n^3 & F_{n,4}/L_n^2 \\ symm. & & & F_{n,2}/L_n \end{bmatrix} \cdot \begin{Bmatrix} w_n \\ \psi_n \\ w_{n+1} \\ \psi_{n+1} \end{Bmatrix} \tag{65}$$

The terms in Equation (65) make use of the following Kolousek's functions [5]:

$$F_{1,n} = -\lambda_n \frac{\sinh \lambda_n - \sin \lambda_n}{\cosh \lambda_n \cos \lambda_n - 1}, \quad F_{2,n} = -\lambda_n \frac{\cosh \lambda_n \sin \lambda_n - \sinh \lambda_n \cos \lambda_n}{\cosh \lambda_n \cos \lambda_n - 1}$$

$$F_{3,n} = -\lambda_n^2 \frac{\cosh \lambda_n - \cos \lambda_n}{\cosh \lambda_n \cos \lambda_n - 1}, \quad F_{4,n} = \lambda_n^2 \frac{\sinh \lambda_n \sin \lambda_n}{\cosh \lambda_n \cos \lambda_n - 1} \quad (66)$$

$$F_{5,n} = \lambda_n^3 \frac{\sinh \lambda_n + \sin \lambda_n}{\cosh \lambda_n \cos \lambda_n - 1}, \quad F_{6,n} = -\lambda_n^3 \frac{\cosh \lambda_n \sin \lambda_n + \sinh \lambda_n \cos \lambda_n}{\cosh \lambda_n \cos \lambda_n - 1}$$

Such a local stiffness matrix corresponds to an interior sub-domain with both ends fixed. A boundary sub-domain can be simplified by methods of degrees of freedom condensation, depending on the boundary conditions. After the global matrix has been assembled, Equation (63) can be substituted and the determinant can be expressed in terms of a single unknown, ω or ω^2. Except for very simple cases, the determinant contains a quite complicated combination of trigonometric and hyperbolic functions, requiring numerical search of the roots. The determinant has many singularities coincident with all natural frequencies of each sub-domain considered separately. In order to avoid special treatment around the singularities, it is possible to solve the roots in the numerator. More details can be found in [13].

4.3 Relation to the infinite beam

The local dynamic stiffness matrix of a single sub-domain can be related to an infinite beam. Let semi-infinite sub-domains be considered first. In this case, only two degrees of freedom must be considered, as exemplified in Figure 8. From the roots specified in Equations (54), the negative-valued ones are used in the positive semi-infinite sub-domains to ensure vanishing of the displacements and rotations for x tending to positive infinity, and *vice versa*.

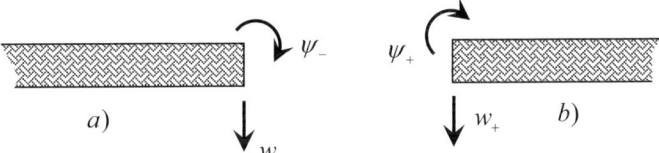

Figure 8: Degrees of freedom of semi-infinite sub-domains: a) negative, b) positive

It yields for the positive and negative semi-infinite sub-domains:

$$\mathbf{K}_+ = \begin{bmatrix} (G\tilde{A} + N + k_p) \dfrac{\hat{s}_4 s_2 - \hat{s}_2 s_4}{\hat{s}_2 - \hat{s}_4} & \dfrac{-(G\tilde{A} + N + k_p)(s_2 - s_4) + G\tilde{A}(\hat{s}_2 - \hat{s}_4)}{\hat{s}_2 - \hat{s}_4} \\ EI \dfrac{\hat{s}_2 \hat{s}_4 (s_2 - s_4)}{\hat{s}_2 - \hat{s}_4} & -EI \dfrac{\hat{s}_2 s_2 - \hat{s}_4 s_4}{\hat{s}_2 - \hat{s}_4} \end{bmatrix} \quad (67a)$$

$$= \frac{EI}{\hat{s}_2 - \hat{s}_4} \begin{bmatrix} -\hat{s}_2 \hat{s}_4 (\hat{s}_2 s_2 - \hat{s}_4 s_4) & \hat{s}_2 \hat{s}_4 (s_2 - s_4) \\ \hat{s}_2 \hat{s}_4 (s_2 - s_4) & -(\hat{s}_2 s_2 - \hat{s}_4 s_4) \end{bmatrix}$$

$$\mathbf{K}_{-} = \begin{bmatrix} \left(G\tilde{A}+N+k_{p}\right)\dfrac{\hat{s}_{1}s_{3}-\hat{s}_{3}s_{1}}{\hat{s}_{1}-\hat{s}_{3}} & \dfrac{\left(G\tilde{A}+N+k_{p}\right)\left(s_{1}-s_{3}\right)-G\tilde{A}\left(\hat{s}_{1}-\hat{s}_{3}\right)}{\hat{s}_{1}-\hat{s}_{3}} \\ -EI\dfrac{\hat{s}_{1}\hat{s}_{3}\left(s_{1}-s_{3}\right)}{\hat{s}_{1}-\hat{s}_{3}} & EI\dfrac{\hat{s}_{1}s_{1}-\hat{s}_{3}s_{3}}{\hat{s}_{1}-\hat{s}_{3}} \end{bmatrix} \quad (67b)$$

$$= \dfrac{EI}{\hat{s}_{1}-\hat{s}_{3}} \begin{bmatrix} \hat{s}_{1}\hat{s}_{3}\left(\hat{s}_{1}s_{1}-\hat{s}_{3}s_{3}\right) & -\hat{s}_{1}\hat{s}_{3}\left(s_{1}-s_{3}\right) \\ -\hat{s}_{1}\hat{s}_{3}\left(s_{1}-s_{3}\right) & \hat{s}_{1}s_{1}-\hat{s}_{3}s_{3} \end{bmatrix}$$

It can be verified simply that $\mathbf{K} = \mathbf{K}_{+} + \mathbf{K}_{-}$ is diagonal and corresponds to the action of a static load on an infinite beam. In order to account for the moving load, the governing Equations (40-41) had to be derived in the moving coordinate system [18]. Then, Equation (51) would include s_{n}^{3}, s_{n} and its coefficients would have the load velocity v implemented. By substitution of $v=0$, the original Equation (51) would be obtained. The full matrix $\mathbf{K}$ would be non-symmetric and $v=v_{crit}$ (the critical velocity) would ensure the nullity of its determinant. A single "mode" shape would define the deflection shape similarly to Equations (55d) and (55c) for subcritical velocities and critical velocities, respectively. Full determination of the quasi-stationary deflection shape would still require the continuity equations at the point of load application. In this case, damping inclusion would cause no problems. The same results as in [6] would be obtained, placing the concept of the dynamic stiffness matrix as a general principle for finite, semi-infinite and infinite beams.

Going back to the stiffness matrixes of the semi-infinite sub-domains (Equations (67)), one can verify that they correspond to the limit of the corresponding terms of the general fixed-fixed or fixed-free sub-domain, when their length tends to infinity. Indeed, for an E-B beam on a Winkler foundation:

$$\mathbf{K}_{+} = EI \begin{bmatrix} -\tilde{\lambda}^{3}(1-i) & i\tilde{\lambda}^{2} \\ i\tilde{\lambda}^{2} & \tilde{\lambda}(1+i) \end{bmatrix}, \quad \mathbf{K}_{-} = EI \begin{bmatrix} -\tilde{\lambda}^{3}(1-i) & -i\tilde{\lambda}^{2} \\ -i\tilde{\lambda}^{2} & \tilde{\lambda}(1+i) \end{bmatrix} \quad (68)$$

where $s_{1} = i\tilde{\lambda}$, $s_{3} = \tilde{\lambda}$, $s_{2} = -i\tilde{\lambda}$, $s_{4} = -\tilde{\lambda}$. Then, $\mathbf{K}_{+}$ corresponds to the limit of the fixed-free (Equation (69)) and fixed-fixed sub-domain (Equation (65-66)) terms, and analogously for $\mathbf{K}_{-}$.

$$K_{11}^{fr} = -\frac{EI\lambda^{3}}{L^{3}} \frac{\cosh\lambda \sin\lambda + \sinh\lambda \cos\lambda}{\cosh\lambda \cos\lambda + 1} \quad (69a)$$

$$K_{12}^{fr} = -\frac{EI\lambda^{2}}{L^{2}} \frac{\sinh\lambda \sin\lambda}{\cosh\lambda \cos\lambda + 1} \quad (69b)$$

$$K_{22}^{fr} = -\frac{EI\lambda}{L} \frac{\cosh\lambda \sin\lambda - \sinh\lambda \cos\lambda}{\cosh\lambda \cos\lambda + 1} \quad (69c)$$

One can see that the stiffness matrix of an E-B, undamped, semi-infinite sub-domain involves complex numbers, which complicates its implementation. These stiffness matrix terms could be interpreted as localized wave-number dependent boundary

Dynamic Analysis of Beam Structures under Moving Loads   121

springs and dampers, but the numerical determination of natural frequencies in such cases is not simple. Semi-infinite elements can work as an absorbing boundary.

4.4 Forced vibrations - generalized displacement

The general solution of Equation (44) is assumed in the form of:

$$w(x,t) = \sum_{j=1}^{\infty} q_j(t) w_j(x) \tag{70}$$

where q_j designates the j-th generalized displacement, but Equation (44) is not convenient for q_j determination, because it has fourth order time derivatives. Generally, the state-space form of the governing equations is more adequate at this point and will be explained later on. Nevertheless, the fundamental issues are more easily explained on a simple case.

Let an E-B beam on a Winkler viscoelastic foundation, with no effect of the normal force, be considered first. Let the localized disturbances (springs, dampers and masses) be only linear. Then, the governing Equations (40-41) or (44) can be simplified as:

$$EI(x)\frac{\partial^4 w}{\partial x^4} + \mu(x)\frac{\partial^2 w}{\partial t^2} + c(x)\frac{\partial w}{\partial t} + k(x)w$$
$$+ \sum_{n=1}^{b+1} \delta(x-x_n)\left(m_n \frac{\partial^2 w}{\partial t^2} + c_{L,n}\frac{\partial w}{\partial t} + k_{L,n}w\right) = p \tag{71}$$

where δ is the Dirac delta function. Here, the piece-wise constant dependence of the beam characteristics is expressed as a function of x, and the localized effects, formerly included in the continuity and boundary Equations (42-43), constitute an integral part of the governing Equation (71). There are two possibilities for determining generalized displacements. They are distinguished by the fact, whether the continuity and boundary conditions (42-43) are included in the natural modes determination or not. In the former case, the undamped vibration modes verify:

$$\left(EI(x)\frac{d^4}{dx^4} + k(x) + \sum_{n=1}^{b+1} \delta(x-x_n) k_{L,n}\right) w_j$$
$$-\omega_j^2 \left(\mu(x) + \sum_{n=1}^{b+1} \delta(x-x_n) m_n\right) w_j = 0, \quad \forall j \tag{72}$$

which means that if

$$\int_{x=0}^{L} w_l \frac{d^4 w_j}{dx^4} dx = \left[w_l \frac{d^3 w_j}{dx^3}\right]_0^L - \left[\frac{dw_l}{dx}\frac{d^2 w_j}{dx^2}\right]_0^L + \left[\frac{d^2 w_l}{dx^2}\frac{dw_j}{dx}\right]_0^L$$
$$- \left[\frac{d^3 w_l}{dx^3} w_j\right]_0^L + \int_{x=0}^{L} \frac{d^4 w_l}{dx^4} w_j dx = \int_{x=0}^{L} \frac{d^4 w_l}{dx^4} w_j dx \tag{73}$$

the system is self-adjoint and the following two orthogonality conditions are fulfilled:

$$\int_0^L \mu(x) w_j w_l dx + \sum_{n=1}^{b+1} m_n w_j(x_n) w_l(x_n) = 0, \quad \forall j \neq l \quad (74a)$$

$$M_j = \int_0^L \mu(x) w_j^2 dx + \sum_{n=1}^{b+1} m_n w_j^2(x_n), \quad \forall j \quad (74b)$$

$$\int_0^L EI(x) \frac{d^4 w_j}{dx^4} w_l dx + \int_0^L k(x) w_j w_l dx + \sum_{n=1}^{b+1} k_{L,n} w_j(x_n) w_l(x_n) = 0, \quad \forall j \neq l \quad (74c)$$

$$\int_0^L EI(x) \frac{d^4 w_j}{dx^4} w_j dx + \int_0^L k(x) w_j^2 dx + \sum_{n=1}^{b+1} k_{L,n} w_j^2(x_n) = S_j, \quad \forall j \quad (74d)$$

where M_j is called the modal mass, but S_j usually does not have a particular designation. The terms in brackets in Equation (73) are zero, because homogeneous boundary conditions are assumed. From the positivity of the potential and kinetic energy, it follows that natural frequencies are real. By substitution of Equation (70) into Equation (71), multiplication by a mode shape function, integration over the full length structure and exploiting Equations (74):

$$S_j q_j + M_j \frac{d^2 q_j}{dt^2} + \int_{x=0}^L \left(c(x) + \sum_{n=1}^{b+1} \delta(x - x_n) c_{L,n} \right) \sum_{l=1}^{\infty} \frac{dq_l}{dt} w_l w_j dx = \tilde{q}_j \quad (75)$$

where

$$\tilde{q}_j = \int_{x=0}^L p(x,t) w_j(x) dx \quad (76)$$

Carrying out the same integration on Equation (72), one obtains:

$$S_j - \omega_j^2 M_j = 0 \quad (77)$$

Joining Equations (75) and (77):

$$M_j \frac{d^2 q_j}{dt^2} + M_j \omega_j^2 q_j + \int_{x=0}^L \left(c(x) + \sum_{n=1}^{b+1} \delta(x - x_n) c_{L,n} \right) \sum_{l=1}^{\infty} \frac{dq_l}{dt} w_l w_j dx = \tilde{q}_j \quad (78)$$

One can see that the damping terms are not uncoupled as in the other ones. Complete uncoupling can only be achieved if the damping is of the Rayleigh type, i.e. proportional to a combination of stiffness and mass terms. α is usually used for the mass coefficient, while β stands for the stiffness coefficient. This is quite a common case, as one can imagine that localized dampers and the continuous damping distribution follow the same variation as the mass terms, thus $\beta=0$. In the general case:

Dynamic Analysis of Beam Structures under Moving Loads

$$M_j \frac{d^2 q_j}{dt^2} + M_j \omega_j^2 q_j + (\alpha M_j + \beta S_j) \frac{dq_j}{dt}$$
$$= M_j \frac{d^2 q_j}{dt^2} + M_j \omega_j^2 q_j + M_j (\alpha + \beta \omega_j^2) \frac{dq_j}{dt} = \tilde{q}_j \quad (79)$$

It is further assumed that $M_j(\alpha + \beta \omega_j^2) = \xi c_{cr} = 2\xi M_j \omega_j$, where ξ is called the damping ratio and $c_{cr} = 2 M_j \omega_j$ the critical damping. In summary:

$$\frac{d^2 q_j}{dt^2} + 2\xi \omega_j \frac{dq_j}{dt} + \omega_j^2 q_j = \frac{1}{M_j} \tilde{q}_j = \tilde{Q}_j \quad (80)$$

The initial conditions also must be expanded in series:

$$w(x,0) = \tilde{w}(x) = \sum_{j=1}^{\infty} q_j(0) w_j, \quad \frac{dw}{dt}(x,0) = \tilde{\tilde{w}}(x) = \sum_{j=1}^{\infty} \frac{dq_j}{dt}(0) w_j \quad (81)$$

which is not a simple task due to the orthogonality conditions. Then, Equations (80-81) can be solved by Laplace transformation (* designates the image):

$$\left(-\frac{dq_j}{dt}(0) - p q_j(0) + p^2 q_j^*(p) \right) +$$
$$2\xi \omega_j \left(-q_j(0) + p q_j^*(p) \right) + \omega_j^2 q_j^*(p) = \tilde{Q}_j^*(p) \quad (82a)$$

$$\Rightarrow q_j^*(p) = \frac{\frac{dq_j}{dt}(0) + p q_j(0) + 2\xi \omega_j q_j(0) + \tilde{Q}_j^*(p)}{p^2 + 2\xi \omega_j p + \omega_j^2} \quad (82b)$$

$$p^2 + 2\xi \omega_j p + \omega_j^2 = (p + \xi \omega_j)^2 + \omega_j^2 (1 - \xi^2) = (p + \xi \omega_j)^2 + \omega_{d,j}^2 \quad (82c)$$

$$q_j^*(p) = \frac{q_j(0)(p + \xi \omega_j)}{(p + \xi \omega_j)^2 + \omega_{d,j}^2} + \left(\xi \omega_j q_j(0) + \frac{dq_j}{dt}(0) \right) \frac{1}{\omega_{d,j}} \frac{\omega_{d,j}}{(p + \xi \omega_j)^2 + \omega_{d,j}^2}$$
$$+ \tilde{Q}_j^*(p) \frac{1}{\omega_{d,j}} \frac{\omega_{d,j}}{(p + \xi \omega_j)^2 + \omega_{d,j}^2} \quad (82d)$$

$$q_j(t) = q_j(0) e^{-\xi \omega_j t} \cos(\omega_{d,j} t) + \frac{q_j(0) \xi \omega_j + \dot{q}_j(0)}{\omega_{d,j}} e^{-\xi \omega_j t} \sin(\omega_{d,j} t) +$$
$$\frac{1}{M_j \omega_{d,j}} \int_{\tilde{t}=0}^{t} \tilde{q}_j(\tilde{t}) e^{-\xi \omega_j (t-\tilde{t})} \sin(\omega_{d,j}(t-\tilde{t})) d\tilde{t} \quad (82e)$$

where Equation (82c) stands for an alternative formulation of the denominator of Equation (82b), and $\omega_{d,j} = \omega_j \sqrt{1-\xi^2}$ is introduced as a damped frequency. However, it has nothing to do with complex frequencies of damped modes. Equation

(82e) is not restricted to subcritical damping ($\xi<1$). For a single constant moving force P (exploiting Equation (76) and the fact that $p(x,t) = P\delta(x-vt)$):

$$\tilde{q}_j(t) = Pw_j(vt)(H(L-vt) - H(-vt)) \tag{83}$$

where the Heaviside function H in the brackets states that this term is only valid when the force is on the beam. For homogeneous initial conditions, Equation (82e) is the same as Duhamel's integral, as expected. Sometimes, it is important to separate the final expression into homogenous and particular solutions of the ordinary differential equation (80). A simple exercise can be done for a simply supported uniform beam with no elastic foundation and homogeneous initial conditions. Then:

$$w_j = \sin\left(\frac{j\pi}{L}x\right) \Rightarrow \tilde{q}_j = P\sin\left(\frac{j\pi}{L}vt\right), \quad M_j = \frac{\mu L}{2} \tag{84a}$$

$$q_{P,j}(t) = q_{AP,j}\sin\left(\frac{j\pi}{L}vt\right) + q_{BP,j}\cos\left(\frac{j\pi}{L}vt\right) \tag{84b}$$

$$q_{AP,j} = \frac{P}{M_j}\frac{1}{\omega_j^2}\frac{1-\Omega_j^2}{(2\xi\Omega_j)^2 + (1-\Omega_j^2)^2} \tag{84c}$$

$$q_{BP,j} = \frac{P}{M_j}\frac{1}{\omega_j^2}\frac{-2\xi\Omega_j}{(2\xi\Omega_j)^2 + (1-\Omega_j^2)^2} \tag{84d}$$

$$q_{H,j}(t) = e^{-\xi\omega_j t}\left(q_{AH,j}\sin(\omega_{d,j}t) + q_{BH,j}\cos(\omega_{d,j}t)\right) \tag{84e}$$

$$q_{AH,j} = \frac{P}{M_j}\frac{\Omega_j}{\omega_j\omega_{d,j}}\frac{\Omega_j^2 + 2\xi^2 - 1}{(2\xi\Omega_j)^2 + (1-\Omega_j^2)^2} \tag{84f}$$

$$q_{BH,j} = \frac{P}{M_j}\frac{1}{\omega_j^2}\frac{2\xi\Omega_j}{(2\xi\Omega_j)^2 + (1-\Omega_j^2)^2} \tag{84g}$$

where $q_{P,j}$ and $q_{H,j}$ define the form of the particular and homogenous solutions, and

$$\Omega_j = \frac{\omega_{f,j}}{\omega_j} = \frac{j\pi v}{L\omega_j} \tag{85}$$

where $\omega_{f,j}$ stands for the forced frequency. It is simple to verify that $q_{P,j}(t) = \tilde{D}(t, \tilde{t}=t)$ and $q_{H,j}(t) = -\tilde{D}(t, \tilde{t}=0)$, where $\tilde{D}(t,\tilde{t})$ is the primitive function of Duhamel's integral. This statement is valid generally.

If the restriction to linear disturbances is removed, the modal mass is defined by:

$$M_j = \int_0^L \mu(x)w_j^2(x)dx + \sum_{n=1}^{b+1}\left(m_n w_j^2(x_n) + J_n\frac{dw_j^2(x_n)}{dx}\right) \tag{86}$$

Regarding the other option mentioned at the beginning of this section, *i.e.* when localized terms are not included in vibration modes determination, it follows:

$$\left(EI(x)\frac{d^4}{dx^4}+k(x)\right)w_j - \omega_j^2 \mu(x)w_j = 0, \quad \forall j \tag{87}$$

meaning that the natural modes and frequencies evaluation is simpler, but final equations will be coupled with no possibility of analytical solution:

$$\mathbf{M}\cdot\ddot{\mathbf{q}}(t)+\mathbf{C}\cdot\dot{\mathbf{q}}(t)+\mathbf{K}\cdot\mathbf{q}(t)=\tilde{\mathbf{q}}(t) \tag{88}$$

where

$$M_{lj} = \delta_{lj}M_j + \sum_{n=1}^{b+1} m_n w_l(x_n)w_j(x_n) + J_n \frac{dw_l}{dx}(x_n)\frac{dw_j}{dx}(x_n) \tag{89a}$$

$$C_{lj} = \int_{x=0}^{L} c(x)w_l w_j dx + \sum_{n=1}^{b+1} c_{L,n} w_l(x_n) w_j(x_n) + c_{LR,n} w_l(x_n)\frac{dw_j}{dx}(x_n)$$
$$+ c_{RL,n}\frac{dw_l}{dx}(x_n)w_j(x_n) + c_{R,n}\frac{dw_l}{dx}(x_n)\frac{dw_j}{dx}(x_n) \tag{89b}$$

$$K_{lj} = \delta_{lj}M_j\omega_j^2 + \sum_{n=1}^{b+1} k_{L,n} w_l(x_n) w_j(x_n) + k_{LR,n} w_l(x_n)\frac{dw_j}{dx}(x_n)$$
$$+ k_{RL,n}\frac{dw_l}{dx}(x_n)w_j(x_n) + k_{R,n}\frac{dw_l}{dx}(x_n)\frac{dw_j}{dx}(x_n) \tag{89c}$$

This formulation includes springs and dampers with the additional specifications *LR* and *RL*, *i.e.* a spring (damper) in which the reaction to applied rotation (rotational velocity) is a force and one in which the reaction to applied displacement (velocity) is a moment. In other words, any additional member characterized by a full, possibly non-symmetric 2x2 matrix, can be added this way. Similar extensions could be done for concentrated masses.

Extending the analysis to the originally assumed T-R beam, it is convenient to write Equations (40-41) in the first order state-space form (similarly as in [22]):

$$\begin{bmatrix} G\tilde{A} - EI\frac{\partial^2}{\partial x^2} & -G\tilde{A}\frac{\partial}{\partial x} & 0 & 0 \\ G\tilde{A}\frac{\partial}{\partial x} & k-(N+k_p)\frac{\partial^2}{\partial x^2}-G\tilde{A}\frac{\partial^2}{\partial x^2} & 0 & 0 \\ 0 & 0 & -\mu r^2 & 0 \\ 0 & 0 & 0 & -\mu \end{bmatrix} \cdot \begin{Bmatrix} \psi(x,t) \\ w(x,t) \\ \vartheta(x,t) \\ \tau(x,t) \end{Bmatrix} +$$
$$+\begin{bmatrix} 0 & 0 & \mu r^2 & 0 \\ 0 & c & 0 & \mu \\ \mu r^2 & 0 & 0 & 0 \\ 0 & \mu & 0 & 0 \end{bmatrix} \cdot \frac{\partial}{\partial t}\begin{Bmatrix} \psi(x,t) \\ w(x,t) \\ \vartheta(x,t) \\ \tau(x,t) \end{Bmatrix} = \begin{Bmatrix} 0 \\ p(x,t) \\ 0 \\ 0 \end{Bmatrix} \tag{90}$$

where:

$$\vartheta(x,t) = \frac{\partial}{\partial t}\psi(x,t), \quad \tau(x,t) = \frac{\partial}{\partial t}w(x,t) \qquad (91)$$

Several simplifications are introduced here. The beam characteristics x-dependence is not written. The localized disturbances could be accounted for by augmented operators, similarly as in [21], but are disregarded for simplicity. The Pasternak damping term is neglected, in order to ensure that the operator matrices in Equation (90) are self-adjoint. Equation (90) allows for recasting the eigenvalues problem into linear equations, but in the following, mostly generalized coordinates solution will be analysed. Two orthogonality relations for damped vibration modes are derived as:

$$\int_{x=0}^{L}\left(\psi_k\left(G\tilde{A}\psi_j - EI\frac{d^2\psi_j}{dx^2} - G\tilde{A}\frac{dw_j}{dx}\right)\right.$$
$$\left. + w_l\left(G\tilde{A}\frac{d\psi_j}{dx} + kw_j - (N+k_p)\frac{d^2w_j}{dx^2} - G\tilde{A}\frac{d^2w_j}{dx^2}\right)\right. \qquad (92a)$$
$$\left. + \omega_l\omega_j\mu r^2\psi_l\psi_j + \omega_l\omega_j\mu w_k w_j\right)dx = \delta_{lj}S_j$$

$$\int_{x=0}^{L}\left(i\mu r^2(\omega_l+\omega_j)\psi_l\psi_j + i\mu(\omega_l+\omega_j)w_k w_j + cw_l w_j\right)dx = \delta_{lj}T_j \qquad (92b)$$

connected by

$$S_j + i\omega_j T_j = 0 \qquad (93)$$

which allows introduction of the modal mass in a generalized form as:

$$S_j = -i\omega_j T_j = 2\omega_j^2 M_j, \quad M_j = -iT_j/(2\omega_j) \qquad (94a)$$

$$M_j = \int_{x=0}^{L}\left(\mu w_j^2 + \mu r^2\psi_j^2(x) + \frac{c}{2i\omega_j}w_j^2\right)dx \qquad (94b)$$

Expanding the unknown functions in modal series:

$$\begin{Bmatrix}\psi(x,t)\\w(x,t)\\\vartheta(x,t)\\\tau(x,t)\end{Bmatrix} = \sum_{j=1}^{\infty}q_j(t)\begin{Bmatrix}\psi_j(x)\\w_j(x)\\\vartheta_j(x)\\\tau_j(x)\end{Bmatrix} \qquad (95)$$

and exploiting Equation (90), one obtains a first order modal equation of motion:

$$S_j q_j + T_j \frac{dq_j}{dt} = \tilde{q}_j \qquad (96)$$

where $\tilde{q}_j$ is given by Equation (83) for a single moving force P. Equation (96) can be normalized into:

$$\frac{d}{dt}q_j - i\omega_j q_j = \frac{-iP}{2\omega_j M_j} w_j(vt) \tag{97}$$

This formulation can be used to show the difference between expansions into damped or undamped modes. But probably, the only analytical solution of complex frequencies that is possible for simply supported uniform E-B beam without any additional effect, is:

$$\omega_{j,l} = i\frac{c}{2\mu} \pm \sqrt{\left(\frac{j\pi}{L}\right)^4 \frac{EI}{\mu} - \left(\frac{c}{2\mu}\right)^2}, \; l=1,2 \tag{98}$$

The solution of Equation (97) can be written as:

$$q_{j,l} = \frac{1}{2i\omega_{j,l} M_{j,l}} \int_{\tilde{t}=0}^{t} \tilde{q}_j(\tilde{t}) e^{i\omega_{j,l}(t-\tilde{t})} d\tilde{t}, \; l=1,2 \tag{99}$$

Exploiting Equations (94b) and (98):

$$2i\omega_{j,1} M_{j,1} = i\mu L \sqrt{\left(\frac{j\pi}{L}\right)^4 \frac{EI}{\mu} - \left(\frac{c}{2\mu}\right)^2} = 2iM_j \omega_{d,j} \tag{100a}$$

$$2i\omega_{j,2} M_{j,2} = -i\mu L \sqrt{\left(\frac{j\pi}{L}\right)^4 \frac{EI}{\mu} - \left(\frac{c}{2\mu}\right)^2} = -2iM_j \omega_{d,j} \tag{100b}$$

where M_j and $\omega_{d,j}$ correspond to the values derived for expansion governed by undamped modes. One can join the solutions into:

$$q_j = q_{j,1} + q_{j,2} = \frac{1}{M_j \omega_{d,j}} \int_{\tilde{t}=0}^{t} \tilde{q}_j(\tilde{t}) e^{-\xi\omega_j(t-\tilde{t})} \frac{e^{\omega_{d,j}(t-\tilde{t})} - e^{-\omega_{d,j}(t-\tilde{t})}}{2i} d\tilde{t}$$
$$= \frac{1}{M_j \omega_{d,j}} \int_{\tilde{t}=0}^{t} \tilde{q}_j(\tilde{t}) e^{-\xi\omega_j(t-\tilde{t})} \sin(\omega_{d,j}(t-\tilde{t})) d\tilde{t} \tag{101}$$

and obtain the previous solution from Equation (82e), as expected. The case presented exhibits only real-valued vibration modes, even for complex frequencies. More complicated cases can be found in [21]. The general case is quite difficult to solve, but when discretization is implemented, the complex frequencies and complex vibration modes can be solved within reasonable difficulty [23].

5   Internal forces, inertial and other effects

Determination of beam bending moment and shear force can be accomplished using Equations (46) and (47). However, when a concentrated moving force is considered, there is a discontinuity in the shear force, *i.e.* a discontinuity in the derivative of the bending moment. This dictates that for reasonable accuracy, many modes, generally more than for the deflection representation, are necessary. An alternative rearrangement derived directly from the governing equations is presented in [24].

Until now, the effect of load inertia was disregarded. If it is employed, the right hand side of the governing Equation (41) must be altered into

$$p - \frac{p}{g}\frac{\partial^2 w}{\partial t^2} \qquad (102)$$

where g is the acceleration of gravity. For a single moving force P, it means that:

$$\left(P - \frac{P}{g}\frac{\partial^2 w}{\partial t^2}\right)\delta(x - vt) \qquad (103)$$

A classical solution to this problem is presented in [25]. An undamped E-B beam with no additional effects is considered. The modal expansion is accomplished over undamped vibration modes determined without the moving mass contribution. The moving mass effect is then moved to the left hand side of the governing equation. This leads to a system of uncoupled equations for the modal coordinates similar to Equations (89). Here it holds:

$$\mathbf{M} \cdot \ddot{\mathbf{q}}(t) + \mathbf{K} \cdot \mathbf{q}(t) = \widetilde{\mathbf{q}}(t) \qquad (104)$$

where

$$M_{lj} = \delta_{lj} M_j + P w_l(vt) w_j(vt) \sqrt{M_l M_j} / g \qquad (105a)$$

$$K_{lj} = \delta_{lj} M_j \omega_j^2 \qquad (105b)$$

Additional effects can be added in the same way as in Equations (89). Extension of this solution to beams on an elastic foundation does not provide correct results, because the terms representing the foundation mass were neglected (see Section 3.1). An approximation can be done by adding a representative soils mass to the beam mass per unit length. Alternative approaches for solving the moving mass problem employ expansion of Equation (103) in series and neglecting insignificant terms.

As a closing remark, it is necessary to highlight that, when the beam deforms, the conventional elastic foundation can sustain both compression and tension. This model was probably motivated more by the desire for mathematical simplicity than by physical reality. The steady state deformation of an infinite beam on a tensionless elastic foundation under a moving load was first studied in [26].

6 Conclusions

In this chapter several aspects related to dynamic analysis of beam structures under moving loads were summarized. Emphasis has been placed on the modal expansion method and a few examples were related to high-speed railways. The concept of the dynamic stiffness matrix was posted as a general principle for finite, semi-infinite and infinite beams. The importance of governing equations reformulations, ensuring the orthogonality of natural vibration modes, has been discussed in detail. Some developments are new; review and summary of published works is far from complete due to a considerable amount of studies that have been published on this subject.

Acknowledgements

The work was partially supported by the project grant PTDC/EME-PME/01419/2008: "SMARTRACK - SysteM dynamics Assessment of Railway TRACKs: a vehicle-infrastructure integrated approach" of Fundação para a Ciência e a Tecnologia of the Portuguese Ministry of Science and Technology.

References

[1] A.N. Krylov, (A.N. Kriloff), "Über die erzwungenen Schwingungen von gleichformigen elastichen Stäben", Mathematische Annalen, 61, 211-234, 1905. (in German)
[2] S.P. Timoshenko, "Forced vibration of prismatic bars", Izvestiya Kievskogo politekhnicheskogo instituta, 1908. (in Russian); Erzwungene Schwingungen prismatischer Stabe, Zeitschrift für Mathematik und Physik, 59(2), 163–203, 1911. (in German)
[3] C.E. Inglis, "A Mathematical Treatise on Vibration in Railway Bridges", The Cambridge University Press, Cambridge, 1934.
[4] A.N. Lowan, "On transverse oscillations of beams under the action of moving variable loads", Philosophical Magazine, Series 7, 19(127), 708–715, 1935.
[5] V. Koloušek, "Dynamics of Civil Engineering Structures—Part I: General problems", 2nd ed.—"Part II: Continuous Beams and Frame Systems", 2nd ed—"Part III: Selected Topics", SNTL, Prague, 1967, 1956, 1961. (in Czech); "Dynamics in engineering structures", Academia, Prague, Butterworth, London, 1973.
[6] L. Frýba, "Vibration of Solids and Structures under Moving Loads", Research Institute of Transport, Prague (1972), 3rd edition, Thomas Telford, London, 1999.
[7] S.P. Timoshenko, "Statical and dynamical stresses in rails", Proceedings of the 2nd International Congress for Applied Mechanics, Zürich (Switzerland), 407-418, 12-17 September 1926.
[8] J.T. Kenney Jr., "Steady-state vibrations of beam on elastic foundation for moving load", ASME Journal of Applied Mechanics, 21, 359-364, 1954.
[9] L. Frýba, "Infinite Beam on an Elastic Foundation Subjected to a Moving Load", Aplikace Matematiky, 2(2), 105-132, 1957. (in Czech)
[10] J.A. Cottrell, A. Reali, Y. Bazilevs, T.J.R. Hughes, "Isogeometric analysis of structural vibrations", Computer Methods in Applied Mechanics and Engineering, 195, 5257-5296, 2006.
[11] Release R2007a Documentation for MATLAB, The MathWorks, Inc., 2007.
[12] Release 11 Documentation for MAPLE, Maplesoft a division of Waterloo Maple, Inc., 2007.

[13] Z. Dimitrovová, "A general procedure for the dynamic analysis of finite and infinite beams on piece-wise homogeneous foundation under moving loads", Journal of Sound and Vibration, 329, 2635–2653, 2010.

[14] J. Máca, "Dynamic analysis of plates on elastic foundation", in H. Grundmann, G.I. Schueller, (Editors), "Proceedings of the 4th International Conference on Structural Dynamics, EURODYN2002", Munich, Germany, 887-892, 2-5 September 2002.

[15] G. Karami, P. Malekzadeh, S.A. Shahpari, "A DQEM for vibration of shear deformable nonuniform beams with general boundary conditions", Engineering Structures, 25, 1169–1178, 2003.

[16] Y-H. Chen, Z-M. Shiu, "Resonant curves of an elevated railway to harmonic moving loads", International Journal of Structural Stability and Dynamics, 4(2), 237-257, 2004.

[17] Y-H. Chen, J-T. Sheu, "Beam length and dynamic stiffness", Computer Methods in Applied Mechanics and Engineering, 129, 311-318, 1996.

[18] Y-H. Chen, Y-H. Huang, "Dynamic stiffness of infinite Timoshenko beam on viscoelastic foundation in moving co-ordinate", International Journal For Numerical Methods In Engineering, 48, 1-18, 2000.

[19] Z. Dimitrovová, J.N. Varandas, "Critical velocity of a load moving on a beam with a sudden change of foundation stiffness: Applications to high-speed trains", Computers & Structures, 87, 1224–1232, 2009.

[20] L. Meirovitch, "Computational methods in structural dynamics", Sijthoff & Noordhoff, 1980.

[21] B. Yang, X. Wu, "Transient response of one-dimensional distributed systems: a closed form eigenfunction expansion realization", Journal of Sound and Vibration, 208(5), 763-776, 1997.

[22] S. Krenk, "Complex modes and frequencies in damped structural vibrations", Journal of Sound and Vibration, 270, 981-996, 2004.

[23] J.C.O. Nielsen, J. Oscarsson, "Simulation of dynamic train–track interaction with state-dependent track properties", Journal of Sound and Vibration, 275, 515-532, 2004.

[24] M. Zehsaz, M.H. Sadeghi, A. Ziaei Asl, "Dynamic response of railway under a moving load", Journal of Applied Sciences, 9(8), 1474-1481, 2009.

[25] J.E. Akin, M. Mofid, "Numerical solution for response of beams with moving mass", ASCE Journal of Structural Engineering, 115, 120–131, 1989.

[26] J. Choros, G.G. Adams, "A steadily moving load on an elastic beam resting on a tensionless Winkler foundation", ASME Journal of Applied Mechanics, 46, 175-180, 1979.

Chapter 5

Virtual Experiments and a Statistically Equivalent Representative Volume Element for Macroscopic Constitutive Laws

M. Šejnoha, J. Vorel, R. Valenta and J. Zeman
Department of Mechanics
Faculty of Civil Engineering
Czech Technical University in Prague, Czech Republic

Abstract

Full scale analysis of complex civil engineering structures often relies on the assumption of material homogeneity, which in turn opens the way to inexpensive phenomenological modeling. However, determining the necessary material parameters from laboratory measurements is generally expensive, tedious and time consuming eventually slowing down the actual design process. Therefore, the present contribution proposes an alternative approach based on a *Virtual testing tool* that integrates small scale experiments and numerical procedures in the well established framework of uncoupled hierarchical modeling to provide the necessary data for a macroscopic material model. The individual steps of this approach are first introduced through potential applications to historical masonry structures, concrete and mastic asphalt mixtures all showing elements of material heterogeneity, anisotropy and structural disorder. Particular attention is then devoted to mastic asphalt to confirm viability of the virtual testing tool to ultimately bring its results to points of practical applications.

Keywords: random composites, statistically equivalent periodic unit cell, homogenization, multiscale modeling, masonry, concrete, mastic asphalt.

1 Introduction

Although often assumed in the micromechanical modeling of materials, the true periodic microstructures can be associated with only a limited number of carefully fabricated material systems. In practice, a majority of construction materials currently used for the design of engineering structures can be classified as random, either natural or artificially made, composites with complex microstructures over a wide range of length scales from nanometers to meters. To span such large size differences, the general scope of hierarchical modeling has been promoted over the last two decades as the main driving force for structural design. On the one hand, such a modeling venture opens the way to the introduction of a fundamentally different theoretical framework for the description of material behavior at a particular scale of interest and, if combined with proper averaging, to address the structural response without the need for calibrating the necessary macroscopic constitutive law at the structural level. On the other hand, one may still offer several arguments as to why this compelling approach is not a practical tool or even admissible for large scale analyses. First, unless restricted to simple academic examples, it is still too computationally demanding. Second, in some modes of deformation such as strain localization, the fully coupled homogenization schemes appear theoretically inadequate. Therefore, even despite tremendous changes in the modeling of details of construction materials including fabrication processes, the large scale analysis of actual engineering structures still relies on the phenomenological description of the macroscopic material behavior. Unfortunately, an automated support for the selection or design of construction materials is currently not available, partially attributed to the fact that specification of material properties and model parameters needed for structural analysis is not an easy task, unable to dispense with laboratory tests at the macrolevel. However, this is often not only expensive but also time-consuming, therefore significantly slowing down the actual design process. It is apparent from the preceding discussion that affording reliable and efficient structural analysis requires supplanting the arduous macroscopic experimental study for deriving the macroscopic model parameters by laboratory measurements performed on finer scales combined with computational micromechanics. A problem of this type fits well within the scope of a virtual testing tool established as an integrated set of models, algorithms and procedures for the prediction of mechanical properties of materials on an arbitrary scale.

The need for the synergy of the combination of experiment and computational modeling already suggested by Drucker [6]: *"...Theory awaits experiment and experiment awaits theory in wide variety of fields. Often two must go hand in hand if significant progress is to be made..."* This precisely reverberates substantiality of the virtual testing tool. It has been the considerable progress in quantification of various microstructural details through high resolution images of material samples or X-ray microtomography together with advances in computational mechanics that enable us to introduce the virtual testing tool in the framework of uncoupled multiscale analysis as a cooperation of the following four tasks:

- Small scale laboratory testing to provide for material data of individual constituents of the *composite*. These micromechanical tests are faster, cheaper and, in combination with a computer model, relatively easy to generalize as they rely on intrinsic properties of composite components.

- Construction of the material database to store both local (phase) and homogenized (macroscopic) data.

- Construction of a certain representative volume element (RVE) of a real material sample used in numerical simulations.

A Statistically Equivalent RVE for Macroscopic Constitutive Laws 133

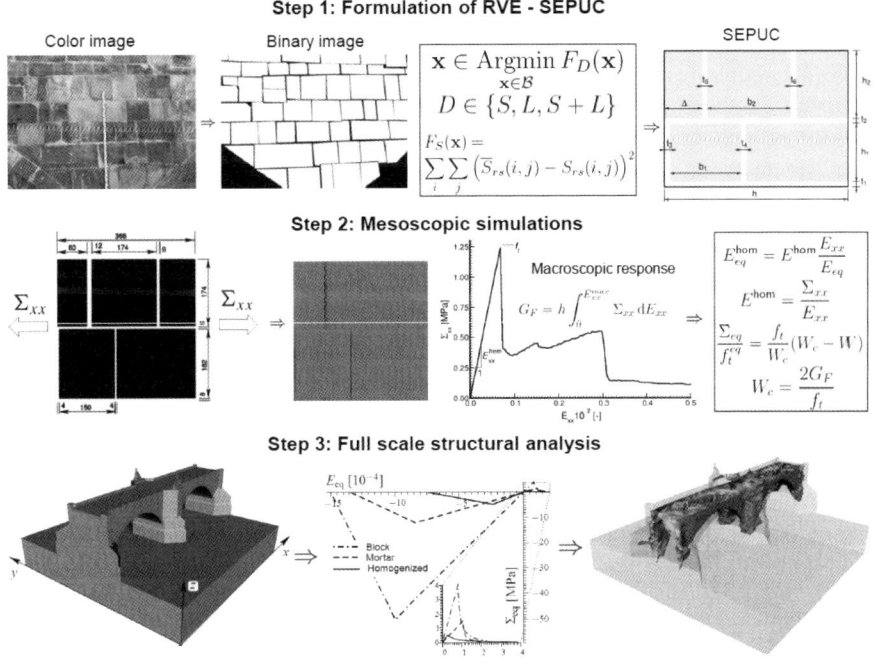

Figure 1: Example of application of virtual laboratory tool

- Numerical simulation of a large scale laboratory experiment exploiting the concept of periodic fields to derive material data needed in the macroscopic analysis.

Such a virtual laboratory tool is then expected not only to greatly reduce the number of laboratory tests at the macrolevel but also to improve the effectiveness of existing laboratory tests and speed up the introduction of new materials. The combination of virtual tests with advanced software for structural analysis may help to define the relation between the material composition and the structural behavior and consequently allow engineers to design a material tailored for the requirements of a particular construction.

However, it goes beyond the present scope of this chapter to provide details on all four tasks mentioned above. Instead, we shall concentrate on the last two items we find essential for the success of this methodology. Although not necessary, we limit our attention to classical civil engineering materials and advocate the generality of the virtual testing tool through the application of this approach to random masonry, concrete and asphalt mixtures - all falling into the category of *random composites*.

The remainder of this chapter essentially will follow the graphical representation of the application of the virtual laboratory tool to the macroscopic analysis of Charles

Bridge in Prague [22] shown in Figure 1. Providing the local constitutive laws for individual components of the composite are known from small scale experiments, the structural analysis splits into three basic steps:

1. It has been shown in a number of our previous works that random composites, generally lacking the geometric periodicity, can be well represented by a statistically equivalent periodic unit cell (SEPUC) derived from a certain minimization problem by matching basic statistical descriptors, *e.g.* the two-point probability function S_{ms}, calculated from binary images of the actual microstructure ($\bar{S}_{ms}$) and SEPUC (S_{ms}). Specific applications to a variety of composite materials can be found in [18, 35, 36, 41–43]. A brief overview of this topic is given in Section 2. Two particular examples of random microstructures of concrete and asphalt mixtures will be addressed to show exploitation of less predicative microstructural information compared to actual microstructure images such as grain size distribution function (concrete) or even a combination of the two (asphalts).

2. Once the basic geometrical data of SEPUC are known the solution proceeds with the prediction of macroscopic or homogenized data needed in the selected macroscopic constitutive equations implemented in the software we wish to use in the subsequent structural analysis. This step is the heartbeat of the virtual testing tool as it replaces the unwanted large scale experiments. As an example we present prediction of some of the basic material parameters requested by the ATENA commercial software [32] such as the homogenized secant modulus of elasticity E^{hom}, tensile strength f_t, or the homogenized fracture energy G_F by loading the unit cell with the macroscopic tensile stress, *e.g.* Σ_{xx}. Recall that the standard definition of G_F yields this quantity as the area under the macroscopic stress-strain curve multiplied by the length of the periodic unit cell h. It is thus analogous to the smeared crack model assuming the macroscopic crack opening displacement W_c being smeared over the periodic unit cell. Further details can be found in [43]. Note that a similar strategy was introduced by Massart *et al.* [19] in a fully coupled multiscale analysis of failure of masonry walls. However, such an approach requires a special treatment of upscaling the local results as the distributed damage developed on a lower scale due to periodicity, see also Figure 1, is essentially localized on an upper scale. The question of applicability of a coupled multiscale analysis to problems involving material softening and strain localization has already been put forward by Bažant [1]. It is interesting to point out that no such obstacle arises within the concept of the, though multiscale but totally uncoupled, virtual testing tool. Furthermore, providing the statistically equivalent *periodic* unit cell is accepted as a sufficiently accurate representative of the real *random* microstructure, the issue of the results being sensitive to small deviations of the microstructural details and as such not admissible as intrinsic material parameters is irrelevant at least theoretically, since *periodic* (no possible deviations) rather than the *random* medium is considered. Owing to space limitation only the mastic asphalt mixture will be

examined again in Section 3 to provide a solid demonstration of this step.

3. The relevant macroscopic data are finally used independently from the processes taking place on lower scales to estimate the response of the structure, which is decisive for an engineer to perform the actual design or to assess the current state of structure, *e.g.* distribution of cracks in the example of Charles Bridge in Figure 1, to propose some rehabilitation measures if needed. Certainly, the reliability of macroscopic predictions strongly depends on the reliability of the macroscopic model adopted. Therefore, some laboratory tests on the macrolevel, despite being greatly reduced compared to those needed for model calibration, are necessary to corroborate the lower scale homogenization step. Section 4 is devoted to this aspect of the proposed three step computational strategy. The example presented is limited to an asphalt mixture touching both the experimental validation and structural use of the corresponding macroscopic constitutive law.

2 Concept of statistically equivalent periodic unit cell

As already indicated in the introductory part, we build upon a fundamental assumption that suggests the possibility of replacing a complex non-periodic microstructure by a certain periodic unit cell, which still optimally resembles the original microstructure in a proper sense. If the periodic unit cell is described by a substantially smaller number of parameters, the construction of SEPUC leads to the best approximation problem. In principle, such a unit cell is found to match the original microstructure as closely as possible in terms of suitable measures. In a purely geometry-based approach, such measures can be represented by various statistical descriptors of the microstructure. Some recent works introduce in addition to material statistics the material physics of local constituents to refine on the microstructure approximation, see *e.g.* [4,24,25,29]. We believe, however, that separation of the geometrical and constitutive descriptions leads to a more flexible and generic framework, leaving aside the computational saving achieved by such a split. Moreover, putting the emphasis on geometrical aspects allows us to exploit advances in reconstruction of random media [27, 39] and maximally utilize the available geometrical data.

A number of statistical measures have been examined in recent decades to address the problem of *random composites* (heterogeneous media with either random distribution of material phases or the presence of various types of material or geometrical imperfections of a more or less random nature). Starting from the characteristic function as the basic descriptor for microstructure quantification, the literature offers a number of derived statistical functions including the radial path function [23], the lineal path function [16] or the general n-point probability function [28]. Among these, the two-point probability function appears as the most prominent one owing to the ease and efficiency of its evaluation particularly when taking advantage of the Fast Fourier transform applied to the binary images of real microstructures [40]. To free the reader from searching a vast body of literature devoted to this function we now briefly summarize its most important features and the essential statistical assumptions allowing its relatively efficient evaluation even for non-periodic media.

2.1 Fundamental function and statistical moments

Consider an ensemble of a two-phase random medium, e.g. the binary image of masonry structure in Figure 1. To provide a general statistical description of such systems it is useful to characterize each member of the ensemble by a stochastic function - characteristic function $\chi_r(\mathbf{x}, \alpha)$, which is equal to one when point $\mathbf{x}$ lies in the phase r of the sample α and equals zero otherwise [2, 28],

$$\chi_r(\mathbf{x}, \alpha) = \begin{cases} 1, & \text{if } \mathbf{x} \in D_r(\alpha), \\ 0, & \text{otherwise,} \end{cases} \qquad (1)$$

where $D_r(\alpha)$ denotes the domain occupied by the r-th phase. Except where noted, composites consisting of a clearly distinguishable continuous matrix phase are considered. Therefore, r is further assumed to take values m for the matrix phase while symbol s is reserved for the second phase. For such a system, the characteristic functions $\chi_m(\mathbf{x}, \alpha)$ and $\chi_s(\mathbf{x}, \alpha)$ are related by

$$\chi_m(\mathbf{x}, \alpha) + \chi_s(\mathbf{x}, \alpha) = 1. \qquad (2)$$

Following [2, 28] we write the ensemble average of the product of characteristic functions

$$S_{r_1,\ldots,r_n}(\mathbf{x}_1, \ldots, \mathbf{x}_n) = \overline{\chi_{r_1}(\mathbf{x}_1, \alpha) \cdots \chi_{r_n}(\mathbf{x}_n, \alpha)}, \qquad (3)$$

where function $S_{r_1,\ldots,r_n}$ referred to as the general n-point probability gives the probability of finding n points $\mathbf{x}_1, \ldots, \mathbf{x}_n$ randomly thrown into a medium located in the phases $r_1, \ldots, r_n$.

2.1.1 Functions of the first and second order

Hereafter, we limit our attention to functions of the order of one and two, since higher-order functions are quite difficult to determine in practice.[1] Therefore, description of a random medium will be provided by the one-point probability function $S_r(\mathbf{x})$

$$S_r(\mathbf{x}) = \overline{\chi_r(\mathbf{x}, \alpha)}, \qquad (4)$$

which simply gives the probability of finding the phase r at $\mathbf{x}$ and by the two-point probability function $S_{rq}(\mathbf{x}_1, \mathbf{x}_2)$

$$S_{rq}(\mathbf{x}_1, \mathbf{x}_2) = \overline{\chi_r(\mathbf{x}_1, \alpha)\chi_q(\mathbf{x}_2, \alpha)}, \qquad (5)$$

---

[1] Note, however, that relatively efficient procedures for approximation of higher-order probability functions for ergodic and statistically isotropic media were proposed in [5].

which denotes the probability of simultaneously finding the phase r at $\mathbf{x}_1$ and the phase q at $\mathbf{x}_2$. In general, evaluation of these characteristics may prove to be prohibitively difficult. Fortunately, a substantial simplification applies when accepting an assumption regarding the material as statistically homogeneous, so that

$$S_r(\mathbf{x}) = S_r, \qquad (6)$$
$$S_{rq}(\mathbf{x}_1,\mathbf{x}_2) = S_{rq}(\mathbf{x}_1 - \mathbf{x}_2). \qquad (7)$$

Further simplification arises when assuming the medium to be statistically isotropic [40]. Then $S_{rq}(\mathbf{x}_1,\mathbf{x}_2)$ reduces to

$$S_{rq}(\mathbf{x}_1 - \mathbf{x}_2) = S_{rq}(\|\mathbf{x}_1 - \mathbf{x}_2\|). \qquad (8)$$

Finally, making an ergodic assumption allows a substitution of the one-point correlation function by its volume average [40], *i.e.* volume concentration or volume fraction of the r-th phase c_r,

$$S_r = c_r. \qquad (9)$$

2.1.2 Limiting values

In addition, the two-point probability function S_{rq} incorporates the one-point probability function S_r for certain values of its arguments such that

$$\text{for } \mathbf{x}_1 = \mathbf{x}_2 \quad : \quad S_{rq}(\mathbf{x}_1,\mathbf{x}_2) = \delta_{rq} S_r(\mathbf{x}_1), \qquad (10)$$
$$\text{for } \|\mathbf{x}_1 - \mathbf{x}_2\| \to \infty \quad : \quad \lim_{\|\mathbf{x}_1-\mathbf{x}_2\|\to\infty} S_{rs}(\mathbf{x}_1,\mathbf{x}_2) = S_r(\mathbf{x}_1) S_q(\mathbf{x}_2), \qquad (11)$$

where symbol δ_{rq} stands for Kronecker's delta. Relation (10) states that the probability of finding two different phases at a single point is equal to 0 (see also Equation (2)) or is given by the one-point probability function if phases are identical. Equation (11) manifests that for large distances points $\mathbf{x}_1$ and $\mathbf{x}_2$ are statistically independent. This relation is often denoted as the no-long range orders hypothesis (see *e.g.* [17, 37]).

Finally, according to Equation (2), we may determine one and two-point probability functions for all phases provided that these functions are given for one arbitrary phase. For one-point probability function of statistically homogeneous and ergodic medium, this relation assumes a trivial form

$$c_m = 1 - c_s. \qquad (12)$$

Relations for the two-point probability functions of statistically uniform and ergodic medium are summarized in Table 1.[2]

Details on numerical evaluation of these functions together with a thorough discussion on the influence of the selected boundary conditions can be found in [8].

---
[2] Note that, by definition (5) and assumption of statistical homogeneity, $S_{rq}(\mathbf{x}) = S_{qr}(\mathbf{x})$.

	Known function		
	$S_{mm}(\mathbf{x})$	$S_{ms}(\mathbf{x})$	$S_{ss}(\mathbf{x})$
$S_{mm}(\mathbf{x})$	$S_{mm}(\mathbf{x})$	$c_m - S_{ms}(\mathbf{x})$	$c_m - c_s + S_{ss}(\mathbf{x})$
$S_{ms}(\mathbf{x})$	$c_m - S_{mm}(\mathbf{x})$	$S_{ms}(\mathbf{x})$	$c_s - S_{ss}(\mathbf{x})$
$S_{ss}(\mathbf{x})$	$c_s - c_m + S_{mm}(\mathbf{x})$	$c_s - c_m - S_{ms}(\mathbf{x})$	$S_{ss}(\mathbf{x})$

Table 1: Relations among two-point probability functions

2.2 Formulation of SEPUC

In general, the statistically equivalent periodic unit cell can be characterized by an N-dimensional vector of parameters x. For example, in the case of random masonry seen in Figure 1 these can be identified with the number and dimensions of stones and the thicknesses of mortar head and bed joints. Once these are specified, the corresponding digital image can be generated and used to determine the statistical descriptors, e.g. the two-point probability function S_{ms}. Then, the following measures of similarity between the original microstructure $\overline{S}_{ms}$ and a periodic unit cell S_{ms} can be introduced

$$F_S(\mathbf{x}) = \sum_i \sum_j \left(\overline{S}_{ms}(i,j) - S_{ms}(i,j)\right)^2. \tag{13}$$

The parameters of SEPUC are then simply found by minimizing the objective function (13). Note that the solution of the introduced optimization problem requires minimization of a multi-dimensional and multi-modal objective function typically exploiting the elements of soft computing [13, 14, 18, to cite a few].

Apart from random masonry structures, several other successful applications of this approach including woven textile composites can be found in [43]. Similar treatment has also been adopted for mastic asphalt [31]. It is shown therein, and will be shortly outlined in Section 2.2.2, that part of this procedure calls for the introduction of a particle size distribution function to select basic geometrical data of the cell to be optimized. Thus before proceeding, we first examine in Section 2.2.1 the potentiality of this function as the only information given (suppose that no microstructural images are available) for the generation of SEPUC to represent the mesoscopic sample of concrete.

2.2.1 Example 1 - concrete

There exists a large group of particle reinforced composites where the particle phase can be considered as random, applying not only to particle position but also to its size and orientation, which projects into material isotropy on the macroscale. Reasonable approximations of real microstructures then can be constructed with limited information such as grading curve (particle size distribution function), particle shape factor and their volume fraction with no need for microstructure images and the tedious minimization step mentioned above. Concrete, the most important construction

A Statistically Equivalent RVE for Macroscopic Constitutive Laws

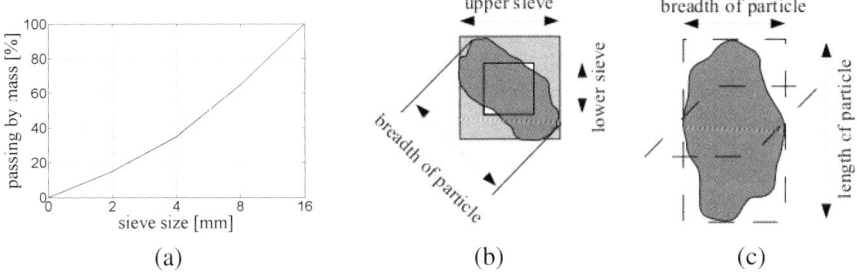

Figure 2: Basic microstructure information: (a) grading curve, (b) particle breadth, (c) particle length

material in civil engineering, is examined here as a particular representative of such systems.

One typical example of the grading curve is plotted in Figure 2(a). Such a curve can be constructed for particle sizes greater than 0.006 mm. Smaller particles must be incorporated directly into the matrix (cement paste) through an independent homogenization step as discussed later in Section 2.2.2. Individual particles are then introduced into the volume of matrix based on the "take and place" method such that their position, size and orientation is generated at random and the final distribution matches the given grading curve. In doing so the entire RVE is subdivided into several subdomains with assigned probability of particle occurrence. Once the particle is introduced in a given subdomain the probability of the particle occurring in that subdomain is reduced, which substantially hastens the generation process.

To generate particles having a realistic shape is not an easy task. At the minimum we should require convexity of particles having the shape that complies with the assumed shape factor. For success of generation the algorithm should be able to control the minimum distance between particles and the overall volume fraction while taking into account periodicity of particles crossing the model boundary. The particle shape is often imagined as spherical or ellipsoidal. Interestingly, more realistic complex shapes of a general polytope type appear more advantageous especially when it comes to the mesh generation step. Note that the breadth of each generated particle must pass through a given upper sieve size and cannot be smaller than the lower sieve size. The particle length is given by the assumed shape factor α being the ratio of length of particle to its breadth, see Figure 2 and [21] for more details. To avoid overlapping of individual particles, the particle size is first slightly enlarged and when successfully positioned within RVE it is shrunk back to maintain the required minimum distance.

Note that considering large range of particle sizes results into complex RVEs with too many particles. Such RVEs are then computationally very expensive particularly in view of complicated numerical simulations of damage processes in concrete. Therefore, only larger particles usually are employed as these are also decisive for the onset and evolution of damage.

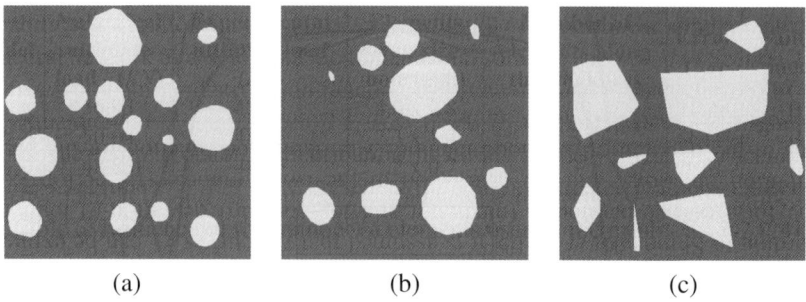

Figure 3: Two-dimensional cuts through three-dimensional samples for various particle shape approximations: (a) sphere, (b) ellipsoid, (c) polytope

Figure 3 shows examples of three particular particle shape approximations for the two largest particle sizes (16–40mm=30%, 8–16mm=70%). The volume of stones for all three samples was considered equal to 15% of the total volume and the shape factor α was set equal to 1.5.

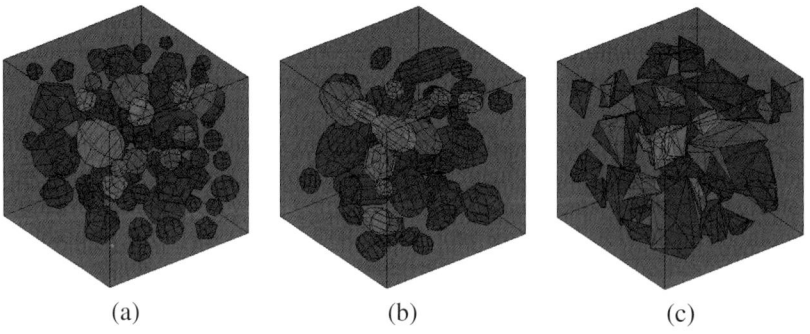

Figure 4: Three-dimensional samples for various particle shape approximations: (a) sphere, (b) ellipsoid, (c) polytope

In [33] the authors performed a set of virtual numerical compression tests to assess the influence of particle shape on the macroscopic response of concrete showing essentially an invariant behavior of samples with respect to this parameter. This, however, might not be true for other modes of loading and requires verification. It is noteworthy that the approximation with polyhedrons, Figure 4(c), required lower number of elements leading to considerable savings in the computational time. Additional information regarding both the RVE formulation and modeling of the influence of aggregate structure on the response of concrete can be found in [9, 11, 38, to cite a few].

2.2.2 Example 2 - mastic asphalt

We choose to inquire here into the study of a mastic asphalt mixture (MAm) as an example of the effective combination of both the already mentioned sources of information, microstructure images and grading curves, exploited in the search for statistically equivalent periodic unit cell. As evident from Figure 5 mastic asphalt belongs to the class of random composites with a relatively complex microstructure.

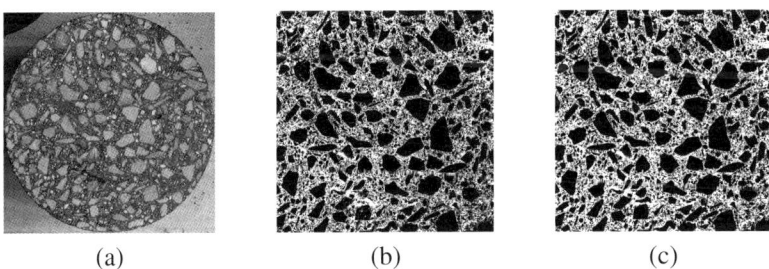

(a)   (b)   (c)

Figure 5: (a) A real microstructure of an asphalt mixture, (b) Original binary image, (c) Improved binary image

In [31] the authors proposed, in the framework of a virtual testing tool, an efficient and yet reasonably accurate procedure for the derivation of the macroscopic constitutive law for MAm that distinguishes three different scales as seen in Figure 6. To identify these scales we first mention the cumulative distribution functions of aggregates plotted in Figure 7(a). The two graphs suggest a relatively large amount of small stones (cumulative distribution function of number of aggregates) which might seem, at least from their volume fraction point of view (cumulative distribution function of volume fraction of aggregates), almost negligible. This suggests neglecting all small aggregates as also assumed for concrete samples in the previous section. One such simplified microstructure corresponding to the removal of all fragments smaller than

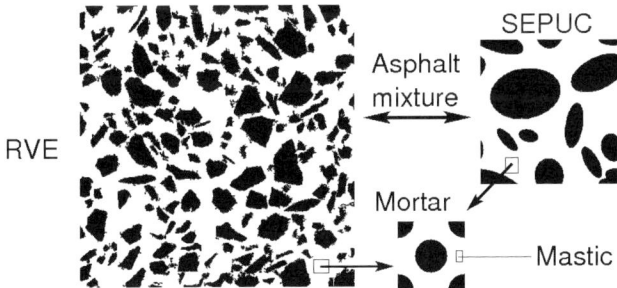

Figure 6: Three distinct scales of Mastic Asphalt mixture

1,200 pixels appears in Figure 7(b). This step reduces the total number of stones by 97% while the original volume fraction of aggregates dropped by only 33%. However, neglecting the small aggregates completely would severely underestimate the final macroscopic predictions. The mesoscopic analysis performed on the level of SEPUC, recall Figure 6, must be preceded by another homogenization step to develop a new binder material as a mixture of mastic matrix and small aggregates. Referring to Figure 6 we denote this scale as mortar. It has been confirmed, see [30,31], that on this level a periodic hexagonal array of aggregates of identical size, see also Figure 6(c), is a reasonable approximation of the true distribution of small particles in the mastic matrix thus giving rise to isotropic homogenized properties of the mortar phase.

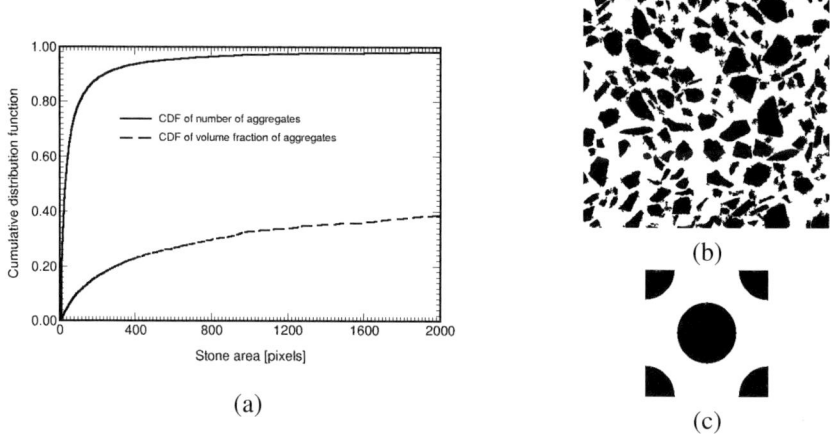

Figure 7: a) Cumulative distribution function, b) binary images of original microstructure after eliminating stone fragments smaller than 1,200px (in area), c) PUC of mortar phase containing stone fragments smaller than 1,200px (in area)

Although considerably simplified when compared to the original microstructure plotted in Figure 5(c) the microstructure in Figure 7(b) is still too complex for any meaningful numerical calculations. In this regard the computational model represented by SEPUC appears as a suitable alternative. It has been demonstrated that SEPUC that statistically resembles the real material system in terms of, *e.g.* the two-point probability function given by Equation (8) can also be expected to deliver a similar mechanical response. In the case of asphalt mixtures such a unit cell can be defined by the following parameters: number of aggregates having elliptical shape, size, position, orientation and aspect ratio of the axes of individual ellipses. The size of stones is based on the cumulative distribution function in Figure 7(a). For example, if 10 stones are selected for a PUC then the smallest stone corresponds to an average size of 10% of the smallest stones determined from the cumulative distribution function. The next stone then reflects the size of the subsequent 10% stones, *etc.* Examples of such unit cells are depicted in Figure 8.

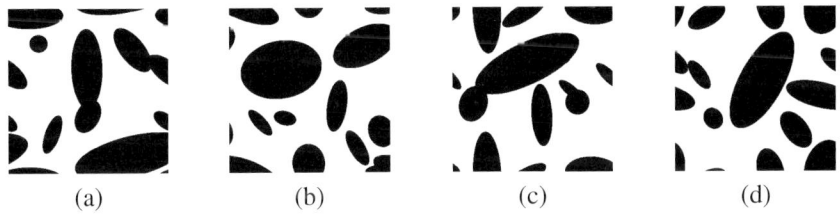

(a)   (b)   (c)   (d)

Figure 8: Examples of SEPUCs corresponding to a binary image in Figure 7(b): (a) SEPUC 6, (b) SEPUC 37, (c) SEPUC 43, (c) SEPUC 48

These parameters are searched by solving the optimization problem (13), recall Section 2.2. Here, the evolutionary algorithm GRADE [13, 14] was adopted to accomplish this task. See also [18, 41] for other applications of genetic algorithm based solution strategies. It is worth mentioning that no interpenetration constraint was introduced as it was naturally enforced through the consistency of volume fractions of aggregates in the SEPUC and targeted microstructure in Figure 7(b). As typical of genetic algorithms based optimization procedures, each run results in a unique SEPUC with a slightly different arrangement of stones, see Figure 8. It therefore remains to confirm that all cells provide the "same" macroscopic response, see the next section.

3 Macroscopic data from numerical simulations

Rendering the desired macroscopic constitutive model that describes the homogenized response of MAm to general loading actions (in the spirit of the virtual testing tool) thus endeavors to formulate a suitable micromechanical model on individual scales and to associated experimental work being jointly the building blocks of the upscaling procedure.

Suppose that the nonlinear viscoelastic generalized Leonov model [15, 35] can be used to represent the matrix behavior on every scale. While the lowest scale, the level of mastic asphalt in Figure 6, is quantified through laboratory measurements [30, 31], the level of mortar and asphalt mixture are treated numerically in the framework of two-step uncoupled homogenization. If limiting our attention to virtual numerical experiments, the literature offers two basic groups of homogenization strategies, one based on the Eshelby solution of an equivalent inclusion problem and the other characterized by detailed finite element simulation of SEPUC exploiting the assumed existence of periodic fields. Applicability of the Mori-Tanaka method as a representative of the first group of methods has been thoroughly scrutinized, *e.g.* in [30] suggesting several limitations of this method particularly when applied to problems with severe localization of strains in the matrix phase. Therefore, we comment here only on the latter approach consistent with the first order homogenization scheme.

Consider a material representative volume element now defined in terms of one of the periodic unit cells in Figure 8. Suppose that the SEPUC is subjected to either boundary displacements Δu_0 resulting in a uniform strain ΔE throughout the body or to uniform boundary tractions Δp_0 (consistent with the macroscopic stresses $\Delta \Sigma$, $\Delta p_0 = \Delta \Sigma \cdot n$, n being the unit outward normal to the boundary of the SEPUC). The strain and displacement fields in the SEPUC then admit the following decomposition

$$\Delta u(\mathbf{x}) = \Delta E \cdot \mathbf{x} + \Delta u^*(\mathbf{x}), \qquad \forall \mathbf{x} \in \Omega, \ u = u_0 \ \forall \mathbf{x} \in S, \qquad (14)$$
$$\Delta \varepsilon(\mathbf{x}) = \Delta E + \Delta \varepsilon^*(\mathbf{x}), \qquad \forall \mathbf{x} \in \Omega. \qquad (15)$$

The first term in Equation (14) corresponds to a displacement field in an effective homogeneous medium with the same overall properties as the composite. The fluctuation part u^* enters Equation (14) as a consequence of the presence of heterogeneities and has to disappear upon volume averaging, see [2] for further discussion. This condition is met for any periodic displacement field with the period equal to the size of the unit cell under consideration, [20, and references therein]. Derivation of the macroscopic response then relies on the application of Hill's lemma [10] written as

$$\langle \delta \varepsilon : \Delta \sigma(\mathbf{x}) \rangle = \delta E : \Delta \Sigma. \qquad (16)$$

On both scales the homogenized creep compliance master curve was derived by performing a set of virtual creep tests at different temperatures by loading the respective RVEs (see Mortar and SEPUC in Figure 6) in shear by the remote stress $\Sigma_{yx} = 1\text{kPa}$. The results for mortar scale are plotted in Figure 9(a). These curves were then horizontally shifted to give the homogenized creep compliance master curve seen in Figure 9(b) used subsequently in the same analysis carried out on the level of SEPUC. The macroscopic compliance function, considered in the actual engineering design (see also the next section), is plotted as the dashed line in Figure 9(c).

For illustration we also present in Figure 9(d) the response of various SEPUCs to macroscopic uniform shear stress rate $\dot{\Sigma}_{xy} = 0.025$ kPa s^{-1}. Inspecting these results suggests that the assumed size of SEPUC is insufficient for the SEPUC to be considered a unique RVE. Selecting the proper size of the SEPUC is crucial for reliable predictions. This, however, is relatively difficult particularly in view of nonlinear analysis, since the SEPUC is derived purely on the basis of geometrical information. Taking mechanical response into consideration when solving the minimization problem thus appears as a reasonable step forward, see *e.g.* [3, 12].

4 Corroboration of macroscopic constitutive law

Since much of the material considered so far has been primarily computational it would be audacious to bring the results of the first three steps of the *Virtual testing tool* directly to points of application without additional experimental validation.

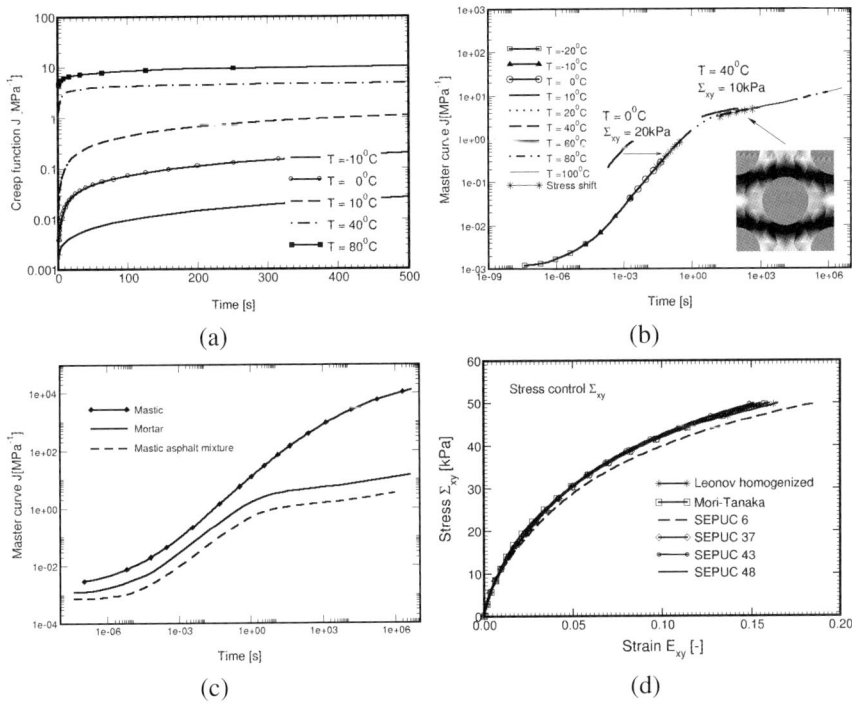

Figure 9: (a) Creep data at different temperatures for reference stress $\Sigma_{xy} = 1$kPa, (b) Master curve for reference temperature $T = 40^0$C, (c) Master curves on individual scales for reference temperature $T = 40^0$C, (d) Macroscopic response for various SEPUCs at $T = 20^0$C under $\dot{\Sigma}_{xy} = 0.025$ kPa s^{-1}

Large scale experiments are therefore needed. Matching these experiments numerically is the necessary condition for the model to be considered in a real engineering design. As an example we present the tracking wheel test as the most common type of standard laboratory test of the hot mix asphalt (HMA).

Wheel tracking is used to assess the resistance of asphaltic materials to rutting under conditions which simulate the effect of traffic. Rutting is defined as the accumulation of small amounts of unrecoverable strain resulting from loads applied to the pavement. The experiment is conducted in such a way that the loaded wheel tracks a sample of asphalt with a given magnitude of load, speed and temperature while the development of the rut is monitored continuously during the test.

The wheel tracker, shown in Figure 10(a), enables us to test a pair of cubical MAm slabs simultaneously. According to European standards the asphalt mixture specimens are 260 mm wide, 300 mm long and 40 mm deep. The wheels are 200 mm in diameter and 50 mm wide. A steel wheel core is covered with a smooth rubber having a stiffness

Figure 10: a) Wheel tracker device, b) sample of MAm after wheel tracking

number of 80 IRHD. The wheel performs 26.5 cycles per minute, one cycle is defined as two passes of the wheel, forward and back. The common amount of executed cycles varies from 10,000 to 20,000. The length of passing is 230 mm. According to European standards one wheel should load the sample with 700 N. Nevertheless, owing to an enormous compliance of the used MAm the load was set equal to 350 N per wheel. Wheel tracking was performed at the prescribed temperature 40^0C.

The rut depth was continuously measured during the experiment at each cycle. Its evolution is continuous due to the averaging along the passing length. The loading sequence was stopped after 2,300 cycles of the wheel tracking due to excessive sample deformations and the failure of separative aluminum sheet visible in Figure 10(b) to prevent a gluing effect between the wheel and the sample surface. These results are rather disturbing and indicate that this particular asphalt mixture is highly inadequate for all practical applications mainly due to the extreme compliance of the mastic binder. Nevertheless, if pursuing only the advisability of the proposed *Virtual testing tool*, the practical applicability of a studied material is not the main concern and the present results, both on micro and macro levels, can still be used in corresponding numerical simulations. Unfortunately, the poor quality of binder caused considerable difficulties in achieving numerical stability of the adopted explicit time integration process.

The experimentally measured rut depth after 2300 cycles was approximately 12 mm. The cross-section of the loaded sample after several month of relaxation is displayed in Figure 11. Both the the rut depth (downward deformation) and shear flow (uplift deformation), increasing the total rutting but not measured in the experiment, are clearly visible.

In an attempt to verify the homogenized macroscopic generalized Leonov model (MGL), recall the homogenization strategy discussed in Sections 2.2.2 and 3, we con-

Figure 11: Cross section of the asphalt sample after wheel tracking

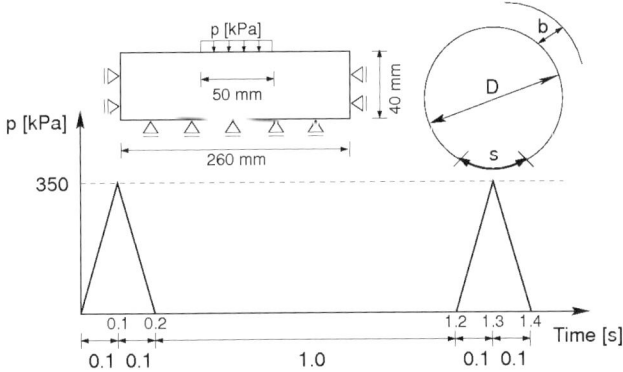

Figure 12: Numerical model scheme of wheel tracking

sidered a two-dimensional plain strain model of asphalt sample with the same cross-section as that used in laboratory measurements. Details regarding the computational model together with the applied load history appear in Figure 12. It was imagined that during the experiment the wheel passing in the direction normal to the cross-section loads a 50 mm wide zone of the cross section for a very short time of 1.2 s (one cycle of two passings lasts approximately 2.4 s). The idea is to model a wheel tracking as short impulses of pressure applied to the loaded zone of the cross section. In analogy with the actual experiment, the temperature was set equal to 40^0C. Judging from the structure geometry and deformed shape of the tested sample, the actual width of contact area s to determine the value of applied pressure was set equal to 20 mm which amounted to the pressure p = 350 kPa. The duration of pressure impulse t_i = 0.2 s was calculated from the wheel speed and the length of contact area s. To reduce the computational time only half of the structure was considered in the analysis due to symmetry conditions.

The actual loading diagram was simplified and assumed to attain a simple triangular shape as evident from Figure 12 to minimize the number of required time steps. Note that for the adopted explicit time integration scheme the stable analysis required at least 20 time steps for each loading branch (loading-unloading sequence). To reproduce the actual experiment stopped after 2,300 cycles of wheel tracking the number of impulses in Figure 12 was set to 4,600 (one impulse represents one passing of the wheel). In addition, a recovery period lasting one day was simulated. To perform numerical calculations the Leonov model was implemented into the commercial finite element software GEO FEM [7]. The results obtained are plotted in Figures 13 and 14.

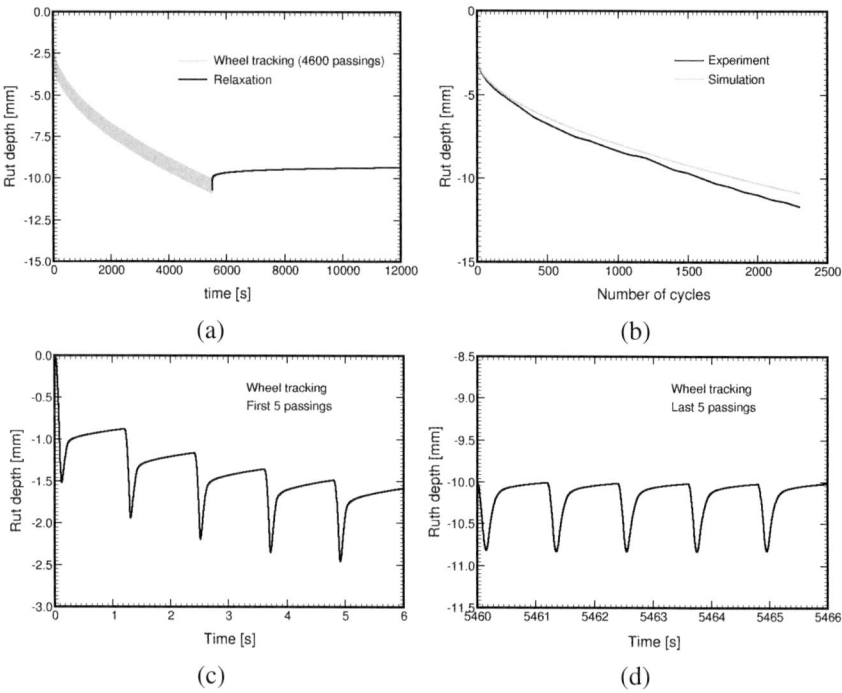

Figure 13: Wheel tracking: (a) numerical simulation - loading and recovery, (b) comparison of maximal rut depth from numerical and experimental results, (c) first 5 passes from numerical simulation, (d) last 5 passes from numerical simulations

In Figure 13(a) the gray curve represents the passing period while the black one represents the recovery period. Note that only two hours of the recovery period are shown for clarity of presentation. As seen, the rut depth after this time remains almost constant. Its value after 24 hours reached 8.1 mm. The shaded region represents all short time impulses of the loading period. To better illustrate the loading sequence we zoomed the first five and the last five passes now clearly visible in Figures 13(c) and (d), respectively. Evidently, the rate of increase of rutting slows down during each pass and appears almost negligible in Figure 13(d) although the rut depth is still increasing as seen in Figure 13(a).

Comparison between experimental data and numerical simulation is depicted in Figure 13(b). Only depression associated with the peak loads of the wheel passing is displayed. Although the experimentally observed response seems slightly more compliant, the numerical predictions appear in the light of the simplified loading diagram rather appealing. Note that more realistic short periods of creep (constant pressure between the loading and unloading branches) were excluded from the numerical simulation as they would cause the analysis to be computationally unfeasible. In every case, the results presented show applicability of the proposed version of the MGL model in the analysis of real engineering problems and confirm the potential of the proposed two step uncoupled homogenization scheme as one particular step of the underlying *Virtual testing tool*.

The deformed shape of the asphalt sample at various stages during the first wheel pass is shown in Figure 14. The three states selected for visualization are labeled in Figure 14(a) also identifying the loading sequence for one wheel pass. In particular, state 1 displayed in Figure 14(b) represents the deformation at the maximum applied pressure, state 2 shown in Figure 14(c) corresponds to the maximum rut depth attained before completing the first pass and finally state 3 represents the sample deformation after first recovery period immediately before the second pass, see Figure 14(d).

Notice that the shape of depression provided by the numerical simulation fully corresponds to the shape of the asphalt sample after one year of recovery visible in Figure 11. The measured rut depth after one year of recovery was approximately 6-7 mm. Recall the value of 8.1 mm calculated one day after terminating the loading sequence.

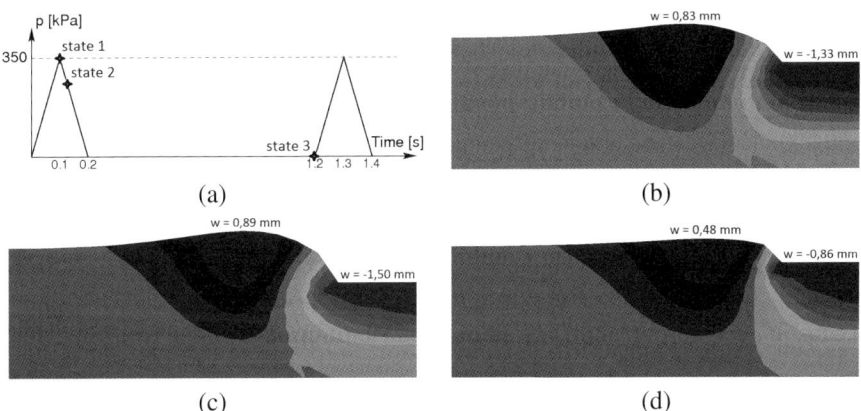

Figure 14: Wheel tracking - first pass: (a) scheme of loading, (b) state 1: depression at maximum pressure, (c) state 2: depression at maximum rut depth, (d) state 3: depression after recovery

A simple example showing application of the MGL model to a real structure as the ultimate goal of the *Virtual testing tool* is presented next. It is not the purpose to analyze one particular structure but rather assess feasibility of the model when combined with inelastic deformation of subsoil. Therefore, only a three-layer structure was considered for simplicity. The soil profile also evident from Figure 15(a) consists of one 100 mm thick layer of asphalt and two 400 mm thick layers of subsoil, the upper one representing a heavily dense gravel overlaying the layer of sand.

Soil	Young mod. E [MPa]	Poisson num. ν [-]	Weight γ [kN/m^3]	Cohesion c [kPa]	Friction angle φ [-]
Gravel	50	0.2	21	5	41.5
Sand	30	0.3	17.5	5	31.5

Table 2: Material properties of subsoil layers

Both subsoil layers were represented by the Drucker-Prager constitutive model with material data listed in Table 2, see *e.g.* [34] for details on even more advanced constitutive models for soils. The same material model as used in the previous paragraphs was employed for the asphalt layer. The essential conclusion of the previous analysis, demonstrating unsuitability of the present asphalt mixture for road applications, required, however, adjusting the material parameters of the MGL model to comply with 0^{0}C temperature in order to arrive at a converged solution even for a relatively short duration of loading.

The wheel load $F = 5$ kN, corresponding to a standing personal vehicle, was assumed to be distributed over the 200 x 200 mm surface area resulting in the prescribed surface pressure $p = 125$ kPa, see Figure 15(a) also showing the boundary conditions and finite element mesh. The analysis was performed again using the GEO FEM [7] finite element code. But even for a 0^{0}C temperature the asphalt material showed enormous susceptibility to creep causing the loaded area essentially to be squashed in all the way to the bottom of the asphalt layer in just a few seconds. Although a converged solution was achieved, the results are not realistic owing to the assumed small strain theory. The results that correspond to an almost instantaneous loading (the total load was applied in 1 s) are shown in Figures 15(b), (d) and (f) whereas Figures 15(c), (e) and (g) were found just after 3 s of the permanent loading after reaching the peak load. In particular, the variation of the vertical displacement appears in Figures 15(b) and (c), the corresponding vertical stress is plotted in Figures 15(d) and (e) and finally the distribution of equivalent deviatoric plastic strain in the subsoil layers is displayed in Figures 15(f) and (g). Increase of the localized plastic deformation due to creep for even such a short duration of permanent loading is evident. Also note a considerable increase of vertical displacement from 2 to 5 mm.

The analysis, although failing to predict the unwanted rutting caused by the long time duration of permanent loading, still supports the possibility of using the proposed MGL model for large scale applications thus supporting the reason for a *Virtual testing tool* and uncoupled multiscale analysis in general.

A Statistically Equivalent RVE for Macroscopic Constitutive Laws 151

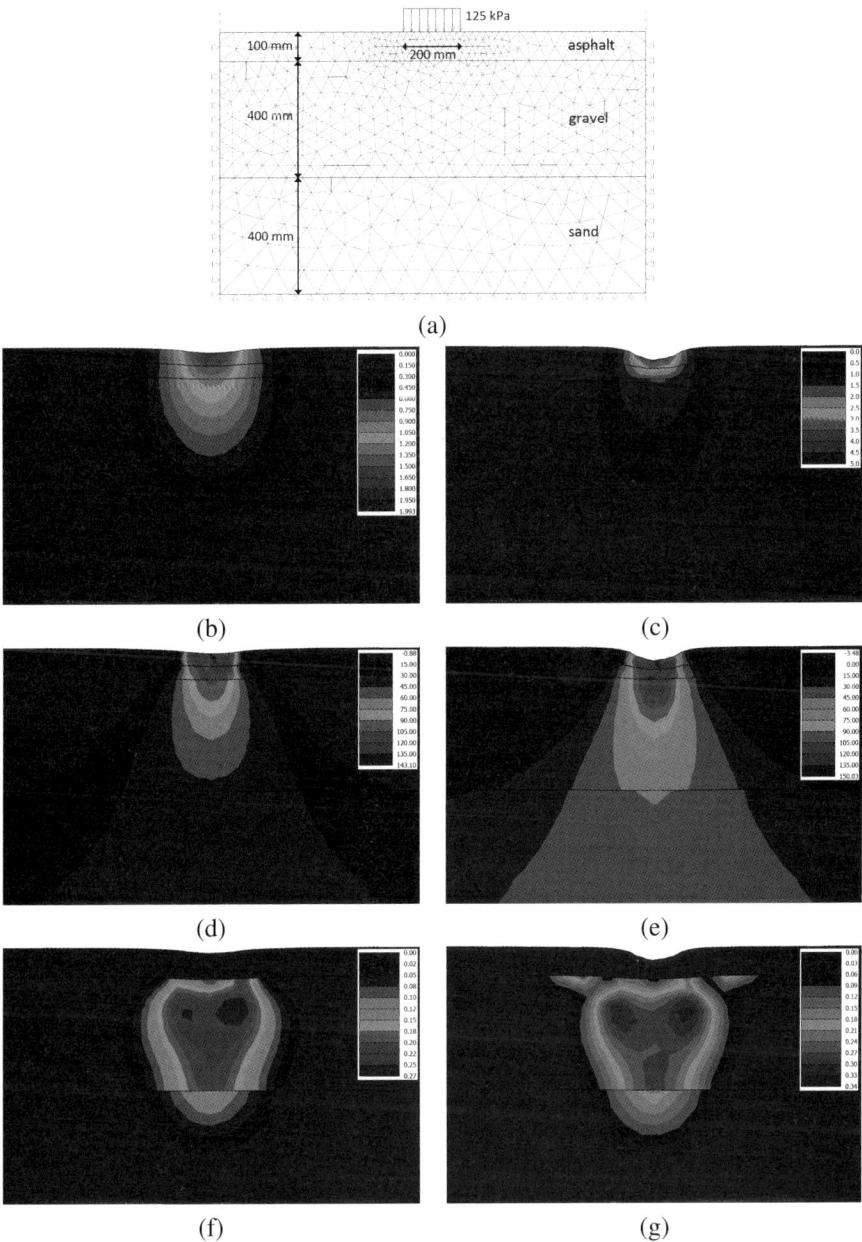

Figure 15: Flexible pavement structure: (a) model scheme, (b)-(c) displacement d_z [mm], (d)-(e) stress σ_z [kPa], (f)-(g) equivalent plastic strain E_d^{pl} [%].

5 Conclusions

The present work provides an overview of theoretical tools for the formulation of macroscopic constitutive law reflecting the confluence of threads coming from experimental work, image analysis, statistical mechanics and traditional disciplines of micromechanics and the first order computational homogenization. Here, the totally uncoupled multiscale modeling approach is favored to enable an efficient analysis of real world large scale structures, which is the principle objective of this chapter.

The leading rule enabling successful achievement of this task is the proposed *Virtual testing tool*, which replaces expensive and arduous large scale experiments with numerical simulations both aiming at the same target - calibration of the macroscopic constitutive law. Prior to its derivation the following issues attracted our attention:

- selection of scales and material models on individual scales,
- selection of experimental program, either laboratory measurements or virtual numerical tests, to homogenize the model on a given scale,
- selection of a suitable homogenization technique,
- selection of full scale laboratory measurements to validate both the numerical predictions and adopted fully uncoupled "bottom-up" multiscale homogenization approach.

Some of these issues were addressed successively in individual sections of this chapter. The concept of a two-step homogenization approach enabling the derivation of macroscopic generalized constitutive model through independent homogenization carried out first at the level of mortar, the matrix reinforced by small stone particles characterized by a greatly simplified micromechanical model, and then on the mesoscale, where the concept of SEPUC should generally be adopted, was advocated for both concrete and asphalt mixtures and can be generalized for many other material systems known from other engineering applications, recall e.g. carbon-carbon textile composites [26].

The message we thus wish to offer is the generality of the *Virtual testing tool* not pertinent to a particular material system or engineering field. One specific example of its application to mastic asphalt mixture was presented to support this final assertion.

Acknowledgements

The financial support of the Ministry of Education, Youth and Sports, project No. 1M0579, within activities of the CIDEAS research centre, and of the GAČR grant 105/11/0224 is gratefully acknowledged.

References

[1] Z.P. Bažant, "Can Multiscale-Multiphysics Methods Predict Softening Damage and Structural Failure?", International Journal for Multiscale Computational Engineering, 8(1), 61–67, 2010.
[2] M.J. Beran, "Statistical Continuum Theories", Monographs in Statistical Physics, Interscience Publishers, 1968.
[3] B. Bochenek, R. Pyrz, "Reconstruction of random microstructures–a stochastic optimization problem", Computational Materials Science, 31(1–2), 93–112, 2004.
[4] F. Cluni, V. Gusella, "Homogenization of non-periodic masonry structures", International Journal of Solids and Structures, 41(7), 1911–1923, 2004.
[5] R. Derr, "Statistical Modeling of Microstructure with Applications to Effective Property Computation in Materials Science", PhD thesis, University of North Carolina, 1999. http://citeseer.nj.nec.com/derr99statistical.html
[6] D.C. Drucker, "Thoughts on the present and future interrelation of theoretical and experimental mechanics", Experimental Mechanics, 97–106, 1968.
[7] Fine-Ltd., "GEO 5 manual", 2011. http://www.fine.cz/
[8] J. Gajdošík, J. Zeman, M. Šejnoha, "Qualitative analysis of fiber composite microstructure: Influence of boundary conditions", Probabilistic Engineering Mechanics, 21, 317–329, 2006.
[9] H. He, Z. Guo, P. Stroeven, M. Stroeven, L. Sluys, "Strategy on simulation of arbitrary-shaped cement grains in concrete", Image Analysis and Stereology, 29, 79–84, 2010.
[10] R. Hill, "Elastic properties of reinforced solids - Some theoretical principles", Journal of the Mechanics and Physics of Solids, 11, 357–372, 1963.
[11] J. Kozicki, J. Tejchman, "Effect of aggregate structure on fracture process in concrete using 2D lattice model", Archives of Mechanics, 59(4-5), 365–384, 2007.
[12] H. Kumar, C. Briant, W. Curtin, "Using microstructure reconstruction to model mechanical behavior in complex microstructures", Mechanics of Materials, Special issue on "Advances in Disordered Materials", 38(8–10), 818–832, 2006.
[13] A. Kučerová, "Identification of nonlinear mechanical model parameters based on softcomputing methods", PhD thesis, Ecole Normale Supérieure de Cachan, Laboratoire de Mécanique et Technologie, 2007.
[14] A. Kučerová, D. Brancherie, A. Ibrahimbegović, J. Zeman, Z. Bittnar, "Novel anisotropic continuum-discrete damage model capable of representing localized failure of massive structures. Part II: identification from tests under heterogeneous stress field", Engineering Computations, 26(1/2), 128–144, 2009.
[15] A.I. Leonov, "Non-equilibrium thermodynamics and rheology of viscoelastic polymer media", Rheol. Acta, 15, 85–98, 1976.
[16] B. Lu, S. Torquato, "Lineal-path function for random heterogeneous materials", Physical Review E, 45(2), 922–929, 1992.

[17] K.Z. Markov, "On the Cluster Bounds for the Effective Properties of Microcraked Solids", Journal of the Mechanics and Physics of Solids, 46(2), 357–388, 1998.
[18] K. Matouš, M. Lepš, J. Zeman, M. Šejnoha, "Applying genetic algorithms to selected topics commonly encountered in engineering practice", Computer Methods in Applied Mechanics and Engineering, 190(13–14), 1629–1650, 2000.
[19] B.C.N. Mercatoris, T.J. Massart, "Assessment of Periodic Homogenization-Based Multiscale Computational Schemes for Quasi-Brittle Structural Failure", International Journal for Multiscale Computational Engineering, 7(2), 153–170, 2009.
[20] J.C. Michel, H. Moulinec, P. Suquet, "Effective properties of composite materials with periodic microstructure: A computational approach", Computer Methods in Applied Mechanics and Engineering, 172, 109–143, 1999.
[21] C. Mora, A. Kwan, H. Chan, "Particle size distribution analysis of coarse aggregate using digital image processing", Cement and Concrete Research, 28(6), 921–932, 1998.
[22] J. Novák, J. Zeman, M. Šejnoha, J. Šejnoha, "Pragmatic multi-scale and multiphysics analysis of Charles Bridge in Prague", Engineering Structures, 30(11), 3365–3376, 2008.
[23] R. Pyrz, "Correlation of microstructure variability and local stress field in two-phase materials", Materials Science and Engineering, A177, 253–259, 1994.
[24] M. Stroeven, H. Askes, L. Sluys, "Numerical determination of Representative Volumes for granular materials", Computer Methods in Applied Mechanics and Engineering, 193(30-32), 3221–3238, 2004.
[25] S. Swaminathan, S. Ghosh, N. Pagano, "Statistically equivalent representative volume elements for composite microstructures, Part I: Without damage", Journal of Composite Materials, 40(7), 583–604, 2006.
[26] B. Tomková, M. Šejnoha, J. Novák, J. Zeman, "Evaluation of Effective Thermal Conductivities of Porous Textile Composites", International Journal for Multiscale Computational Engineering, 6(2), 153–168, 2008.
[27] S. Torquato, Random heterogeneous materials: Microstructure and macroscopic properties, Springer-Verlag, 2002.
[28] S. Torquato, G. Stell, "Microstructure of two-phase random media.I. The n-point probability functions", Journal of Chemical Physics, 77(4), 2071–2077, 1982.
[29] D. Trias, J. Costa, A. Turon, J. Hurtado, "Determination of the critical size of a Statistical Representative Volume Element (SRVE) for carbon reinforced polymers", Acta Materialia, 54(13), 3471–3484, 2006.
[30] R. Valenta, Micromechanical modeling of asphalt mixtures, PhD thesis, Czech Technical University in Prague, Faculty of Civil Engineering, 2011.
[31] R. Valenta, M. Šejnoha, J. Zeman, "Macroscopic constitutive law for Mastic Asphalt Mixtures from multiscale modeling", International Journal for Multiscale Computational Engineering, 8(1), 2010.
[32] V. Červenka, L. Jendele, J. Červenka, "ATENA Program Documentation – Part I : Theory", Červenka Consulting Company. http://www.cervenka.cz

[33] J. Vorel, A. Kučerová, V. Šmilauer, Z. Bittnar, "Virtual testing of concrete", in "Engineering Mechanics 2011", Instititue of Thermomechanics, Academy of Sciences of the Czech Republic, 2011, Presented at the 17th International conference Engineering Mechanics 2011.

[34] M. Šejnoha, "Finite element analysis in geotechnical design", Saxe-Coburg Publications, Stirlingshire, Scotland, UK, 2011. (in print)

[35] M. Šejnoha, R. Valenta, J. Zeman, "Nonlinear Viscoelastic Analysis of Statistically Homogeneous Random Composites", International Journal for Multiscale Computational Engineering, 2(4), 645–673, 2004.

[36] M. Šejnoha, J. Zeman, "Overall viscoelastic response of random fibrous composites with statistically quasi uniform distribution of reinforcements", Computer Methods in Applied Mechanics and Engineering, 191(44), 5027–5044, 2002.

[37] J.R. Willis, "Bounds and self-consistent estimates for the overall properties of anisotropic composites", Journal of the Mechanics and Physics of Solids, 25, 185–202, 1977.

[38] R. Xu, X. Yang, A. Yin, S. Yang, Y. Ye, "A three-dimensional aggregates generation and packing alghorithm for modeling asphalt mixture with graded aggregates", Journal of Mechanics, 26(2), 165–171, 2010.

[39] C.L.Y. Yeong, S. Torquato, "Reconstructing random media", Physical Review E, 57(1), 495–506, 1998.

[40] J. Zeman, "Analysis of composite materials with random microstructure", CTU Reports, 7, CTU in Prague, 2003.

[41] J. Zeman, M. Šejnoha, "Numerical evaluation of effective properties of graphite fiber tow impregnated by polymer matrix", Journal of the Mechanics and Physics of Solids, 49(1), 69–90, 2001.

[42] J. Zeman, M. Šejnoha, "Homogenization of balanced plain weave composites with imperfect microstructure: Part I – Theoretical formulation", International Journal for Solids and Structures, 41(22–23), 6549–6571, 2004.

[43] J. Zeman, M. Šejnoha, "From random microstructures to representative volume elements", Modelling and Simulation in Materials Science and Engineering, 15(4), S325–S335, 2007.

Chapter 6

Applications of Graph Products and Canonical Forms in Structural Mechanics: A Review

A. Kaveh[1] and H. Rahami[2]
[1] School of Civil Engineering
  Iran University of Science and Technology, Tehran, Iran
[2] School of Engineering Science, College of Engineering
  University of Tehran, Iran

Abstract

In this paper different canonical forms which have applications in engineering problems are studied. These matrices are in block forms and the arrangement of block lead to different canonical forms changing the solution of the problem.

Here first we use Schur's method and Form I and Form II are introduced as special cases. Then Form II is generalized to Forms F and G. In all cases the aim is to calculate the eigenvalues and eigenvectors followed by finding their inverse. Then the graph products are introduced. Then Form III is introduced and it is transformed to Form II. Parallel to this the transformation of a special case of F to G is performed. Finally the circulant matrices as a special class of block matrices are discussed.

Keywords: canonical forms, graph products, eigensolution, diagonalization, decomposition, circulant.

1   Introduction

In spite of considerable advances in the computational capability of computers in recent years, efficient methods for more time saving solutions of structures are of great interest. Large problems arise in many scientific and engineering problems. While the basic mathematical ideas are independent of the size of the matrices, the numerical determination of the displacement and internal forces become more complicated as the dimensions of matrices increase and their sparsity decreases. The use of prefabrication in industrialized building construction, often results in structures with regular patterns of elements exhibiting symmetry of various types, and special methods are beneficial for the efficient solution of such problems.

Well established techniques exist for the eigensolution of bilateral symmetry in the work of Kaveh and Sayarinejad [1,2], Kaveh [3], Kaveh and Salimbahrami [4].

Other eigensolution methods are also available for cyclically symmetric structures in Thomas [5], Williams [6-7], Hasan and Hasan [8], and Aghayere [9] among many others. The history of the developments in symmetry and application of different mathematical tools can be found in the excellent review paper of Kangwai et al. [10].

Mathematicians have also contributed a great deal to the canonical forms and graph products. Block matrices can be found in the work of [11-12]. For eigensolution and inversion of block circulant matrices one can refer to [13-17]. The inverse of tri-diagonal and penta-diagonal block matrices are extensively discussed in References [18-22].

In this chapter different canonical forms which have applications in engineering problems are studied. These matrices are in block forms and the arrangement of block lead to different canonical forms changing the solution of the problem.

Here first we use Schur's method and Form I and Form II are introduced as special cases. Then Form II is generalized to Forms F and G. In all cases the aim is to calculate the eigenvalues and eigenvectors followed by finding their inverse. Then the graph products are introduced. Then Form III is introduced and it is transformed to Form II. Parallel to this the transformation of a special case of F to G is performed. Finally the circulant block matrices as a special class of block matrices is discussed.

2 Different Matrix Forms

Considering the decomposition of Schur, the determinant of a 2-by-2 block matrix can be written as

$$\mathbf{M} = \begin{bmatrix} A & B \\ C & D \end{bmatrix}; \quad \det(M) = \begin{cases} \det(D)\det(A - BD^{-1}C); & \det(D) \neq 0 \\ \det(A)\det(D - CA^{-1}B); & \det(A) \neq 0 \end{cases} \tag{1}$$

Therefore the eigenvalues can be obtained easily. In this case all the submatrices are selected as square.

Two special cases of this decomposition are as follows [23]:

1. If $D = A$ and $B = C = 0$ we will have Form I.
2. If $D = A$ and $C = B$ we will obtain Form II.

It should be noted that in the Schur's method we need to find the inverse of a matrix, while form II can be written as

$$M = \begin{bmatrix} A & B \\ B & A \end{bmatrix} \approx \begin{bmatrix} A+B & 0 \\ 0 & A-B \end{bmatrix} \tag{2}$$

Therefore the eigenvalues of M will be the union of the eigenvalues of $A + B$ and $A - B$. It should be noted here that the union means that if an eigenvalue is repeated twice or more, all these repeated elements will be present in the union.

It should be noted that the Kronecker product of two matrices A and B will be represented as $S = A \otimes B$ in which the ij th entry of A is replaced by $a_{ij} B$. Therefore a Form II matrix can be written as

$$M = I_2 \otimes A + \begin{bmatrix} 0 & 1 \\ 1 & 0 \end{bmatrix} \otimes B \tag{3}$$

To generalize this form F is defined as follows:

$$F_n(A_m, B_m, C_m, D_m) = \begin{bmatrix} A_m & B_m & D_m & & & & \\ B_m & C_m & B_m & D_m & & & \\ D_m & B_m & C_m & . & . & & \\ D_m & . & . & . & D_m & & \\ & . & . & C_m & B_m & D_m \\ & & D_m & B_m & C_m & B_m \\ & & & D_m & B_m & A_m \end{bmatrix}_n \tag{4}$$

It should be mentioned that the Form G block matrices are exactly the same as F with the difference of having B_m submatrices in their corners. As can be seen the matrices F and G in general have peta block diagonal form. If this matrix is formed with three arguments, then $D_m = 0$. The blocks can be matrices or simply numbers. With this definition the generalized Form II matrix will have the following form:

$$M = F_n(A_m, B_m, A_m) = I_n \otimes A_m + F_n(0,1,0) \otimes B_m \tag{5}$$

As is shown in Reference [24], the determinant of this matrix will be

$$\det(M) = \sum_{n=0}^{[\frac{n}{2}]} (-1)^n \binom{n-i}{n} A^{(n-2i)} B^{2i} = \sum_{i=1}^{n} (A + \alpha_i B) \tag{6}$$

Where

$$\alpha_i = 2 \cos \frac{i\pi}{n+1} \quad ; \quad i = 1 : n \tag{7}$$

For evaluating the eigenvalues of M one can use the above relationships. Equivalent to this is employing the following theorems:

First we assume M can be expressed as the sum of two Kronecker products as

$$M = A_1 \otimes B_1 + A_2 \otimes B_2 \tag{8}$$

Now if u is an orthogonal matrix which can diagonalize A_1 and A_2, then as is shown in [24], the matrix $U = u \otimes I$ will diagonalize the matrix M, i.e. $U'MU$ will be a block diagonal. Thus M is block diagonalized and for calculating its eigenvalues, we need to calculate the eigenvalues of the blocks on the diagonal. This is because if U is orthogonal, then M and $U'MU$ will be similar matrices. Now the question which arises is whether the assumption considered at the beginning is feasible or not. This means can we find an orthogonal matrix u which can simultaneously diagonalize A_1 and A_2? For diagonalizing a matrix it is sufficient for the matrix to be Hermischian. It should be mentioned that if the matrix is not Hermischian, it is only sufficient to have independent eigenvectors. However, for two matrices to be diagonalizable the necessary and sufficient conditions are given in the following theorem:

Theorem 1: The necessary and sufficient condition for simultaneous diagonalization of two Hermishian matrices A_1 and A_2, using an orthogonal matrix, is that:

$$A_1 A_2 = A_2 A_1 \tag{9}$$

If this property holds, then considering the similarity of M and $U'MU$ we can write:

$$\lambda_M = \bigcup_{i=1}^{n} eig(M_i); \; M_i = \lambda_i(A_1)B_1 + \lambda_i(A_2)B_2 \tag{10}$$

In this relationship the dimensions of the matrices A_1 and A_2 is n, and the dimensions of B_1 and B_2 are equal to m.

As a special case, if $A_1 = I_2$ and $A_2 = F_2(0,1,0)$, since $I_2 A_2 = A_2 I_2$ we will have

$$\lambda_M = \bigcup_{i=1}^{n} [eig(B_1 + \lambda_i(A_2)B_2)] \;, \; \lambda_i(A_1) = 1 \;, \; \lambda_i(A_2) = \{1,-1\} \tag{11}$$

Therefore

$$\lambda_M = \cup [eig(A+B), eig(A-B)] \tag{12}$$

This is the same result as that of the Form II.

Now we calculate the eigenvectors. We suppose the condition (9) to hold. Then both matrices A_1 and A_2 can simultaneously be diagonalized by vector u. First we assume $A_1 = I$ to hold. In this case the eigenvalues will be as follows:

$$\lambda_M = \bigcup_{i=1}^{n} eig(M_i); M_i = B_1 + \lambda_i(A_2)B_2 \tag{13}$$

If μ is the eigenvalue and v is the eigenvector, then $M_i = B_1 + \lambda_i(A_2)B_2$. Then

$$(B_1 + \lambda B_2)v = \mu v \tag{14}$$

In the following we show that $u \otimes v$ will be the eigenvector of M. Since

$$(A \otimes B)(C \otimes D) = AC \otimes BD \tag{15}$$

Therefore

$$(A_1 \otimes B_1 + A_2 \otimes B_2)(u \otimes v) = (A_1 u) \otimes (B_1 v) + (A_2 u) \otimes (B_2 v) \tag{16}$$

Since we have assumed $A_1 = I$, therefore

$$A_1 u = u \quad ; \quad A_2 u = \lambda u \tag{17}$$

In this manner we will finally have

$$(A_1 \otimes B_1 + A_2 \otimes B_2)(u \otimes v) = u \otimes (B_1 v) + \lambda u \otimes (B_2 v) = u \otimes (B_1 + \lambda B_2)v = \mu(u \otimes v) \tag{18}$$

This relationship shows that $u \otimes v$ is an eigenvector of M.

In case $A_1 \neq I$, then the proof will stay valid. Since we have assumed the Equation (9) to hold, in this case, using the QZ decomposition one can find Q and Z such that:

$$QA_1 Z = I \quad ; \quad QA_2 Z = D \tag{19}$$

Where D is a diagonal matrix and can transform A_1 to I, as shown in Reference [25].

Now we suppose M is the sum of three Kronecker products, *i.e.*

$$M = \sum_{i=1}^{3} A_i \otimes B_j \tag{20}$$

In this case if we want to block diagonalize such a matrix, similar to the way we described for two matrices, we should show that each pair of A_i should commute with respect to multiplication, *i.e.*

$$A_i A_j = A_j A_i \qquad i, j = 1:3 \qquad i \neq j \qquad (21)$$

It can be seen that the important problem is to recognize whether they have the commutativity property with respect to multiplication. The following theorem provides a simple means for such recognition.

Theorem 2: Two penta-diagonal (and special cat tri-diagonal) matrices of the form $A_i = F(a_i, b_i, c_i, d_i)$ satisfy $A_i A_j = A_j A_i$ when $t_1 = \dfrac{a_i - c_i}{d_i}$ and $t_2 = \dfrac{a_i - c_i + d_i}{b_i}$ are identical for the two matrices.

It should be noted that for two tri-diagonal matrices t_1 are identical for the two matrices and only the equality of t_2 should be controlled.

3 Inverse of a Block Matrix

In this section we want to solve a set of algebraic equations with the corresponding coefficient matrix being in a block form. As an example, consider the following equations:

$$Mx = B \qquad (22)$$

It is obvious that after calculating the eigenvalues and eigenvectors of M one can easily solve the equations. If λ_i and $\{\varphi\}_i$ for different values of i these are the eigenvalues and eigenvectors of M, then introducing $B_j = \{\varphi\}_j^t B$ we will have

$$y_j = \frac{B_j}{\lambda_j} \Rightarrow \{x\}_n = \sum_{i=1}^{n} \{\varphi\}_i y_i = \sum_{i=1}^{n} \{\varphi\}_i \frac{B_i}{\lambda_i} = \sum_{i=1}^{n} \frac{\{\varphi\}_i \{\varphi\}_i^t}{\lambda_i} B \qquad (23)$$

If none of the A_is and B_is have the commutativity property, and $k = 2$, then using QZ transformation the set of equations can be solved. This transformation and its applications are introduced in [25]. In this approach there is no need to inverse any matrix.

If the aim is to find M^{-1} having the eigenvalues and eigenvectors, this can easily be done. Let V be the matrix of eigenvalues and D be a diagonal matrix containing the eigenvalues as its diagonal entries, then $M = VDV^t$. Since the eigenvalues of M^{-1} are the inverse of the eigenvalues of M and the eigenvectors are identical, therefore:

$$M^{-1} = VD^{-1}V^t = V \begin{bmatrix} 1/\lambda_1 & & & 0 \\ & 1/\lambda_2 & & \\ & & \ddots & \\ 0 & & & 1/\lambda_{m*n} \end{bmatrix} V^t \qquad (24)$$

Where D^{-1} can be obtained by simply inverting the entries of the main diagonal of D. Also the eigenvectors of such a matrix will be as $u \otimes v$, **where** u is a vector which simultaneously diagonalizes both matrices A_1 and A_2, and v is the eigenvector of $M_i = \sum_{j=1}^{k}(\lambda_i(A_j)B_j)$.

4 Graph Products

Before introducing the graph products, it is necessary to define two important matrices associated with graphs.

The adjacency matrix of a graph (A) is a square matrix with dimension as the number of nodes of the graph. The *ij*th entry is zero, unless the two nodes i and j are connected in the graph, and such an entry is equal to 1.

The Laplacian matrix is defined as $L = D - A$, where D is a diagonal matrix and the *ii*th entry is the degree of the node i. The degree of a node is the number of edges connected to that node.

A set of nodes connected to each other sequentially is a path and will be denoted by P_n. If the first and last nodes of a path coincide, then it is called a cycle and is denoted by C_n.

Now we briefly introduce four types of graph products.

4.1 Cartesian product

The simplest Boolean operation on the graph, is the Cartesian product $S = K \times H$. The Cartesian product is a Boolean operation $S = K \times H$ in which for any two nodes $u = (u_1, u_2)$ and $v = (v_1, v_2)$ in $N(K) \times N(H)$, the edge uv is in $M(S)$ whenever,

$$\begin{aligned} u_1 = v_1 \text{ and } u_2 v_2 \in M(H) \\ \text{or} \\ u_2 = v_2 \text{ and } u_1 v_1 \in M(K) \end{aligned} \qquad (25)$$

This definition means that two nodes in the Cartesian product are connected to each other when their first components are identical and the second components in the second graph are connected, and vice versa

4.2 Direct product

The direct product of two graphs K and H is denoted by $S = K * H$. For two nodes $u = (u_1, u_2)$ and $v = (v_1, v_2)$ in $N(K) \times N(H)$, we have an edge uv in $M(S)$ if the following condition is satisfied:

$$u_1 v_1 \in M(K) \quad \text{and} \quad u_2 v_2 \in M(H) \tag{26}$$

This means that two nodes in the direct product are connected to each other when their first components in the first graph and their second components in the second graph are connected.

4.3 Strong Cartesian product

The strong Cartesian product of two graphs K and H is denoted by $H \boxtimes S = K$. For two nodes $u = (u_1, u_2)$ and $v = (v_1, v_2)$ in $N(K) \times N(H)$, we have an edge uv in $M(S)$ if the following condition is satisfied:

$$\begin{array}{c} u_1 = v_1 \quad \text{and} \quad u_2 v_2 \in M(H) \\ \text{or} \\ u_2 = v_2 \quad \text{and} \quad u_1 v_1 \in M(K) \\ \text{or} \\ u_1 v_1 \in M(K) \quad \text{and} \quad u_2 v_2 \in M(H) \end{array} \tag{27}$$

This means that this product is in fact the union of the previous two products.

4.4 Lexicographic product

The lexicographic product of two graphs K and H is denoted by $S = K o H$. For two nodes $u = (u_1, u_2)$ and $v = (v_1, v_2)$ in $N(K) \times N(H)$, we have an edge uv in $M(S)$ if the following condition is satisfied:

$$\begin{array}{c} u_1 v_1 \in M(K) \\ \text{or} \\ u_1 = v_1 \quad \text{and} \quad u_2 v_2 \in M(H) \end{array} \tag{28}$$

In fact the graph H is copied on each node of K, and if two nodes i and j of the graph K are connected to each other, then all the nodes of the copied graph H in these two nodes are connected to each other.

With these definitions and the above mentioned theorems, methods for calculating the eigenvalues of the Laplacian matrices of these product graphs are presented.

It should be noted that considering the relationships presented in [26], the eigenvalues of path and cycle graphs will be as follows:

For adjacency matrices

$$\lambda_m^P = 2\cos\frac{k\pi}{m+1} \quad : \quad k = 1:m$$
$$\lambda_m^C = 2\cos\frac{2k\pi}{m} \quad : \quad k = 1:m \tag{29}$$

In this case it is obvious that $\lambda_m \in \{-2, 2\}$

For Laplacian matrices

$$\lambda_m^P = 2 + 2\cos\frac{k\pi}{m} \quad : \quad k = 1:m$$
$$\lambda_m^C = 2 - 2\cos\frac{2k\pi}{m} \quad : \quad k = 1:m \tag{30}$$

In this case it is obvious that $\lambda_m \in \{0, 4\}$. Also we have

$$\lambda_2^P = (2\sin\frac{\pi}{2m})^2$$
$$\lambda_2^C = (2\sin\frac{\pi}{m})^2 \tag{31}$$

The first eigenvalue of both is zero.

Now using the above relationships the eigenvalues and eigenvectors of the Cartesian product can be obtained. These calculations become more important for the second eigenvalues, since these can be utilized in nodal ordering for profile reduction, and bisection for parallel computing.

In the following the results of block diagonalization of adjacency and Laplacian matrices are discussed for four types of product graphs.

4.5 Block diagonalization of adjacency matrices

For the adjacency matrices of three types of product graph (Cartesian, strong Cartesian and direct) we have:

$$M_{mn} = F_n(A_m, B_m, A_m) = I_n \otimes A_m + F_n(0,1,0) \otimes B_m \tag{32}$$

Since Equation (6) holds, thus it can be diagonalized and we have

$$eig(M_{mn}) = \bigcup_{i=1}^{n} \{eig(A_m + \lambda_i(T_n)B_m)\}; \lambda_i(I_n) = 1; \ i = 1:n \tag{33}$$

For the lexicographic product the adjacency matrix can be expressed in terms of the Kroneker product as the following:

$$A_{mn} = I_n \otimes H_m + G_n \otimes O_m \tag{34}$$

where O_m is a matrix with unit entries. Since $A_{mn} = A_1 \otimes B_1 + A_2 \otimes B_2$ and $A_i A_j = A_j A_i$, therefore it is diagonalizable and we can write:

$$eig(A_{mn}) = \bigcup_{i=1}^{n} \{eig(H_m + \lambda_i(G_n)O_m)\} \tag{35}$$

In this way instead of calculating the eigenvalues of a matrix of dimension $mn \times mn$, we need only to calculate the eigenvalues of a matrix of dimension m.

4.6 Block diagonalization of Laplacian matrices

In the Cartesian product this matrix is in the following form:

$$M_{mn} = F_n(A_m, B_m, C_m) = I_n \otimes (A+B)_m + F_n(1,-1,2) \otimes (-B_m) \tag{36}$$

Similar to the previous case this matrix is also diagonalizable and we have

$$eig(M_{mn}) = \bigcup_{i=1}^{n} \{eig[(A+B)_m - \lambda_i(T_n)B_m]\} \tag{37}$$

However for direct product and strong Cartesian product, though one can write them as the sum of two Kronecker products, however, Equation (9) does not hold. For direct product we will have:

$$M_{mn} = F_n(0,1,0) \otimes F_m(0,-1,0) + F_n(1,0,2) \otimes F_m(1,0,2) \tag{38}$$

And for Cartesian product we have:

$$M_{mn} = F_n(1,1,1) \otimes F_m(-1,-1,-1) + F_n(2,0,3) \otimes F_m(2,0,3) \tag{39}$$

Both are in the form $A_1 \otimes B_1 + A_2 \otimes B_2$. However in both cases $A_1 A_2 \neq A_2 A_1$ and therefore we cannot block diagonalize with the same method as we presented. In [27] members are added to the edges. In fact doing this A_2 and B_2 in the direct product are changed to $2I$ and in the strong Cartesian product these are changed to $3I$, and thus decomposition has become feasible. However, it can be observed that there is no need to change B_2 and altering A_2 is sufficient to make it block diagonalizable.

In these products if we add members only to two opposite edges (say top and bottom) in place of all around the product graph, then M_{mn} can be written similar to the Cartesian product as:

$$M_{mn} = I_n \otimes (A+B)_m + F_n(1,-1,2) \otimes (-B)_m \qquad (40)$$

Here as well M_{mn} is diagonalizable and we have:

$$eig(M_{mn}) = \bigcup_{i=1}^{n} \{ eig[A_m - (1 + 2\cos\frac{i\pi}{n})B_m]\} \qquad (41)$$

Since in this case edges are added only to one side of the graph, hence compared to the case where members were added all around the graph, the results are more accurate and the error is less, and using the Rayleigh quotient method only once the accuracy can be increased.

In Section 5 a different method for the decomposition of the Laplacian matrices of these two products into two blocks will be presented.

In order to calculate the eigenvalues of the Laplacian matrices of the Lexicographic products we have to present the following:

In some graphs it is possible not to have the commutativity property of the matrices A_i s. In another theorem we saw that if the B_i s commute pairwise, then the matrix can be block diagonalized. In this case, we will have the following theorem:

Theorem 3: two matrices $M = \sum_{i=1}^{k}(A_i \otimes B_i)$ and $N = \sum_{i=1}^{k}(B_i \otimes A_i)$ are similar.

The simplest result of this theory is that according to Theorem 1, the condition for M to be block diagonalizable is $A_i A_j = A_j A_i$. Now considering Theorem 3, if $B_i B_j = B_j B_i$ also holds, one can make the make a block diagonal one. The difference is that in the first case the matrix will have m blocks and in the second case it will have n blocks. The graph interpretation of this is the numbering of $G \circ H$ starting with nodes of G or H which lead to these two cases.

Considering the above cases we can replace L_{mn} with its similar one L_{nm}.

$$L_{nm} = H_m \otimes I_n + O_m \otimes G_n + K_m \otimes DG_n = \sum_{i=1}^{3}(B_i \otimes A_i) \qquad (42)$$

One can easily show that $B_i B_j = B_j B_i$ and in this way we can write:

$$eig(L_{nm}) = \bigcup_{i=1}^{m}[eig(\lambda_i(H_m)I_n + \lambda_i(O_m)G_n + \lambda_i(K_m)DG)_n)] \qquad (43)$$

After diagonalization of H_m, O_m and K_m simultaneously, the eigenvalues of these matrices will appear on their main diagonals and will be used in the above relationship. Then the eigenvalues of these matrices will be in the following form:

$$eig(L_{nm}) = \bigcup_{i=1}^{m}\left[eig\left(\begin{bmatrix}0\\\lambda_2\\\cdot\\\cdot\\\lambda_m\end{bmatrix}I_n + \begin{bmatrix}m\\0\\\cdot\\\cdot\\0\end{bmatrix}G_n + \begin{bmatrix}0\\m\\\cdot\\\cdot\\m\end{bmatrix}DG_n\right)\right] \qquad (44)$$

or

$$eig(L_{nm}) = \bigcup_{i=1}^{m}[eig(mDG_n + \lambda_i(H_m)I_n)] \cup eig(mG_n) \qquad (45)$$

However $mDG_n + \lambda_i(H_m)I_n$ is diagonal and the eigenvalues are the numbers on the main diagonal, thus instead of forming the matrix we form only its main diagonal. Since from this relationship for $i = 2 : m$ and for each i we obtain m numbers, it means, similar to Cartesian product, the vector on the diagonal of the matrices mDG_n and $\lambda_i(H_m)I_n$ can be added. It is obvious that the numbers on the diagonal of DG_n are the degrees of the nodes of the G_n (i.e. $d(G_n)$. The above steps can be summarized as:

1. Calculate the eigenvalues of G_n and H_m and delete the zero from the eigenvalues of the H_m.

2. Multiply the vector $d(G_n)$ by m and sum it with the vector on the diagonal $\lambda_i(H_m)I_n$ similar to the Caresian product two by two. We call these as the primary group of answers.

3. Multiply the eigenvalues of G_n by m to obtain the second group of answers.

In this way we need to calculate only the eigenvalues of G_n and H_m, and the produced diagonal blocks are fortunately diagonal themselves and their eigenvalues appear on the main diagonal.

As an example of Theorem 3, we want to block diagonalise the Laplacian matrix of the strong Cartesian product. For the case $P \square C$, we want to find the eigenvalues in terms of its blocks. In this case

$$L_{mn} = F_n(A_m, B_m, C_m)$$
$$A_m = G_m(5,-1,5), \quad B_m = G_m(-1,-1,-1) \text{ and } C_m = G_m(8,-1,8)$$

Thus

$$A_m = 6I_m + B_m, \quad C_m = 9I_m + B_m$$

Therefore

$$L_{mn} = F(6I_m + B_m, B_m, 9I_m + B_m) - 3F_n(2,0,3) \otimes I_m + F_n(1,1,1) \otimes B_m$$

Since

$$eig(B_m) = -(1 + 2\cos\frac{2k\pi}{m}) \quad k = 1:m$$

Using Theorem 3 we have

$$eig(\sum(A_i \otimes B_i)) = eig(\sum(B_i \otimes A_i))$$

Therefore

$$eig(L_{mn}) = eig\{I_m \otimes 3F_n(2,0,3) + B_m \otimes F_n(1,1,1)\}$$

Since I_m and B_m have commutativity property, thus

$$eig(L_{mn}) = \bigcup_{k=1}^{m}[eig\{3F_n(2,0,3) - (1 + 2\cos\frac{2k\pi}{m})F_n(1,1,1)\}]$$

In this way the Laplacian matrix becomes a block form without the need for adding members to the edges of the graph. The only important point is that the use of Theorem 3 made the matrix diagonal in n blocks, with the performed replacement m blocks are positioned on the diagonal and the eigenvalues of L are obtained in terms of the eigenvalues of each block independently.

Instead of each calculation one can use a weighted graph which will produce equivalent results [28]. As an example if we assume $m=5$ and $n=4$, it is sufficient first to obtain the eigenvalues (equivalent to $eig(B_m)$) for the graph shown in Figure 1(a), and then substitute each magnitude in the weighted graph (Figure 1(b)), and find its eigenvalues.

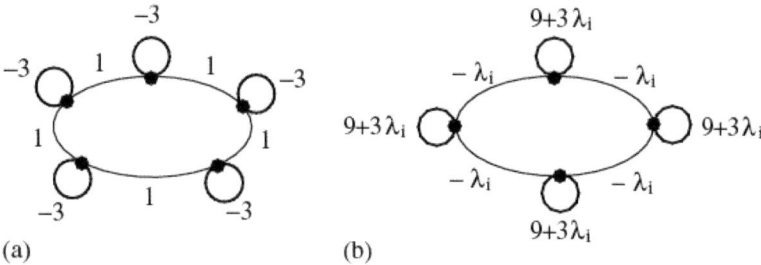

Figure 1: Weighted graphs for finding eigenvalues

5 Decomposition of the Matrices of the Form $F_n(A_m, B_m, C_m)$

We found that the form $L_{mn} = F_n(A_m, B_m, C_m)$ if the condition (9) does not hold, then decomposition cannot be performed. In [29] it is shown that in such a case at least we can decompose it into two blocks. As an example we can apply this to the Laplacian matrices of the strong Cartesian product and direct product. The relationships for both even and odd n are presented in the following:

A. If n is even (here it is $n = 4$), we interchange the rows and columns 3 and 4 block wise, then the eigenvalues will stay the same and L_{mn} will be as the following:

$$L_{m \times 4} = \begin{bmatrix} A_m & B_m & 0 & 0 \\ B_m & C_m & 0 & B_m \\ 0 & 0 & A_m & B_m \\ 0 & B_m & B_m & C_m \end{bmatrix} = \begin{bmatrix} M_{m \times 2} & N_{m \times 2} \\ N_{m \times 2} & M_{m \times 2} \end{bmatrix} \qquad (46)$$

This is in fact the Form II and its eigenvalues are obtainable from $M_{m \times 2} \pm N_{m \times 2}$. The generalization of this work leads to the following two matrices ($n = 2k$):

$$M_{mk} = \begin{bmatrix} A_m & B_m & & & 0 \\ B_m & C_m & \cdot & & \\ & B_m & \cdot & \cdot & \\ & & \cdot & \cdot & \cdot \\ & & & \cdot & B_m \\ 0 & & & B_m & C_m \end{bmatrix}_k, \quad N_{mk} = \begin{bmatrix} 0 & 0 & & & 0 \\ 0 & 0 & \cdot & & \\ & 0 & \cdot & \cdot & \\ & & \cdot & \cdot & \cdot \\ & & & \cdot & 0 \\ 0 & & & 0 & B_m \end{bmatrix}_k \qquad (47)$$

B. When n is odd (e.g. $n = 3$ in here), then one zero block row and block column are added between block 2 and block 3 followed by the previous operations. After some row and column operations, since we have an even number of blocks, the rows and columns 3 and 4 are interchanged block wise and we obtain the following form:

$$L_{m \times 4} = \begin{bmatrix} A_m & B_m & 0 & B_m \\ B_m & C_m & B_m & C_m \\ \overline{2} & \overline{2} & \overline{2} & \overline{2} \\ 0 & B_m & A_m & B_m \\ B_m & C_m & B_m & C_m \\ \overline{2} & \overline{2} & \overline{2} & \overline{2} \end{bmatrix} = \begin{bmatrix} M_{m \times 2} & N_{m \times 2} \\ N_{m \times 2} & M_{m \times 2} \end{bmatrix} \tag{48}$$

This means we will have again a Form II matrix.
The generalization results in the following two matrices M_{mk} and N_{mk}. For $n = 2k - 1$ we have

$$M_{mk} = \begin{bmatrix} A_m & B_m & & & & 0 \\ B_m & C_m & \cdot & & & \\ & B_m & \cdot & \cdot & & \\ & & \cdot & \cdot & \cdot & \\ & & & B_m & C_m & B_m \\ 0 & & & & \frac{B_m}{2} & \frac{C_m}{2} \end{bmatrix}_k, \quad N_{mk} = \begin{bmatrix} 0 & 0 & 0 & & & 0 \\ 0 & 0 & \cdot & & & \\ & & 0 & \cdot & \cdot & \\ & & & \cdot & \cdot & 0 \\ & & & & \cdot & B_m \\ 0 & & & 0 & \frac{B_m}{2} & \frac{C_m}{2} \end{bmatrix}_k$$
(49)

Obviously after calculating the eigenvalues $M_{m \times 2} \pm N_{m \times 2}$ we will have m extra zeros which correspond to the addition of the zero rows and columns which we have added and should now be omitted.

6 Conditions for decomposibility of block matrices

In some cases, matrix calculations can be reduced to numerical calculations (matrices of dimension 1). If both A_i s and B_i s fulfill the commutativity conditions in multiplication, then the calculations will be simplified. In such a case the eigenvalues of the sum of some matrices will reduce to the sum of their eigenvalues The following theorem clarifies this problem:

Theorem 4: One condition for

$$eig(\sum_{i=1}^{n} A_i \otimes B_i) = \sum_{i=1}^{n} eig(A_i \otimes B_i) \tag{50}$$

to hold is that

$$A_i A_j = A_j A_i, B_i B_j = B_j B_i; \quad i,j = 1:n \tag{51}$$

For calculating the eigenvectors it is sufficient for v to be an eigenvector which diagonalizes the two matrices B_1 and B_2 simultaneously. In this case, if μ is an eigenvalue corresponding to the eigenvector, after using QZ transformation and transforming B_2 to I, we will have:

$$B_2 v = v \quad ; \quad B_1 v = \mu v \tag{52}$$

In this way we have

$$(A_1 \otimes B_1 + A_2 \otimes B_2)(u \otimes v) = u \otimes \mu v + \lambda u \otimes v = (\mu + \lambda)(u \otimes v) \tag{53}$$

In this way $u \otimes v$ will be the eigenvector of M with the difference that here v is an eigenvector which diagonalizes the two matrices B_1 and B_2 simultaneously, while in the previous part only A_is were commutative with respect to multiplication and v was assumed to be the eigenvector of $M_i = B_1 + \lambda_i (A_2) B_2$.

As an example of Theorem 4, the eigenvalues of C☐C are calculated in the following:

$$L_{mn} = F(A_m, B_m, C_m)$$
$$A_m = G_m(8,-1,8), \quad B_m = G_m(-1,-1,-1) \text{ and } C_m = A_m$$

Therefore $A_m = 9I_m + B_m$ and thus

$$L_{mn} = G_n(9I_m + B_m, B_m, 9I_m + B_m) = 9I_n \otimes I_m + G_n(1,1,1) \otimes B_m$$

Finally we have

$$eig(L_{mn}) = \bigcup_{k=1}^{m} [eig\{9I_n - (1 + 2\cos\frac{2k\pi}{m})G_n(1,1,1)\}]$$
$$= 9 - (1 + 2\cos\frac{2k\pi}{m})(1 + 2\cos\frac{2k'\pi}{n})$$

Since the two matrices I_n and G_n commute, therefore using Theorem 4 decomposes the matrix. Similar operations can be repeated for direct products.

Here, similar to the previous case, introducing

$$\lambda_1 = -(1 + 2\cos\frac{k\pi}{m}) \quad ; \quad \lambda_2 = 1 + 2\cos\frac{k'\pi}{n}$$

The corresponding graph is as shown in Figure 2.

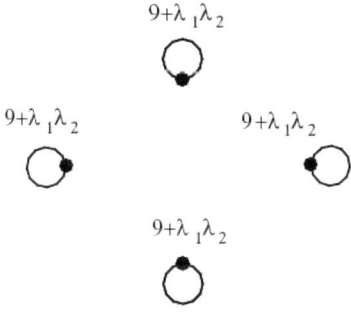

Figure 2: The weighted graph for finding the eigenvalues

7 Transformation of Form *III* to Form *II* and its generalization

As is shown in Reference [23], Form *III* is as the following:

$$\begin{bmatrix} A & B & P \\ B & A & P \\ Q & H & R \end{bmatrix} \quad (54)$$

As can be seen here we have one additional row and column compared to the Form *II*. The aim is to transform this into a Form II matrix. Then we consider an additional row and column for the matrix $G_n(A_m, B_m, A_m)$ and the operations on it are also generalized.

First we add an arbitrary zero block of row and column as the following:

$$\begin{bmatrix} A & B & P \\ B & A & P \\ Q & H & R \end{bmatrix} \Rightarrow \begin{bmatrix} A & 0 & B & P \\ \frac{H-Q}{2} & 0 & \frac{Q-H}{2} & 0 \\ B & 0 & A & P \\ Q & 0 & H & R \end{bmatrix} \quad (55)$$

Half of the column 4 is added to column 2, and a similar operation is performed on the corresponding rows:

$$\begin{bmatrix} A & \frac{P}{2} & B & P \\ \frac{H}{2} & \frac{R}{4} & \frac{Q}{2} & \frac{R}{2} \\ B & \frac{P}{2} & A & P \\ Q & \frac{R}{2} & H & R \end{bmatrix} \Rightarrow \begin{bmatrix} A & P & B & \frac{P}{2} \\ \frac{H}{2} & \frac{R}{2} & \frac{Q}{2} & \frac{R}{4} \\ B & P & A & \frac{P}{2} \\ Q & R & H & \frac{R}{2} \end{bmatrix} \qquad (56)$$

Since columns 2 and 4 are proportional these can be interchanged, column 4 is multiplied by 2 and row 4 is divided by 2. Ultimately we get:

$$\begin{bmatrix} A & P & B & P \\ \frac{H}{2} & \frac{R}{2} & \frac{Q}{2} & \frac{R}{2} \\ B & P & A & P \\ \frac{Q}{2} & \frac{R}{2} & \frac{H}{2} & \frac{R}{2} \end{bmatrix} \Rightarrow \begin{bmatrix} M & N \\ N & M \end{bmatrix} \qquad (57)$$

The generalization of the Form III for matrix $G_n(A_m, B_m, A_m)$ will be as follows (for $n = 5$):

$$L_{5m+k} = \begin{bmatrix} A_m & B_m & Z_m & Z_m & Z_m & B_m & P_{mk} \\ B_m & A_m & B_m & Z_m & Z_m & Z_m & P_{mk} \\ Z_m & B_m & A_m & B_m & Z_m & Z_m & P_{mk} \\ Z_m & Z_m & B_m & A_m & B_m & Z_m & P_{mk} \\ Z_m & Z_m & Z_m & B_m & A_m & B_m & P_{mk} \\ B_m & Z_m & Z_m & Z_m & B_m & A_m & P_{mk} \\ P_{mk}^t & P_{mk}^t & P_{mk}^t & P_{mk}^t & P_{mk}^t & P_{mk}^t & R_k \end{bmatrix}_{5m+k} \qquad (58)$$

Where Z_m is a zero matrix of dimension m. Similar to the above operations the process can be summarized as [30]:

1. First we add $n-1$ rows and columns of zero blocks similar to the one we did for the case $n = 2$. In this way these rows and columns are considered in a distance of each row and column of the initial matrix. It should be mentioned that in the considered matrix Q and H are both P^t.

2. Then the last column is multiplied by $\frac{1}{n}$ and the result is added to all the columns. Similar operations are performed for the rows. In fact the last column which is an additional column is divided among the remaining columns.

3. The even columns are interchanged and at the end the last column is multiplied by n and the last row is multiplied by $\frac{1}{n}$.

As an example for $n = 5$ the final matrix will be in the following form:

$$L_{5(m+k)} = \begin{bmatrix} M_{m+k} & N_{m+k} & W_{m+k} & W_{m+k} & N_{m+k} \\ N_{m+k} & M_{m+k} & N_{m+k} & W_{m+k} & W_{m+k} \\ W_{m+k} & N_{m+k} & M_{m+k} & N_{m+k} & W_{m+k} \\ W_{m+k} & W_{m+k} & N_{m+k} & M_{m+k} & N_{m+k} \\ N_{m+k} & W_{m+k} & W_{m+k} & N_{m+k} & M_{m+k} \end{bmatrix} \quad (59)$$

Where

$$M_{m+k} = \begin{bmatrix} A_m & P_{mk} \\ \frac{1}{n}P_{mk}^t & \frac{1}{n}R_k \end{bmatrix} \; ; \; N_{m+k} = \begin{bmatrix} B_m & P_{mk} \\ \frac{1}{n}P_{mk}^t & \frac{1}{n}R_k \end{bmatrix} \; ; \; W_{m+k} = \begin{bmatrix} Z_m & P_{mk} \\ \frac{1}{n}P_{mk}^t & \frac{1}{n}R_k \end{bmatrix} \quad (60)$$

It should be noted that all the matrices which were zero block (matrix Z), now have k additional rows and columns which are non-zero (matrix W). Another important issue is that performing these operations increases the dimension of the matrix to $n(m+k)$ and since this is done by the addition of rows and columns to the matrix, thus the rank of each submatrix is increased by k. This means that the eigenvalues of $n-1$ submatrices will have k additional zero eigenvalues which should be omitted from the answer. It should be noted that the above obtained matrix is in fact a circulant matrix which will be introduced in the subsequebnt section.

The Laplacian matrix can be expressed as

$$L_{n(m+k)} = \sum_{i=1}^{3}(A_i \otimes B_i) = I_n \otimes M_{m+k} + G_n(0,1,0) \otimes N_{m+k} + K_n \otimes W_{m+k} \quad (61)$$

where

$$K_n = O_n - I_n - G_n(0,1,0) \; ; \; O_n = ones(n) \quad (62)$$

It can simply be seen that Equation (21) holds and thus it can be block diagonalized. According to Theorem 2 we have

$$\lambda_L = \bigcup_{i=1}^{n}\{eig\,[L_i]\} \; ; \; L_i = \sum_{j=1}^{3}\{\lambda_j(A_i)B_i\} = M_{m+k} + \lambda(G_n(0,1,0))N_{m+k} + \lambda(K_n))W_{m+k} \quad (63)$$

From the results of Ref. [26] we will have

$$\lambda(G_n(0,1,0)) = 2\cos\frac{2k\pi}{n} \quad ; \quad k = 1:n \tag{64}$$

For calculating the eigenvalues of K_n we can write:

$$K_n = O_n - I_n - G_n(0,1,0) = \sum_{i=1}^{3} C_i \tag{65}$$

Since

$$C_i C_j = C_j C_i \quad i,j = 1:3, \ i \neq j \tag{66}$$

Therefore

$$eig(K_n) = eig(O_n) - eig(I_n) - eig(G_n(0,1,0)) \tag{67}$$

Or

$$eig(K_n) = \begin{bmatrix} 0 \\ 0 \\ \cdot \\ \cdot \\ \cdot \\ 0 \\ n \end{bmatrix} - \begin{bmatrix} 1 \\ 1 \\ \cdot \\ \cdot \\ \cdot \\ 1 \\ 1 \end{bmatrix} - \begin{bmatrix} 2\cos(2\pi/n) \\ 2\cos(4\pi/n) \\ \cdot \\ \cdot \\ \cdot \\ 2\cos(2(n-1)\pi/n) \\ 2 \end{bmatrix} \tag{68}$$

In this way we get

$$eig(K_n) = \{\bigcup_{k=1}^{n-1}[-(1+2\cos\frac{2k\pi}{n})]\} \cup \{n-3\} \tag{69}$$

Finally

$$\lambda_L = \{\bigcup_{i=1}^{n-1} eig\,(M_{m+k} + 2\cos\frac{2i\pi}{n}(N_{m+k} - W_{m+k}) - W_{m+k})\} \cup \{eig(M_{m+k} + 2N_{m+k} + (n-3)W_{m+k})\} \tag{70}$$

It should be mentioned that one can perform similar operations on the form $F_n(A_m, B_m, A_m)$. However the difficulty with this form is that it does not satisfy the Equation (9). Of course one can relate the two forms F and G in a special case. We know

$$\begin{aligned} eig(F_n(a,b,a)) &= a + 2b\cos\frac{k\pi}{n+1} \quad ; \quad k = 1:n \\ eig(G_m(a,b,a)) &= a + 2b\cos\frac{2k'\pi}{m} \quad ; \quad k' = 1:m \end{aligned} \tag{71}$$

Graph Products and Canonical Forms in Structural Mechanics

In the simplest way by equating these two we have

$$k = k' \quad ; \quad \cos\frac{k\pi}{n+1} = \cos\frac{2k\pi}{m} \Rightarrow m = 2n+2 \tag{72}$$

This means that a path of length n is transformed into a cycle of length $2n+2$. This is equivalent to putting an additional path next to the existing one and from each side connecting them with an auxiliary node.

Considering the above relationship for G it can be realized that in a cycle all the eigenvalues are repeated twice except the maximum and minimum eigenvalues ($a+2b$ and $a-2b$) which correspond to the two added auxiliary nodes. All that is needed is to delete these two values and the repeated eigenvalues.

8 Circulant block matrices

A matrix of the following form is known as a block circulant matrix:

$$C = \begin{bmatrix} A_1 & A_2 & \cdots & A_{n-1} & A_n \\ A_n & A_1 & \cdots & A_{n-2} & A_{n-1} \\ \cdot & & & & \cdot \\ \cdot & & & & \cdot \\ A_3 & A_4 & \cdots & A_1 & A_2 \\ A_2 & A_3 & \cdots & A_n & A_1 \end{bmatrix} \tag{73}$$

As can be seen, the repeated blocks in the first row are circulated in the other rows. Suppose P_i be a unit matrix which is shifted i columns. In such a case C can be expressed as

$$C = \sum_{i=1}^{n}(P_i \otimes A_i) \tag{74}$$

In Reference [31] the matrix function $H(x)$ is defined as follows:

$$\begin{cases} H : C \to C^2 \\ H(x) = \sum_{i=1}^{n}(x^{i-1} \otimes A_i) \end{cases} \tag{75}$$

The eigenvalues of the matrix P_i of dimension n are as follows:

$$eig(P_i) = \omega^k \quad ; \quad \omega = e^{\frac{2\pi i}{n}} \quad ; \quad k = 0 : n-1 \qquad (76)$$

In this case it was observed that $H(\omega^i)$ are the eigenvalues of C. Also if u and v are the eigenvectors of $H(\omega^i)$ and P_i respectively, then $v \otimes u$ will be the eigenvectors of C.

This form appears in those models which contain a cycle. In fact $G_n(A_m, B_m, A_m)$ is a special case of C.

9 Complementary examples

In this Section three examples are included to illustrate some applications of the developed theorems and algorithms. In Example 1 the critical load of a plate is calculated [24]. Example 2 deals with static analysis of a circular structure taken from [32]. Finally in the third example, the frequencies and vibration natural modes of a structure are calculated [31].

Example 1: Consider a simply supported a×a square plate, as shown in Figure 3. The load is applied in x direction. Using the finite difference approach, the critical load of the plate is calculated.

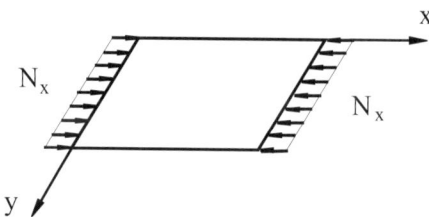

Figure 3: A simply supported plate

Considering the governing differential equation as

$$\nabla^4 w + \frac{N_x}{D} \frac{\partial^2 w}{\partial x^2} = 0,$$

and employing the finite difference method, the matrix **M** is obtained in the following form:

Graph Products and Canonical Forms in Structural Mechanics

$$\mathbf{M} = \mathbf{I} \otimes \mathbf{A} + \mathbf{T} \otimes \mathbf{B} + \mathbf{S} \otimes \mathbf{I} = \sum_{i=1}^{3} \mathbf{A}_i \otimes \mathbf{B}_i$$

where

$$\mathbf{B}_1 = F(18-2x, x-8, 19-2x, 1), \mathbf{B}_2 = F(-8, 2, -8), \mathbf{A}_2 = F(0,1,0), \mathbf{A}_3 = F(0,0,1,1).$$

Since $\mathbf{A}_i \mathbf{A}_j = \mathbf{A}_j \mathbf{A}_i$ for each pair of i and j, using Equation (26) we obtain:

$$\lambda_\mathbf{M} = \bigcup_{j=1}^{n} [\text{eig} \sum_{i=1}^{3} \lambda_j(\mathbf{A}_i)\mathbf{B}_i)]; \quad \lambda_{\mathbf{A}_1} = 1, \lambda_{\mathbf{A}_2} = 2\cos\frac{k\pi}{n+1}, \lambda_{\mathbf{A}_3} = 1 + 2\cos\frac{2k\pi}{n+1}$$

Once $\lambda_\mathbf{M}$ is calculated, it can be observed that it contains diagonal blocks and each block has the form **F**. Thus, the diagonalization is performed once again for each block, since $\mathbf{A}_i \mathbf{A}_j = \mathbf{A}_j \mathbf{A}_i$ still holds for these blocks.

For critical load (k=1), we have

$$18 - 2x - 16\cos\frac{\pi}{m} + 1 + 2\cos\frac{2\pi}{m} + 2\cos\frac{\pi}{m}(x - 8 + 4\cos\frac{\pi}{m}) + 1(1 + 2\cos\frac{2\pi}{m}) = 0; m = n+1$$

$$\Rightarrow x = 4(3 - 4\cos\frac{\pi}{m} + \cos\frac{2\pi}{m})/(1 - \cos\frac{\pi}{m})$$

In this relationship, $m \to \infty$ leads to an accurate value of the critical load as $x = \frac{4\pi^2}{m^2}$, where $1 - \cos\alpha \cong \frac{\alpha^2}{2}$ when $\alpha \to 0$. This result is in good agreement with the exact value, which is

$$N_{cr} = \frac{xD}{(\frac{a}{m})^2} = \frac{4\pi^2 D}{a^2}$$

Example 2: Suppose we want to study a structure in the form of the strong Cartesian product of P_2 by C_5. Three dimensional and two dimensional configurations of this structure are shown in Figure 4. In these figures the geometric properties and loading are specified. In fact this structure is formed by two equilateral polygons with five edges, where the distance of the five external and internal nodes from their centers are 3 and 1.5, respectively. The external nodes 2, 4, 6, 8 and 10 are at a height of 1.5m. All the cross sections areas are 5cm² and the elastic modulus is taken as 200kN/mm². The loads P_1 at node 2 is 30kN and the load P_2 at node 6 is 20kN. It can be seen that the loading is non-symmetric.

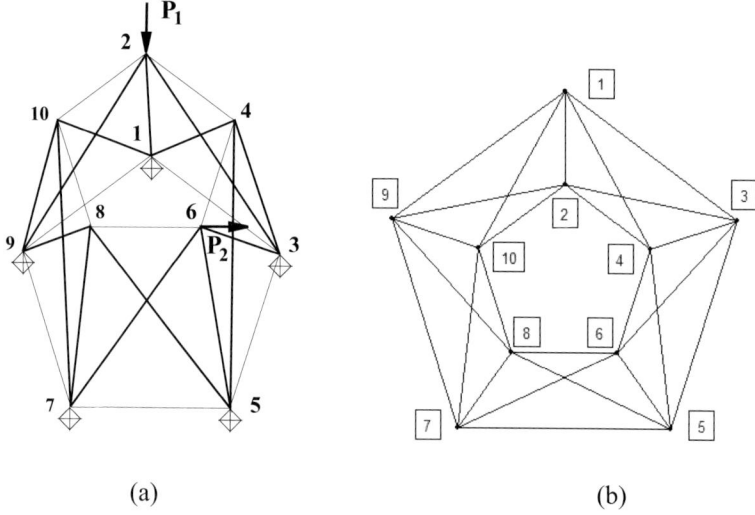

Figure 4: Three and two dimensional views of a structure

If we form the stiffness matrix of this structure in the Cartesian coordinate, we will find out that it does not obey the pattern of the repetitive form (Equation (8)). However, forming this matrix in the cylindrical coordinate system, the reduced stiffness matrix will have the following form:

$$M = A_1 \otimes B_1 + A_2 \otimes B_2 + A_3 \otimes B_3 \quad ; \quad A_1 = I \quad ; \quad B_3 = B_2^t$$

$$A_3 = A_2^t = \begin{bmatrix} 0 & 1 & & & & 0 \\ & 0 & 1 & & & \\ & & \cdot & \cdot & & \\ & & & \cdot & \cdot & \\ & & & & 1 & \\ & & & & 0 & 1 \\ 1 & & & & & 0 \end{bmatrix}_n$$

Where

$$B_1 = 10^5 \begin{bmatrix} 0.6463 & 0 & -0.1867 \\ 0 & 1.2063 & 0 \\ -0.1867 & 0 & 0.3639 \end{bmatrix} \text{ and } B_2 = 10^4 \begin{bmatrix} 1.9593 & 2.6967 & 0 \\ -2.6967 & -3.7117 & 0 \\ 0 & 0 & 0 \end{bmatrix}$$

The corresponding eigenvalues of A_3 (and A_2) are provided by :

$$\det(A_3 - \lambda I) = 0 \Rightarrow \lambda^n = 1 \Rightarrow \lambda = \{1, e^{\frac{2\pi i}{n}}, e^{\frac{4\pi i}{n}}, ..., e^{\frac{2(n-1)\pi i}{n}}\}$$

Graph Products and Canonical Forms in Structural Mechanics

In this example, both matrices have the dimension equal to 5.

$$eig(A_2) = \{1, e^{\frac{2\pi i}{5}}, e^{\frac{4\pi i}{5}}, e^{\frac{6\pi i}{5}}, e^{\frac{8\pi i}{5}}\}$$
$$= \{1, 0.3090 + 0.9511i, -0.8090 + 0.5878i, -0.8090 - 0.5878i, 0.3090 - 0.9511i\}$$

The eigenvalues of the matrix M can be found as follows:

$$eig(M) = \bigcup_{i=1}^{5}[eig\{B_1 + \lambda_i(A_2)B_2 + \lambda_i(A_2')B_2'\}_3]$$

As an example the biggest eigenvalue is calculated as $\lambda_{max} = 1.0864e5$

The eigenvectors of this matrix are $u \otimes v$ where both u and v are introduced before. As an example the eigenvectors corresponding to the above given eigenvalue are as

$$\{\varphi\} = [1 \quad 1 \quad 1 \quad 1 \quad 1]^t \otimes [0.4330 \quad 0 \quad -0.1119]^t.$$

It should be noted that for a structure the governing equation is $K\Delta = P$. In this equation K is the reduced cylindrical stiffness matrix of the structure and P is the force vector which is expressed in the cylindrical coordinate system. Using Equation (23), the displacements are found in the cylindrical coordinate system (ρ, θ, z) as

$$\Delta = \theta \begin{matrix} \rho \\ \theta \\ z \end{matrix} \begin{bmatrix} 0.5330 & -0.2351 & 0.4221 & -0.2549 & -0.1770 \\ 0.0028 & 0.0924 & 0.2169 & 0.1621 & -0.1254 \\ 1.0978 & -0.1206 & 0.2166 & -0.1308 & -0.0908 \end{bmatrix}$$

and after transformation, the displacements are obtained in the Cartesian coordinate system (x, y, z) as

$$\Delta = \begin{matrix} x \\ y \\ z \end{matrix} \begin{bmatrix} -0.0028 & -0.2521 & 0.4236 & 0.2810 & 0.2071 \\ 0.5330 & 0.0152 & -0.2140 & 0.1110 & 0.0646 \\ 1.0978 & -0.1206 & 0.2166 & -0.1308 & -0.0908 \end{bmatrix}$$

Here the columns 1 to 5 contain the displacements of the nodes 2, 4, 6, 8 and 10 in the specified directions.

Example 3: In this example a structure is considered as shown in Figure 5. We want to calculate the frequencies and the natural modes of this structure under its self-weight. The area for all the cross sections are considered as 5cm^2, the elastic modulus is taken as 200kN/mm^2 (MPa). For all the elements $\rho = 78.5$ KN/m^3.

Similar to Reference [32] first the stiffness and mass matrices are constructed in a cylindrical coordinate system. Here, the stiffness and mass matrices will have a similar pattern to that of the Laplacian matrix of the previous example, with the only difference being that the dimensions will be three times larger. These matrices have the following patterns:

$$K = \begin{bmatrix} A_m & B_m & 0 & B_m^t & P_{mk} \\ B_m^t & A_m & B_m & 0 & P_{mk} \\ 0 & B_m^t & A_m & B_m & P_{mk} \\ B_m & 0 & B_m^t & A_m & P_{mk} \\ P_{mk}^t & P_{mk}^t & P_{mk}^t & P_{mk}^t & R_k \end{bmatrix} \quad ; \quad M = \begin{bmatrix} A_m & B_m & 0 & B_m^t & P_{mk} \\ B_m^t & A_m & B_m & 0 & P_{mk} \\ 0 & B_m^t & A_m & B_m & P_{mk} \\ B_m & 0 & B_m^t & A_m & P_{mk} \\ P_{mk}^t & P_{mk}^t & P_{mk}^t & P_{mk}^t & R_k \end{bmatrix}$$

Where $m = 18$, $n = 4$ and $k = 3$. It can be observed that here the last block column has three numerical columns in place of a single one. Therefore, in performing the algorithm for decomposing this block column to other blocks, we consider three additional zero columns next to each block matrix. In this case, the stiffness and mass matrices will have the following form after deleting the rows and columns corresponding to the support nodes.

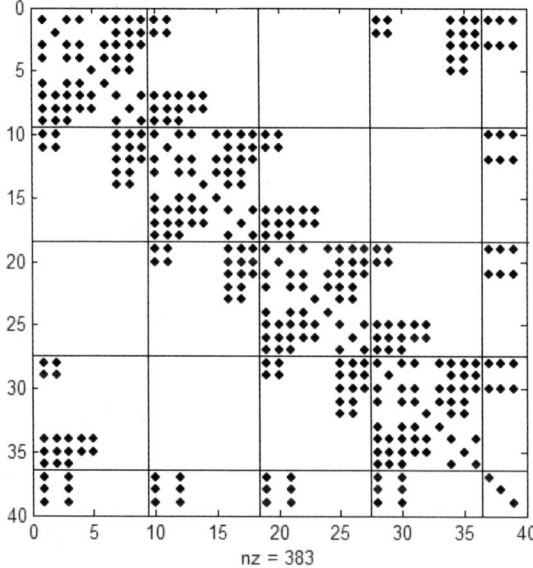

Figure 5: The patterns of the stiffness and mass matrices of Example 3

It can be seen that the mass matrix is not diagonal, and one cannot calculate the eigenvalue for the free vibration merely by considering the stiffness matrix. Thus the $K - M\omega^2$ matrix is constructed as

$$K - M\omega^2 = \begin{bmatrix} M_K - M_M\omega^2 & N_K - N_M\omega^2 & W_K - W_M\omega^2 & S_K - S_M\omega^2 \\ S_K - S_M\omega^2 & M_K - M_M\omega^2 & N_K - N_M\omega^2 & W_K - W_M\omega^2 \\ W_K - W_M\omega^2 & S_K - S_M\omega^2 & M_K - M_M\omega^2 & N_K - N_M\omega^2 \\ N_K - N_M\omega^2 & W_K - W_M\omega^2 & S_K - S_M\omega^2 & M_K - M_M\omega^2 \end{bmatrix}$$

This is also a circulant matrix. Therefore $H(x)$ can be written as

$$H(x) = x^0 \otimes (M_K - M_M\omega^2) + x^1 \otimes (N_K - N_M\omega^2) + x^2 \otimes (W_K - W_M\omega^2) + x^3 \otimes (S_K - S_M\omega^2)$$

We also have

$$\omega^k = (e^{2\pi i/4})^k = \{\pm 1, \pm i\} \quad ; \quad k = 0:3$$

For all values of x, the submatrices $H(x)$ are obtained in terms of ω, and equating them to zero, the corresponding eigenvalues are calculated. The largest period of this structure is obtained as $T_1 = 0.0826$. For the construction of the eigenvectors, as we mentioned before, u and v should be calculated. Then we obtain the eigenvectors which can be calculated as

$$\omega = e^{2\pi i/4} = i \Rightarrow v = \{1, \omega, \omega^2, ..., \omega^{(n-1)}\}^t = \{1, i, -1, i\}^t$$

For all values of u, the sunbmatrices $H(\omega^i)$ are calculated, and finally the vibrating modes $v \otimes u$ are obtained. As an example, for the 6th period ($T_6 = 0.0456$), the 6th natural mode is shown in Figure 6.

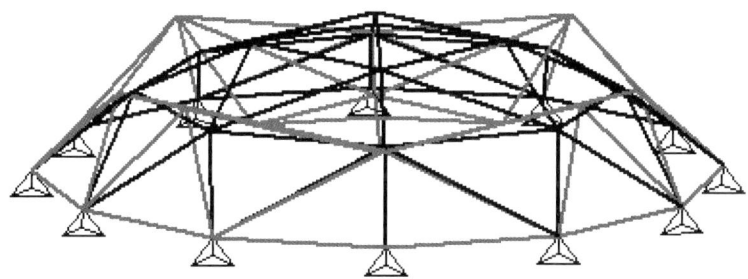

Figure 6: The 6th natural mode of the structure in Example 1

10 Conclusions

In this chapter different canonical forms are studied. Depending on the layout of the constituting sub-matrices of the matrix different methods should be used for calculating the eigenvalues and eigenvectors, and finding their inverse. Such matrices arise in the adjacency and Laplacian matrices of the product graphs. The stiffness matrices of repetitive structures in block form may also be formed in canonical forms using appropriate coordinate systems. Here conditions required for the formation of such matrices are studied and methods are presented for their decomposition into smaller sub-matrices. In certain cases some row and column operations alter the layout of the block matrices and make them suitable for decomposition and block diagonalization. In certain matrices the circulant nature helps to achieve the decomposition by introducing a matrix function to calculate the eigenvalues and eigenvectors. For each case examples are presented for clarification.

References

[1] A. Kaveh, M.A. Sayarinejad, "Eigensolutions for matrices of special patterns", Communications in Numerical Methods in Engineering, 19, 125-136, 2003.
[2] A. Kaveh, M.A. Sayarinejad, "Eigenvalues of factorable matrices with form IV symmetry", Communications in Numerical Methods in Engineering, 21, 269-278, 2005.
[3] A. Kaveh, "Structural Mechanics: Graph and Matrix Methods", 3rd ed. Research Studies Press, Somerset, UK, 2004.
[4] A. Kaveh, B. Salimbahrami, "Eigensolutions of symmetric frames using graph factorization", Communications in Numerical Methods in Engineering, 20, 889-910, 2004.
[5] D.L. Thomas, "Dynamics of rotationally periodic structures", International Journal for Numerical Methods in Engineering, 14, 81-102, 1979.
[6] F.W. Williams, "An algorithm for exact eigenvalue calculations for rotationally periodic structures", International Journal for Numerical Methods in Engineering, 23, 609-622, 1986.
[7] F.W. Williams, "Exact eigenvalue calculations for rotationally periodic sub-structures", International Journal for Numerical Methods in Engineering, 23, 695-706, 1986.
[8] M.A. Hasan, J.A.K. Hasan, "Block eigenvalue decomposition using nth roots of identity matrix", 41st IEEE Conference on Decision and Control, 2, 2119-2124, 2002.
[9] A.O. Aghayere, "Structural Systems with Polar Symmetry: Solution by Quasi-circulant Matrices", M.Sc. thesis, Massachusetts Institute of Technology, March 1983.
[10] R.D. Kangwai, S.D. Guest, S. Pellegrino, "An introduction to the analysis of symmetric structures", Computers and Structures, 71, 671-688, 1999.
[11] Y. Vanberghen, R. Vandebril, M. Van Barel, "A QZ-method based on semi-separable matrices", Journal of Computational and Applied Mathematics, 218, 482-491, 2008.
[12] G.A.A. Kahou, L. Grigori, M. Sosonkina, "A partitioning algorithm for block-diagonal matrices with overlap", Parallel Computing, 34, 332–344, 2008.

[13] M. Abreu, D. Labbate, R. Salvi, N. Zagaglia Salvi "Highly symmetric generalized circulant permutation matrices", Linear Algebra and its Applications, 429, 367–375, 2008.

[14] N.L. Tsitsas, E.G. Alivizatos, G.H. Kalogeropoulos, "A recursive algorithm for the inversion of matrices with circulant blocks", Applied Mathematics and Computation, 188, 877–894, 2007.

[15] S.M. El-Sayed, "A direct method for solving circulant tridiagonal block systems of linear equations", Applied Mathematics and Computation, 165, 23–30, 2005.

[16] M. Batista, "A cyclic block-tridiagonal solver", Advances in Engineering Software, 37, 69–74, 2006.

[17] M. Abreu, E.G. Alivizatos, G.H. Kalogeropoulos, "Highly symmetric generalized circulant permutation matrices", Linear Algebra and its Applications, 429, 367–375, 2008.

[18] R. Sh. Ran, T. Zhu Huang, "The inverses of block tridiagonal matrices", Applied Mathematics and Computation, 179, 243–247, 2006.

[19] M.H. Koulaei, F. Toutounian, "Factored sparse approximate inverse of block tridiagonal and block pentadiagonal matricies", Applied Mathematics and Computation, 184, 223–234, 2007.

[20] M.A. El-Shehawey, Gh.A. El-Shreef, A.Sh. Al-Henawy, "Analytical inversion of general periodic tridiagonal matrices", J. Math. Anal. Appl., 345, 123–134, 2008.

[21] H.B. Li, T.Z. Huang, H. Li, "On some new approximate factorization methods for block tridiagonal matrices suitable for vector and parallel processors", Mathematics and Computers in Simulation, 79, 2135–2147, 2009.

[22] M. Batista, "The use of the Sherman–Morrison–Woodbury formula to solve cyclic block tri-diagonal and cyclic block penta-diagonal linear systems of equations", Applied Mathematics and Computation, 210, 558–563, 2009.

[23] A. Kaveh, M.A. Sayarinejad, "Eigensolutions for factorable matrices of special patterns", Communications in Numerical Methods in Engineering, 20, 133-146, 2004.

[24] A. Kaveh, H. Rahami, "Compound matrix block diagonalization for efficient solution of eigenproblems in structural mechanics", Acta Mechania, 188(3-4), 155-166, 2007.

[25] A. Kaveh, H. Rahami, "Special decompositions for eigenproblems in structural mechanics", Communications in Numerical Methods in Engineering 22(9), 943-953, 2006.

[26] W.C. Yueh, "Eigenvalues of several tri-diagonal matrices", Applied Mathematics, 5, 66–74, 2005.

[27] A. Pothen, H. Simon, K.P. Liou, "Partitioning sparse matrices with eigenvectors of graphs", SIAM Journal of Matrix Analysis and Applications, 11, 430-452, 1990.

[28] A. Kaveh, H. Rahami, "Factorization for efficient solution of eigenproblems of adjacency and Laplacian matrices for graph products", International Journal for Numerical Methods in Engineering, 75(1), 58–82, 2007.

[29] A. Kaveh, H. Rahami, "Block diagonalization of adjacency and Laplacian matrices for graph product; applications in structural mechanics", International Journal for Numerical Methods in Engineering, 68(1), 33-63, 2006.
[30] A. Kaveh, H. Rahami, "Eigenvalues of the adjacency and Laplacian matrices for modified regular structural models", International Journal for Numerical Methods in Biomedical Engineering, 26, 1836–1855, 2010.
[31] A. Kaveh, H. Rahami, "Block circulant matrices and applications in free vibration analysis of cyclically repetitive structures", Acta Mechanica, 217, 51–62, 2011.
[32] A. Kaveh, H. Rahami, "An efficient analysis of repetitive structures generated by graph products", International Journal for Numerical Methods in Engineering, 84(1), 108-126, 2010.

Chapter 7

Numerical and Experimental Assessment of Stainless and Carbon Bolted Tensioned Members

P.C.G. da S. Vellasco[1], L.R.O. de Lima[1], J. de J. dos Santos[2]
A.T. da Silva[2], S.A.L. de Andrade[3] and J.G.S. da Silva[1]
[1] Structural Engineering Department
 State University of Rio de Janeiro (UERJ), Brazil
[2] Post Graduate Program in Civil Engineering (PGECIV)
 State University of Rio de Janeiro (UERJ), Brazil
[3] Civil Engineering Department
 Pontifical Catholic University of Rio de Janeiro (PUC-Rio), Brazil

Abstract

Changes in attitudes associated with the building construction industry and a global transition to a sustainability development reduction in environmental impacts has caused a boost in stainless steel use. Despite this fact, the current stainless steel design codes, Eurocode 3, part 1.4, 2003 are still largely based on carbon steel structural behaviour analogies. However, considering that these codes represented one of the first attempts to produce specific stainless steel structural design rules, the idea of using similar rules to the ones adopted for carbon steel, enabled engineers to perform a smooth transition to the stainless steel design. The present investigation presents experimental and numerical studies aimed at evaluating the tension capacity of stainless and carbon steel bolted structural elements performed over the last few years. One of the most significant conclusions from these studies was that, due to the large strain observed in these structures, additional criteria based on deformation limits also should be used in stainless steel design.

Keywords: steel structural, beam-to-column joints, extended endplate joints, semi-rigid joints, experimental analysis, component method.

1 Introduction

Stainless steel has been used in various types of construction due to its main characteristics associated with high corrosion resistance, durability, fire resistance 0, ease of maintenance, appearance and aesthetics. The development of the construction process and the new tendencies adopted in the architecture design concept, highlighted the need for materials that can combine versatility with durability. Once again stainless steel presents itself as a promising material for constructions that require these characteristics, mainly related to high corrosion strength avoiding the need for short period maintenance.

Stainless steel is indicated as a structural element in construction for multiple reasons. Its high ductility allows its use in structures subjected to cyclic loadings, enabling the dissipation of the energy associated with this type of load, through load redistributions before the structural collapse. The cost reduction achieved with less need for structural maintenance, and the increase in its capacity to dissipate impact loads, also enhanced the stainless steel structure reliability.

Changes in attitudes associated with the building construction industry and a global transition to a sustainability development reduction in environmental impacts has caused an increase in stainless steel use as can be observed in the Figure 1. Despite these facts, current stainless steel design codes like the Eurocode 3, part 1.4 [2], are still largely based on carbon steel structural analogies [2]. This strategy was used as a first attempt to produce specific stainless steel structural design rules, enabling engineers to perform a smooth transition for stainless steel design.

Figure 1: Sá Ferreira airport, Porto - Portugal

The structural joints perform a fundamental role in the global response to steel and composite structures. The search for a broader understanding of the actual behaviour of stainless steel bolted joints has motivated several investigations to be performed in various research centres like: Burgan *et al.* [4], Kouhi *et al.* [5], Van Den Berg [6], Gardner and Nethercot [7], Graham [8] and Bouchaïr *et al.* [9]. The main motivation for these studies was the search for the most cost-effective structure resulting from an optimum joint design, as well as an improvement in the joint fabrication and assembly costs. A complete understanding of the stainless steel bolted joints response will also contribute to the production of more accurate design equations that will probably be incorporated into future editions in the stainless steel design standards.

An experimental investigation of the behaviour of riveted joints in shear was performed by D'Aniello *et al.* [10]. The results indicated the proportionality of the relation between the force needed to install the rivet and the connection strength.

The magnitude variation of this assembly force led to different forms of joint collapses and large differences in joint load distributions. The tests performed on riveted joints collapsed at loads 20% larger than the Eurocode 3 - EN 1-8 [11] provisions, with the failure occurring in the plate net sections demonstrating the conservative aspect attributed to these design formulae. The investigation also pointed to the need for a better assessment of the riveted joint design in shear and the associated plate net area failure without losing the simplicity associated with the current formulae present in the Eurocode standard.

Welded joints in shear were also investigated, by Oosterhof and Driver [12], focusing on their geometry and weld arrangements. The experimental investigation validated a finite element model that was later used in a parametric study centred on the assessment of a different weld arrangement and joint geometries. The results confirmed the use a single parameter $U_t=1.25$, that can be utilized in the joint conception. The authors concluded that joints presenting longitudinal fillet weld (Type B) performed satisfactorily and the addition of a transversal fillet weld (Type A) did not significantly enhance the joint shear capacity, as presented in Figure 2.

(a) Type A: transversal and longitudinal fillet weld

(b) Type B: longitudinal fillet weld

Figure 2: Welded joints type [12]

The performance and response of joints with filler plates under shear was investigated by Borello et al. [13], as illustrated in Figure 3. The results indicated large slips and ruptures associated with premature bolt collapses due to stress concentrations. These facts indicate that some bolts were overloaded while others were subjected to minimal or even no load at all. The authors also observed that adding new filler plates led to a reduction of the joint slip resistance and that this issue was more significant to the joint slip performance than other variables like: plate thickness or even the bolt hole size. The authors also concluded that bolt shear capacity reduces with the simultaneous reduction of the plate thickness.

Denavit et al. [14] also investigated the response of bolted joints with filler plates under shear indicating the various mechanisms that influence the structural joint response. A stochastic analysis was conducted to identify the main effects and variables related to the joint slip performance focusing on the extra layers created by the addition of multiple filler plates to the joint. The authors main conclusions were: the reduction in the joint slip resistance by increasing the number of fillet plates; extending the joint to enable the use of additional bolts in the joint can balance the detrimental effects of having additional fillet plates (depending of course on how and where to use the new bolts).

Figure 3: Bolted joint with filler plates [13]

Može and Beg [15] performed an experimental investigation on bolted high strength steel in double shear as depicted in Figure 4, and concluded that friction played an important role in the joint collapses. The authors observed that the bolts load carrying capacity was reached with an associated deformation of $d_0/6$. Numerical models developed by Kim and Yura [17] and Aalberg and Larsen [18], [18] indicated that the joint capacity increases with an increase in the bolt distance up to a value of $3d_0$, a fact that was guaranteed by the e_1 and p_1 values. The authors observed two types of bearing force distributions with the various bolt load distributions and concluded that the Eurocode provisions for the joint bearing capacity are conservative.

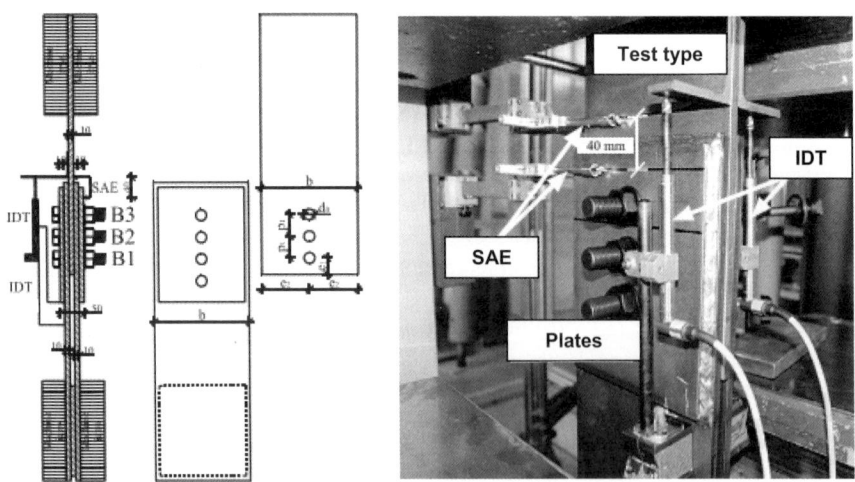

Figure 4: Bolted joints in shear details [15]

Aiming to improve joint ductility, knowledge of the bearing force distribution proves to be a key issue for the identification of the actual joint collapse mechanism. One of the possible solutions for the correct prediction of the bearing forces induced by the bolts was presented by Može and Beg [15]. They proposed a modified version of the bearing load design equation that is suitable for various collapse types. This equation was based on the edge distances ratio (e_1/e_2) and could be used with a larger number of bolts in the load bearing direction by the addition of extra coefficients. The main advantage of this approach is that the sum of the individual bearing forces in the bolts leads to the joint capacity, also enabling the designer to choose the joint failure mode.

A study of the bearing collapse of austenitic and ferritic stainless steel joints, in single and double shear, with thin and thick plate, as illustrated in Figure 5, was made by Salih et al. [20] focusing on the variables e_1/d_0 and e_2/d_0. The authors evaluated the equivalent strain percentages that can happen at various angles of the bolt holes and plate arrangements. The stainless steel stress *versus* strain curves is different from the carbon steel stress vs. strain curves presenting a higher ultimate stress capacity and a nonlinear relationship. These differences pointed to a better assessment of the carbon steel design rules to be adapted to a consistent stainless steel design.

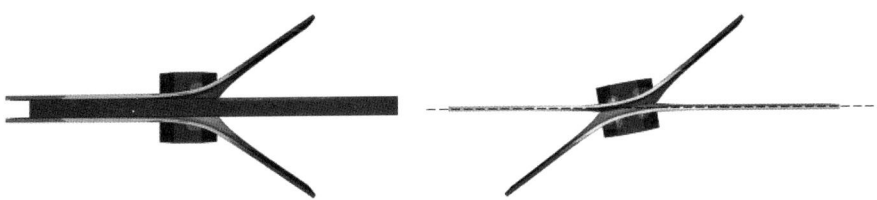

Figure 5: Numerical tests [20]

Experimental studies indicated that different types of collapses, especially due to serviceability limitations, were observed in stainless steel joints with thin and thick plates under shear. Numerical studies also pointed out that the stainless steel ultimate stress f_u could be adopted, instead of the reduced stainless steel ultimate stress $f_{u,red}$, proposed by the Eurocode in the joint shear design formulae. The net section rupture represents one of the ultimate limit states usually verified for structural elements submitted to tension normal stress. The present chapter presents an experimental numerical investigation aiming to evaluate the tension capacity of carbon and stainless steel bolted structural elements. The results are discussed and compared in terms of the stress distribution (that detects, for instance, first yield), and force-displacement curves, among others. The assessment of the results was made by comparisons with the Eurocode 3 provisions for carbon and stainless steels. The investigation indicated that when stainless steel is used in certain structural engineering applications like joints under shear forces, the current design criteria based on deformation limits needs to be re-evaluated especially due to the differences in the yield to ultimate deformation and stress ratios.

2 Eurocode 3 Provisions

The current investigation uses the European design code for stainless steel elements - Eurocode 3, part 1.4 [2]. In this design standard, the failure modes for a plate with holes under tension axial forces are governed by two ultimate limit states: the gross area yield and the net area tension rupture. The presence of staggered holes in the transversal section, Figure 6, makes an immediate identification of the plate critical net section difficult. This process is not new since in 1922, Cochrane [21], performed one of the first attempts to characterize staggered bolted connection failure modes by the use of Equation (1). This expression adds a term to the original net width to obtain the final net section area and is present in major steel design codes all over the world.

$$b_n = b - d_b + \frac{s^2}{4p} \quad (1)$$

where b is the plate width, d_b is the bolt diameter, s and p represent the staggered centre to centre hole distances measured parallel and perpendicular to the member axis. The Eurocode 3, part 1.4 (2003) [2], establishes the guidelines for the stainless steel plate design submitted to axial tension forces. The structure failure is associated with the smallest tension axial force obtained considering two limit states: gross cross-section plastic resistance given by Equation (2), or the ultimate net cross-section tension rupture expressed by Equation (3).

$$N_{pl,Rd} = \frac{A_g \cdot f_y}{\gamma_{M0}} \quad (2)$$

where $N_{pl,Rd}$ is the tension design plastic resistance,
 A_g is the plate gross area,
 f_y is the steel yielding stress,
 γ_{M0} is the partial safety factor, in this case equal to 1.

$$N_{u,Rd} = \frac{k_r \cdot A_n \cdot f_u}{\gamma_{M2}} \quad \text{with} \quad k_r = \left(1 + 3r(d_0/u - 0.3)\right) \quad (3)$$

where A_n is the net cross-section plate area,
 f_u is the steel tension rupture stress,
 k_r is obtained from Equation (3),
 γ_{M2} is the partial safety factor, in this case equal to 1.25,
 r is the ratio between the number of bolts at the cross-section and the total joint bolt number,
 d_0 is the hole diameter,
 u = 2.e_2 but u ≤ p_2 where e_2 is the edge distance measured from the bolt hole centre to its adjacent edge, in the direction perpendicular to the load transfer direction and p_2 is the hole centre-to-centre distance measured perpendicular to the load axis.

Figure 6: Cover plate joint detail and strain gauges location

The tension joint design also has some additional recommendations:
a) in bolted joints, the hole width should be considered $2mm$ larger than the nominal bolt diameter, perpendicular to the applied force direction;
b) in the case of staggered holes, when a diagonal direction to the load axis or zigzag is considered, the net width should be calculated first deducing from the initial gross width, all the holes present in it, and after that adding for each staggered hole a value equal to $s^2/4p$, where s and g, represent the considered longitudinal and traverse hole spacing;
c) the bolted joint critical net width is the smallest evaluated net width for all the different net rupture possibilities;
d) for angles, the dimension p of opposite leg holes is equal to the sum of the dimensions, measured from the angle corner, minus its thickness;
e) the net cross-section area for joints with fillet or spot welds present in the holes should not consider the weld metal area;
f) joints without holes should be evaluated considering that the net area is equal to the gross area, $A_{net} = A_g$.

3 Experimental Investigations

Kim and Kuwamura [22] performed a series of experiments on austenitic stainless steel SUS304, bolted joints and compared their results with bi- and three-dimensional finite element models using contact elements, Figure 7. The authors specified that three possible failure modes, associated with ultimate limit states, ULS, have to be checked: tension and shear stresses (CF1), equivalent stress (CF2) and equivalent plastic strain (CF3). The net area rupture under tension ULS (A) occurs in the cross section area passing through bolt holes, while the plate tearing ULS (B) is caused by high shear stress concentrations in the region between the bolt holes and the plate edge. Finally the shear block ULS (C) is characterized by yield or rupture of the cross section area perpendicular to the load direction and by the shear rupture of the cross section area parallel to the load direction.

Figure 7: Kim and Kuwamura experiments [22]

The ultimate joint load in the analysis corresponds to the maximum load obtained from the load vs. displacement curve. The above mentioned failure criteria are summarized in the following equations:

1) CF1 - tension and shear stresses

$$\frac{\sigma_{ii}}{\sigma_{t\max}} \geq 1.0 \qquad (4)$$

$$\frac{\tau_{ij}}{\tau_{t\max}} \geq 1.0 \qquad (5)$$

2) CF2 - equivalent stress

$$\frac{\sigma_{eq}}{\sigma_{t\max}} \geq 1.0 \qquad (6)$$

3) CF3 - equivalent plastic strain

$$\frac{\varepsilon_{eq}^{p}}{\varepsilon_{t\max}} \geq 1.0 \qquad (7)$$

where,
- σ_{ij} are the normal stresses;
- τ_{ij} are the shear stresses;
- $\sigma_{t\,max}$ is the maximum true normal stress;
- $\tau_{t\,max}$ is the maximum true shear stress;
- σ_{eq} is the maximum equivalent Von Mises stress;
- ε_{eq}^{p} is the equivalent total plastic strain;
- $\varepsilon_{t\,max}$ is the ultimate strain.

The experimental programme performed by Kim and Kuwamura [22] involved three different bolt layout configurations, as illustrated in Figure 8. The single plane shear experiments were made on Austenitic stainless steel SUS304 plates with 1.5mm and 3mm thicknesses and 12mm bolts (A2-50 SUS bolts and 10T-SUS HFSG bolts). The plate specimen ends were attached to an Amsler universal test machine that gradually applied the load to the tested plates.

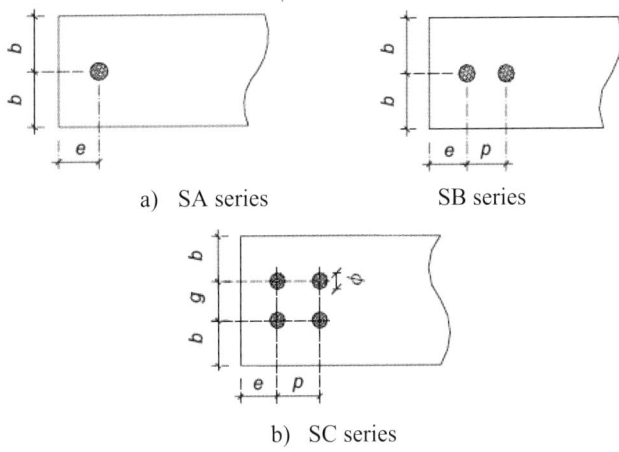

Figure 8: Kim and Kuwamura joint layouts [22]

The experimental results can be depicted in Table 1, with their respective dimensions for each test and its associated failure mode. The nominal stress vs. strain curve of the SUS304 stainless steel is illustrated in Figure 9, as well as the true stress vs. true strain curve obtained using Equations (8) and (9), respectively where σ_t, ε_t, f_y, and ε_n represent the true stress, the true strain, the yield stress and original measured strain, respectively. Table 2 presents the measured mechanical properties determined in the tension test made to characterize the material constitutive law.

Series	Specimen	Thickness (mm)	e (mm)	b (mm)	g (mm)	p (mm)	Failure mode	Ultimate load (kN)
SA (1 hole)	SA1-1	1.5	12	25	-	-	B	12.28
	SA2-2	3.0	18	25	-	-	B	48.05
SB (2 holes)	SB1-4	1.5	60	25	-	30	A	43.34
	SB2-4	3.0	60	25	-	30	A	85.62
SC (3 holes)	SC1-4	1.5	60	55	30	30	C	79.53
	SC2-1	3.0	12	55	30	30	B-C	115.62
	SC2-3	3.0	30	55	30	30	C	162.34
	SC2-4	3.0	60	55	30	30	C	163.30

Table 1: Summary of experimental results [22]

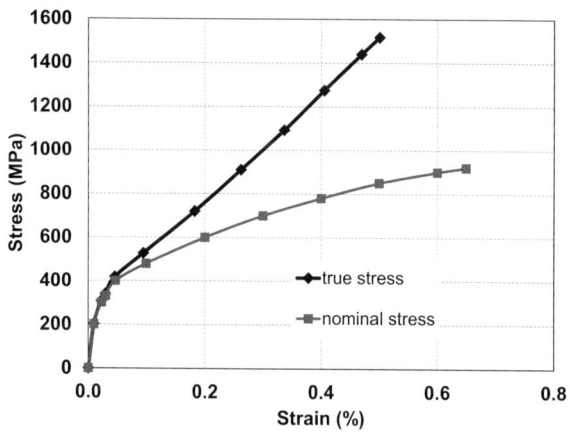

Figure 9: SUS304 stainless steel stress vs. strain curves [22]

$$\sigma_t = f_y(1+\varepsilon_n) \quad (8)$$

$$\varepsilon_t = ln(1+\varepsilon_n) \quad (9)$$

Material	Elasticity modulus (GPa)	Yield stress (MPa)	Ultimate stress (MPa)	Strain limit (%)
SUS304 Stainless steel	204	288	842.35	63.0

Table 2: SUS304 stainless steel mechanical properties [22]

An innovative experimental program was used to evaluate the tension capacity of carbon and stainless steel plates with staggered bolts [23]. The experiments involved bolted cover plate joints made of stainless steel A304 and carbon steel USI300 denominated EX_INOX_Y and EX_CARBON_Y [23], respectively. All the geometrical properties for the tests are presented in Table 3. The bolted joints were made of two 3*mm* thick stainless and carbon steel plates and two 15*mm* thick carbon steel plates used to transfer load to the 3*mm* plate with a 5*mm* gap. The horizontal bolt pitch, s, was modified in each test and the vertical bolt pitch, p, was 55*mm* (see Figure 10). The bolted cover plate joint tests were carried out on a 600*kN* Universal Lousenhausen test machine, see Figure 10. The data acquisition in terms of strains, displacements and applied load was performed using the National Instruments system NI-PXI-1050. The strain measurements were performed using linear strain gauges located in both stainless steel plates named SG, Figure 6.

ID	s (*mm*)	p (*mm*)	e1 (*mm*)	e2 (*mm*)	d0 (*mm*)	STEEL	t_{base} (*mm*)	bolts
E3_CARB_S50	50	55	40	17.6	14.7	carbon	15	6
E4_CARB_S30	30	55	40	17.6	14.7	carbon	15	6
E5_STAIN_S50	50	55	40	17.6	14.7	stainless	15	6
E6_CARB_S30_P10	30	55	40	17.6	14.7	carbon	10	6
E7_STAIN_S30	30	55	40	17.6	14.7	stainless	15	6
E8_CARB_S50_P8	50	55	40	17.6	14.7	carbon	8	6
E9_STAIN_S23	23	55	40	17.6	14.7	stainless	15	6

Table 3: Summary of experimental tests [23]

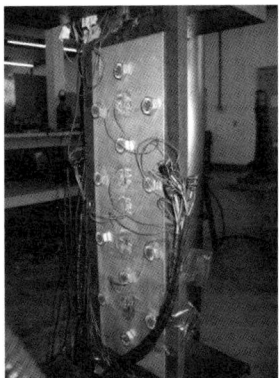

(a) Universal test machine, 600*kN*   (b) cover plate joint detail

Figure 10: Santos et al test layout [23]

The tensile coupons test curves presented a nonlinear expected behaviour, mainly for the stainless steel are shown in Figure 11. The stainless steel yield stress was determined using a straight line parallel to the initial stiffness at a 0.2% deformation, leading to a value equal to 350.6MPa while the ultimate tension stress was 710.7MPa. For the carbon steel, these values were equal to 386.8MPa and 478.7MPa for the yield and ultimate stress, respectively. These results are summarised in Table 4. Figure 11 also present the results of a true stress *versus* true strain curve obtained using Equations (8) and (9), respectively. This curve was used in the finite element modelling due to the large strain and stresses associated with the problem investigated.

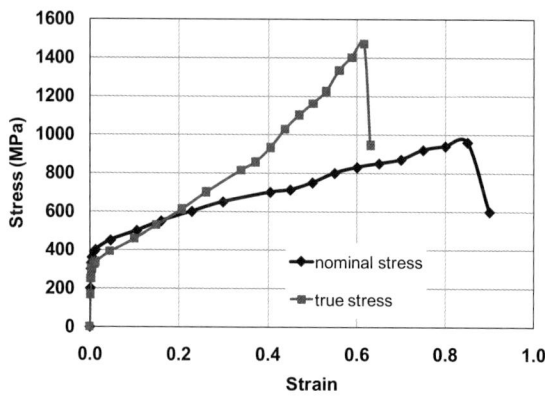

Figure 11: Stress *versus* strain curves for the stainless steel A304

Carbon coupons	f_y (MPa)	f_u (MPa)	Stainless coupons	f_y (MPa)	f_u (MPa)
AC 1	388.97	485.97	I 1	352.30	713.10
AC 2	383.77	478.43	I 2	345.70	699.00
AC 3	348.29	450.55	I 3	352.20	730.50
AC 4	404.39	495.91	I 4	347.80	692.80
AC 5	401.31	488.81	I 5	349.60	725.80
AC 6	394.01	472.43	I 6	356.10	703.10
Mean	386.79	478.68	Mean	350.62	710.72
Standard Deviation	20.34	16.02	Standard Deviation	3.70	15.11

Table 4: Summary of tensile coupons tests [23]

Figure 12 presents the comparison between the results from tests E3_CARB_S50 and E5_STAIN_S50 in terms of the load *versus* axial displacement curves. In this figure, are depicted the experimental ultimate loads of 310.0kN and 469.4kN for E3_CARB_S50 and E5_STAIN_S50 tests, respectively. According to the Eurocode 3 [2,3], Equations (2) and (3), for the E3_CARB_S50 test, the design resistances were 337.0*kN* for gross cross-section plastic resistance and 298.7*kN* for ultimate net cross-section tension rupture (section with three holes). While for the E5_STAIN_S50 test, these values were 305.5*kN* and 810.8*kN* (net section with three holes), respectively. The partial safety factor was taken equal to 1.0. Figure 12 also indicates that in both tests, the test rupture occurred in the section represented by two holes near the joint symmetry axis. Despite the fact that for the carbon steel test E3_CARB_S50, the theoretical and experimental values presented a good agreement, the stainless steel test E5_STAIN_S50 presented a larger difference in terms of the ultimate design equation and test loads.

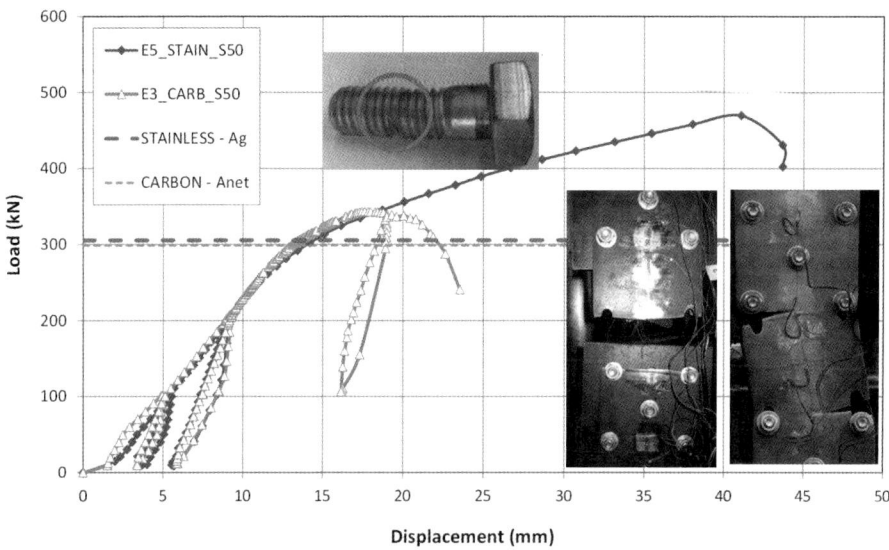

Figure 12: Load *versus* displacement - E3_CARBON_S50 & E5_STAIN_S50 [23]

Figure 13 presents the comparison between the results from tests E4_CARB_S30 and E7_STAIN_S30 in terms of the load *versus* displacement curves. In this figure, it may be depicted the experimental ultimate loads of 303.6*kN* and 545.8*kN* for E4_CARB_S30 and E7_STAIN_S30 tests, respectively. According to the Eurocode 3 [2, 3] , Equations (2) and (3), for the E4_CARB_S30 test, the design resistances were 337.0*kN* for gross cross-section plastic resistance and 291.7*kN* for ultimate net cross-section tension rupture (section with three holes). In the case of the E5_STAIN_S50 test, these values were 305.5*kN* and 791.9*kN* (net section with three holes), respectively. Figure 13 indicates that in both tests, the rupture occurred in the section represented by two holes near the joint symmetry axis. Once again the carbon steel test E3_CARB_S50 theoretical and experimental values presented a good agreement while the stainless steel test E5_STAIN_S50, still shown a non-negligible difference between the ultimate design equation and test loads.

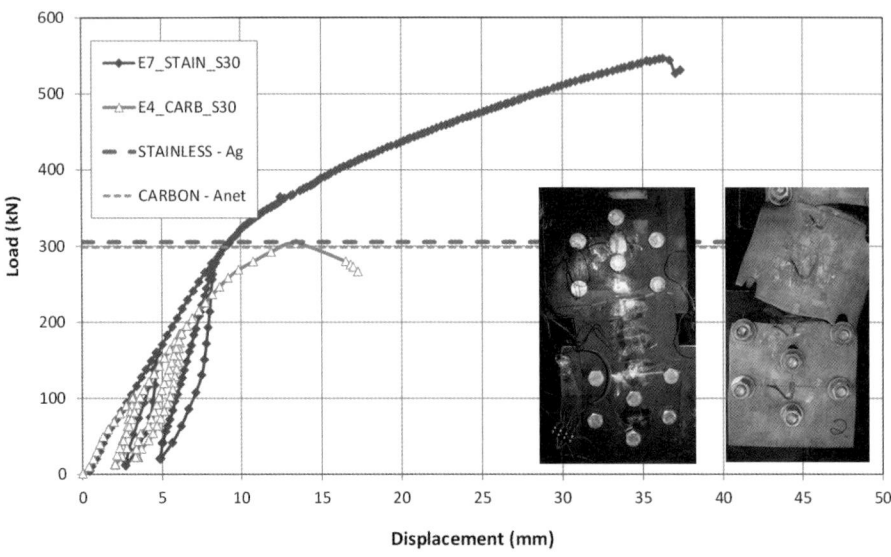

Figure 13: Load *versus* displacement - E4_CARBON_S30 & E7_STAIN_S30 [23]

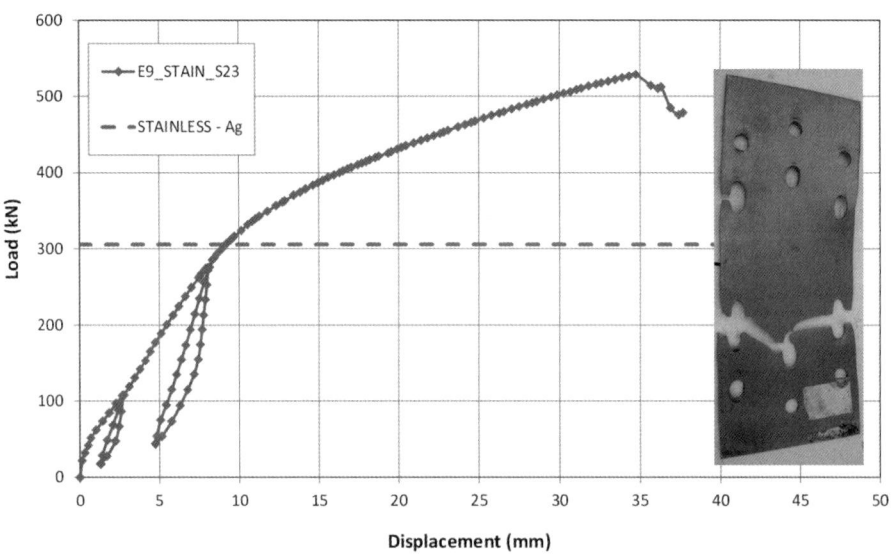

Figure 14: Load *versus* displacement - E9_STAIN_S23 [23]

Considering the difference between the failure modes for the two stainless steel joints presented before, a third test, E9_STAIN_S23, was performed to investigate this issue where the horizontal bolt pitch was taken equal to 23*mm*. This value was considered to increase the difference between the load failure in sections with two and three bolts, respectively. According to Eurocode 3 [2,3], for this test, the design resistances were 305.4*kN* for gross cross-section plastic resistance and 787.7*kN* for ultimate net cross-section tension rupture (section with three holes). Figure 14 presents the load *versus* displacement for this test where the ultimate experimental load was equal to 526.5*kN*. It can also be observed that the joint rupture occurred in the net section passing through three bolts, in agreement with the Eurocode 3 provision [2,3]. Despite this fact, the design equation and the experimental ultimate loads still presented a large difference.

A summary of these results is presented in Table 5. It may be concluded that for carbon steel joints, good agreement was observed comparing the theoretical and experimental results. Alternatively, for the stainless steel joints, larger differences were found in terms of ultimate (rupture) loads.

ID	Experiment Failure Mode	Experiment Ultimate Load (kN)	EC3 Failure Mode	EC3 Ultimate Load (kN)	% $\dfrac{EXP}{EC3}$
E3_CARB_S50	2H	310.0	3H	298.3	3.9
E4_CARB_S30	2H	296.0	3H	282.5	4.8
E5_INOX_S50	2H	480.0	AB	302.9	58.5
E6_CARB_S30_P10	3H	309.5	3H	282.5	9.6
E7_INOX_S30	2H	459.0	AB	302.9	51.5
E8_CARB_S50_P8	2H	326.0	2H	282.5	13.3
E9_INOX_S23	3H	436.0	AB	302.9	43.9
2H: two hole net rupture; 3H: three hole net rupture and AB: gross section yielding					

Table 5: Summary of experimental tests [23]

Another key issue also was investigated in the current study and involved the assessment of the influence of the load application plate thickness, adopted initially equal to 15*mm* (E3_CARB_S50 and E4_CARB_S30). Two other tests were performed, E6_CARB_S30_P10 (load plate thickness equal to 10*mm*) and E8_CARB_S50_P8 (load plate thickness equal to 8*mm*). Comparing the E3, E8 and E4 tests with E6, Figure 15, it may be concluded that the load application plate thickness significantly alters the joint response in terms of the ultimate load and associated failure mode.

4 Numerical investigations

Finite element numerical analyses provide a relatively inexpensive and time efficient alternative to physical experiments. Despite this fact, due to their nature these numerical simulations [24] have to be properly calibrated against experimental test results. If the validity of FE analysis is assured, it is possible to investigate the structural behaviour against a wide range of parameters with the FE model [25].

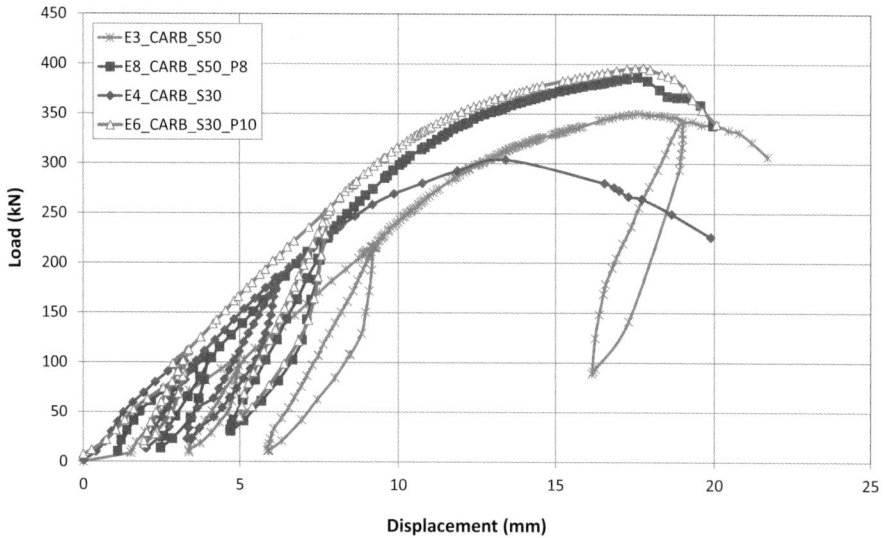

Figure 15: Load *versus* displacement – load plate thickness variation [23]

A finite element model was developed to investigate the tension capacity of cover plate joints with the aid of the ANSYS 11 finite element package [26]. The numerical model adopted solid elements (SOLID45) defined by eight nodes with three degrees of freedom per node: translations were in the nodal x, y and z directions. The adopted mesh was chosen so that the elements had a proportion and size to avoid numerical problems.

Contact elements (CONTA174 and TARGE170) presented in the ANSYS Element Library [26] were considered between the plates and between the holes and the bolt shanks. The load was applied by means of axial displacements in the load plate as presented in Figure 16. In this figure, it is also possible to observe that the bolt head and nuts were simulated through UZ displacement restraints at the hole adjacent area.

The numerical model validation was made with the aid of specimen SA2-2, that presented an ultimate experimental load of de 48.05kN. The best performance of the numerical model proposed by Kim and Kuwamura [22] in terms of the collapse load was 44.85kN, Figure 17. The present investigation concluded that there was no influence on the plate length when a value greater than or equal to 150mm was used. This was the reason for adopting a 150mm plate length in the present model. The load was incrementally applied to the models by means of displacements, divided into sub-steps adopting a tolerance convergence criteria in terms of the standard square root of the displacement metric variation of 0.01.

Figure 17 depicts the experiment and numerical results in terms of load vs. displacement curves. It is possible to note that models 2 and 3 presented a similar response. Alternatively, model 1 indicated the same trend as the others up to a 5mm displacement. From this point on the first model performed differently, leading to a

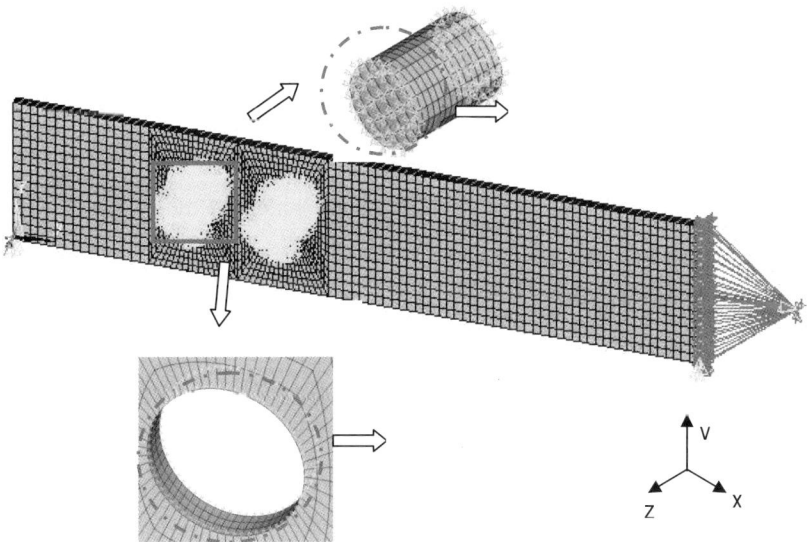

Figure 16: Finite element model and contact elements [24]

substantially larger ultimate load than its experimental counterpart. This result could have been influenced by the mesh discretization adopted in the first model that presented larger elements than the other two subsequent models, being a less accurate representation of the regions with high stress concentrations. Table 6 and Table 7 depict these experimental and numerical results for the Kim and Kuwamura [22] test and numerical programme.

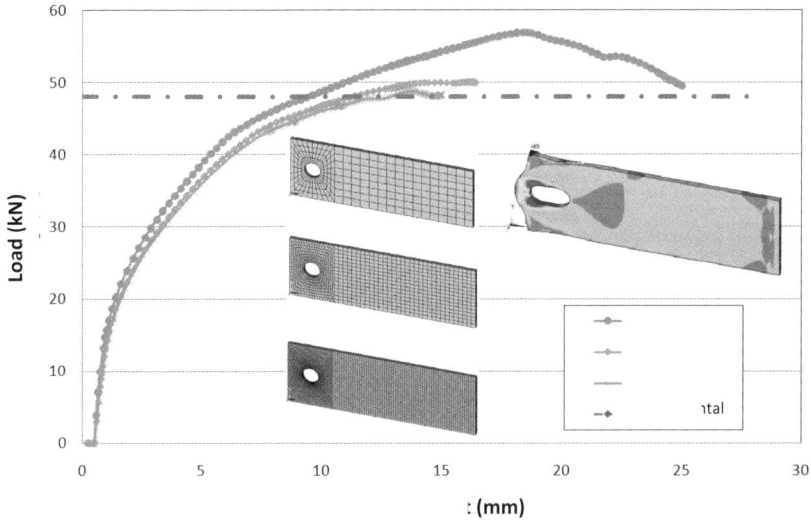

Figure 17: Load *versus* displacement curves - SA2-2 specimen [24]

Model	Numerical ultimate load - (kN)	Experimental ultimate load (kN)	Elements mean size - hole (mm)	Number of elements along the thickness	Error (%)	Failure mode
1	56.93	48.05	4.6 x 2.8	3	18.48	B
2	50.04	48.05	2.3 x 1.4	3	4.14	B
3	48.83	48.05	1.2 x 0.7	3	1.61	B

Table 6: Influence of the number of elements used in the mesh - SA2-2 [24]

Model	Numerical ultimate load - (kN)	Numerical load - Kim & Kuwamura (kN)	Error - Kim & Kuwamura (%)	Error (%)
1	56.93	44.85	6.60	18.48
2	50.04	44.85	6.60	4.14
3	48.83	44.85	6.60	1.61

Table 7: Comparison varying the number of elements used in the mesh - SA2-2 [24]

An inspection of the results indicates that the third model presented an ultimate load closer to the experimental collapse load, with the same collapse mode numerically and experimentally found by Kim and Kuwamura [22]. It is also interesting to observe that the proposed numerical model ultimate load was closer to the experimental value than the numerical load proposed by Kim and Kuwamura [22].

The influence of the number of finite elements used along the plate thickness was also investigated. The best performing finite element mesh (three elements along the plate thickness, model 3) when compared to the experiments was used to create two further models with two (model 3_R2) and four (model 3_R1) elements along the plate thickness, as can be seen in Figure 18 and in Table 8. The best results were obtained with model 3_R1, that presented a larger number of elements, but the difference in the model performance was less than 1%.

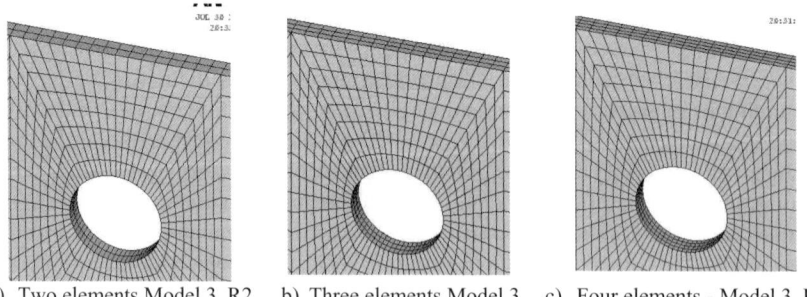

a) Two elements Model 3_R2   b) Three elements Model 3   c) Four elements - Model 3_R1

Figure 18: Models varying the number of elements along the plate thickness [24]

Model	Numerical ultimate load (*kN*)	Experimental ultimate load (*kN*)	Elements mean size - hole (*mm*)	Number of elements along the thickness	Error (%)	Failure mode
3	48.83	48.05	1.2 x 0.7	3	1.61	B
3_R1	48.81	48.05	1.2 x 0.7	4	1.58	B
3_R2	49.02	48.05	1.2 x 0.7	2	2.00	B

Table 8: Variation of the number of elements along the thickness - SA2-2 [24]

The numerical model validation proceeded with specimen SA2-4, that presented an ultimate experimental load of de $85.62kN$. The best performance of the numerical model proposed by Kim and Kuwamura [22] in terms of the collapse load was $86.81kN$, Figure 19. Kim and Kuwamura [22] indicated that there was no influence on the plate length when a value greater than or equal to $150mm$ was used. This was the reason for adopting a $230mm$ plate length in the present model.

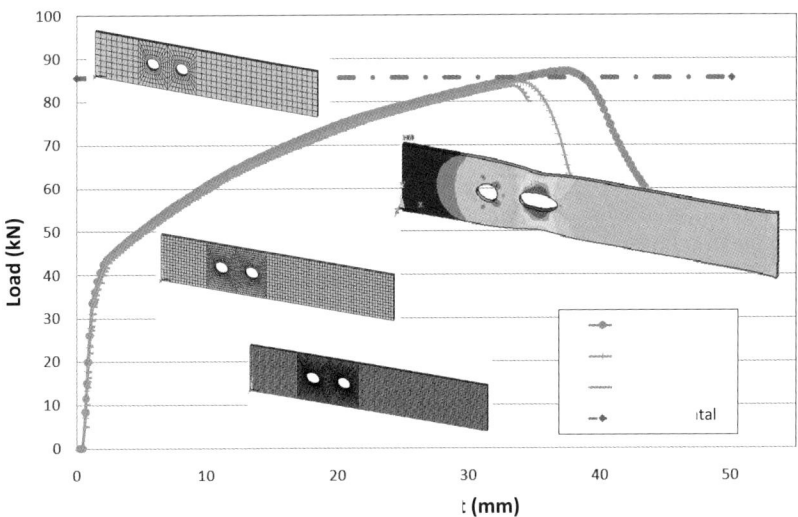

Figure 19: Load *versus* displacement curves - SA2-4 [24]

Figure 19 depicts the experiment and numerical results in terms of load vs. displacement curves where it is possible to observe that all the models presented a similar response up to a $30mm$ displacement, but the first model reached the maximum load with the largest associated displacement, while the smallest associated displacement was reached with the third model. This result again could have been influenced by the mesh discretization adopted in the first model that presented larger elements than the other two subsequent models, being a less accurate representation of the regions with high stress concentrations.

Tables 9 and 10 depict these experimental and numerical results for the Kim and Kuwamura [22] test and numerical programme. An inspection of the results indicates that the second model presented an ultimate load closer to the experimental collapse load, with the same collapse mode numerically and experimentally found by Kim and Kuwamura [22].

Model	Numerical ultimate load - (kN)	Experimental ultimate load - (kN)	Elements mean size – hole (mm)	Number of elements along the thickness	Error (%)	Failure mode
1	87.09	85.62	4.6 x 2.4	3	1.71	A
2	84.46	85.62	2.4 x 1.2	3	1.36	A
3	83.58	85.62	1.2 x 0.6	3	2.38	A

Table 9: Influence of the number of elements used in the mesh - SA2-4 [24]

Model	Numerical ultimate load - (kN)	Numerical load - Kim & Kuwamura (kN)	Error – Kim & Kuwamura (%)	Error (%)
1	87.09	86.81	1.4	1.71
2	84.46	86.81	1.4	1.36
3	83.58	86.81	1.4	2.38

Table 10: Variation of the number of elements used in the mesh - SA2-4 [24]

The influence of the number of elements used along the plate thickness was again verified. The best performing finite element mesh (three elements along the plate thickness, model 3) when compared to the experiments was used to create two further models with two (model 2_R1) and four (model 3_R1) elements along the plate thickness, as can be seen in Figure 19 and in Table 11. The best results were obtained with model 3_R1, that presented a larger number of elements, but the difference in the model performance was around 1%.

Model	Numerical ultimate load (kN)	Experimental ultimate load (kN)	Elements mean size - hole (mm)	Number of elements along the thickness	Error (%)	Failure mode
2	84.46	85.62	2.4 x 1.2	3	1.36	A
2_R1	84.47	85.62	2.4 x 1.2	4	1.34	A
2_R2	84.32	85.62	2.4 x 1.2	2	1.51	A

Table 11: Comparison varying the number of elements used in the mesh - SB2-4 [24]

The numerical model validation for the joint with four bolts was made with specimen SC2-4, that presented an ultimate experimental load of de 160.32kN. The best performance of the numerical model proposed by Kim and Kuwamura [22] in terms of the collapse load was 152.79kN, Figure 20. For the same reason as the previous two models the present model has a 240mm plate length.

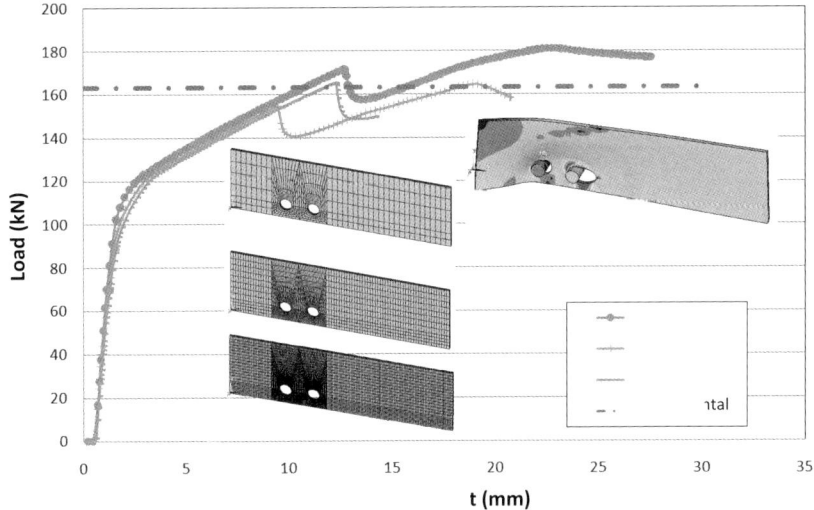

Figure 20: Load *versus* displacement curves - SC2-4 [24]

Figure 20 illustrates the experiment and numerical results in terms of load vs. displacement curves where it is possible to observe again that all the models presented a similar response up to a 10mm displacement, but the second model presented the best performance.

Table 12 and Table 13 depict these experimental and numerical results for the Kim and Kuwamura [22] test and numerical programme. An inspection of the results indicates that the second model presented an ultimate load closer to the experimental collapse load, with the same collapse mode numerically and experimentally found by Kim and Kuwamura [22]. Analysing the curves present in Figure 20 and the deformed shape of the plate it was possible to identify the curling effect, which is associated with the deformation, in the direction out of the plane, at the plate edge. This effect occurs when the distance from the bolt hole to the edge is too large. An inspection of the collapse load reached by the models indicated no influence of the curling effects over these values because the curling effect happened after the peak load was reached.

It could also be noticed that the proposed model was able to reproduce the experimental failure mode of Kim and Kuwamura [22] with an ultimate load close to its experimental counterpart.

Table 14 presents the results of the influence of the number of elements used along the plate thickness. For this specimen the model with three element along the thickness presented the best performance. The accurate performance achieved with the Kim and Kuwamura [22], experiments motivated the subsequent phase of this investigation that involved the modelling of the experiments made by Santos [23].

Model	Numerical ultimate load (kN)	Experimental ultimate load (kN)	Elements mean size - hole (mm)	Number of elements along the thickness	Error (%)	Failure mode
1	181.11	162.32	12.0 x 4.0	3	11.58	C
2	164.60	162.32	6.0 x 2.0	3	1.40	C
3	165.24	162.32	3.0 x 1.0	3	1.80	C

Table 12: Influence of the number of elements used in the mesh [24]

Model	Numerical ultimate load (kN)	Numerical load - Kim & Kuwamura (kN)	Error - Kim & Kuwamura (%)	Error (%)
1	181.11	152.79	5.9	11.58
2	164.60	152.79	5.9	1.40
3	165.24	152.79	5.9	1.80

Table 13: Variation of the number of elements used in the mesh - SC2-4 [24]

Model	Numerical ultimate load (kN)	Experimental ultimate load (kN)	Elements mean size - hole (mm)	Number of elements along the thickness	Error (%)	Failure mode
2	164.60	162.32	6.0 x 2.0	3	1.40	C
2_R1	164.94	162.32	6.0 x 2.0	4	1.62	C
2_R2	160.03	162.32	6.0 x 2.0	2	1.41	C

Table 14: Variation of the number of elements used in the mesh - SC2-4 [24]

Figure 21 presents a typical mesh configuration for the complete model. It is emphasized here that only half of the model was considered using the symmetry boundary conditions, being sufficient to characterize the joint ultimate limit states.

The adopted material properties were: Young's modulus of $210 GPa$ (see Figure 11) and a Poisson's coefficient of 0.3. As previously mentioned, stainless steel true stress *versus* true strain curves with a nonlinear behaviour were adopted using data from the tensile coupons tests Figure 11.

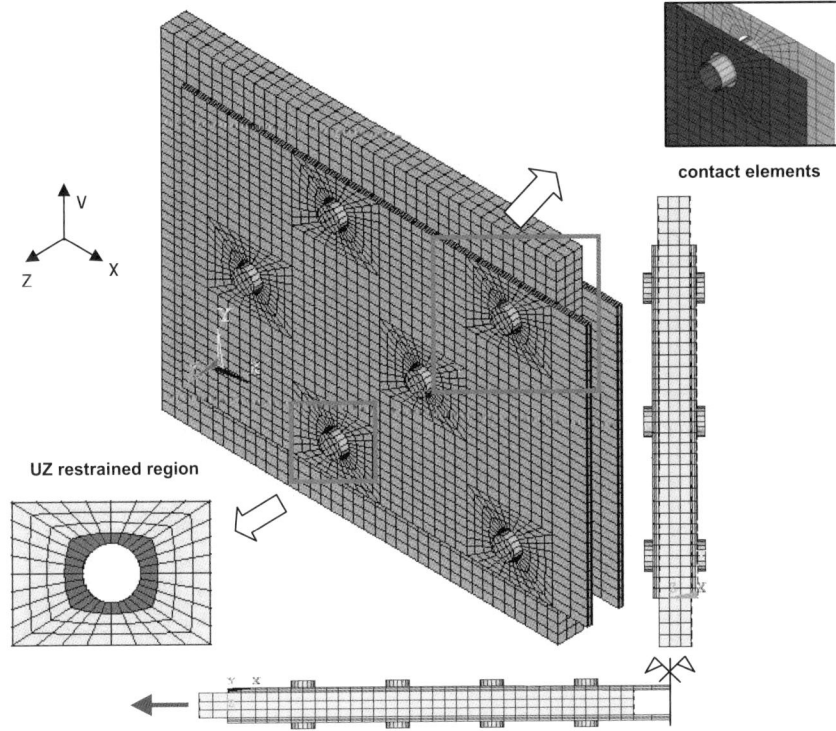

Figure 21: Finite element model and contact elements [24]

A full nonlinear analysis was performed for the developed numerical model. The material non-linearity was considered using a Von Mises yield criterion associated with a multi-linear stress-strain relationship and isotropic hardening response. The geometrical non-linearity was introduced into the model by using an updated Lagrangean formulation. This procedure represents the full structural assessment of the analysed bolted joints, and may be summarized in several outputs, namely the stress distribution (that detects, among other data, first yield), or the force-displacement curve for any node within the connection.

Figure 22 presents the load *versus* displacements curves for each individual test, where it can be observed that the ultimate load of experiments E5_INOX_S50, E7_INOX_S30 e E9_INOX_S23 were: 389kN, 389kN and 385kN, respectively. All the numerical model loads were situated in an interval between the experimental loads and the Eurocode 3 (Part 1.4 2003) estimated values.

Figure 23 depicts the Von Mises stress distributions for the three numerical models, where it could be noticed that all models presented high stress concentrations in the region between the bolt holes and the plate edge. In the numerical modelling corresponding to the E5_INOX_S50 test the stress distribution indicates a possible rupture in the stainless steel plate net area passing through two bolt holes, a mode of failure that also occurred in the experiment. In the numerical

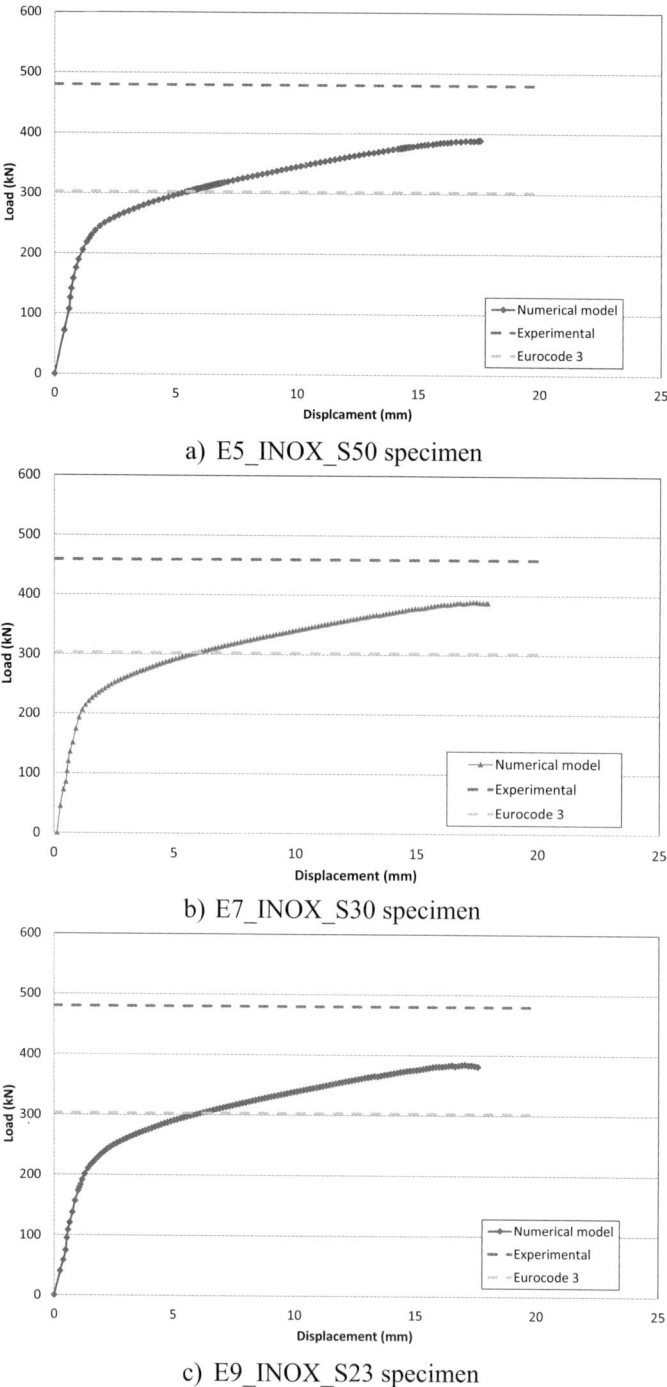

a) E5_INOX_S50 specimen

b) E7_INOX_S30 specimen

c) E9_INOX_S23 specimen

Figure 22: Stainless steel load *versus* displacement curves [24]

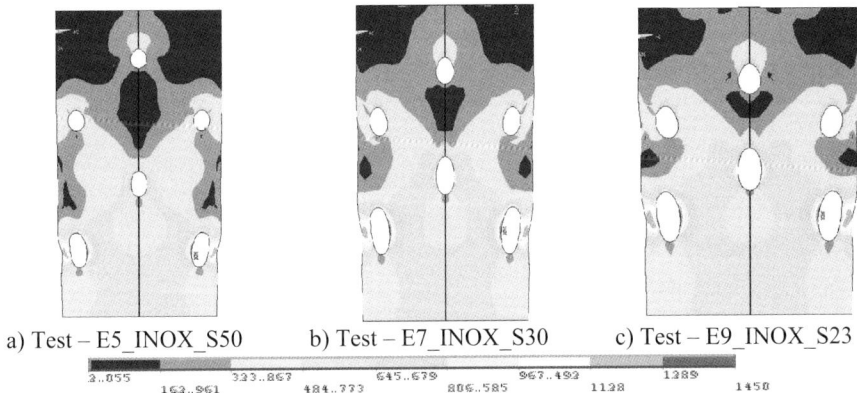

a) Test – E5_INOX_S50   b) Test – E7_INOX_S30   c) Test – E9_INOX_S23

Figure 23: Von Mises stress distribution (in MPa) at the ultimate load

modelling corresponding to E7_INOX_S30, the section in which the net area rupture occurs is not clear, *i.e.* it could be located in the plate region passing through two or three bolt holes. In the numerical modelling corresponding to the E9_INOX_S23 test the stress distribution indicates a possible rupture in the stainless steel plate net area passing through three bolt holes, a failure mode that also was identified in the experiment.

In order to validate the proposed numerical model, a series of comparisons will be presented in terms of load vs. deflection curves for specific points that were measured in the experimental programme. Figure 24 presents numerical and experimental strain comparisons for the E5_INOX_S50 specimen, measured in the region close to the joint centre line and to the symmetry axes. It can be observed that a similar response was found in the numerical and experimental curves, for strain gauges 2(4) and 3(8). However, the numerical model presented, in the plastic range, a stiffness value lower than their experimental counterparts, for both investigated points. Figure 25 depicts load vs. strain curves for the E7_INOX_S30 test where it can be observed that the numerical model results were not as good as the previous tests, presenting, for all the load levels, a lower stiffness and achieving larger strains than their experimental counterparts. Alternatively, the results of the E9_INOX_S23 tests, Figure 26, indicated good similarity between the numerical and experimental evidence.

The plate section at which the stainless steel net area rupture failure mode occurred was determined with the aid of Figure 27 where load vs. strain curves are illustrated for a point located at the plate cross section with two bolt holes at the horizontal symmetry axes. From this graph it can be observed that as the horizontal distance between two bolt holes increases the magnitude of the stresses on the bolt present in adjacent sections diminishes. For example, for a $250kN$ load level, the left curve is associated to a strain level lower than the rest highlighting the net area rupture failure occurring in a plate section with three bolt holes. On the other hand, for this load level, E5_INOX_S50 numerical model highlights that the failure mode is associated with the net area rupture failure occurring in a plate section with two bolt holes.

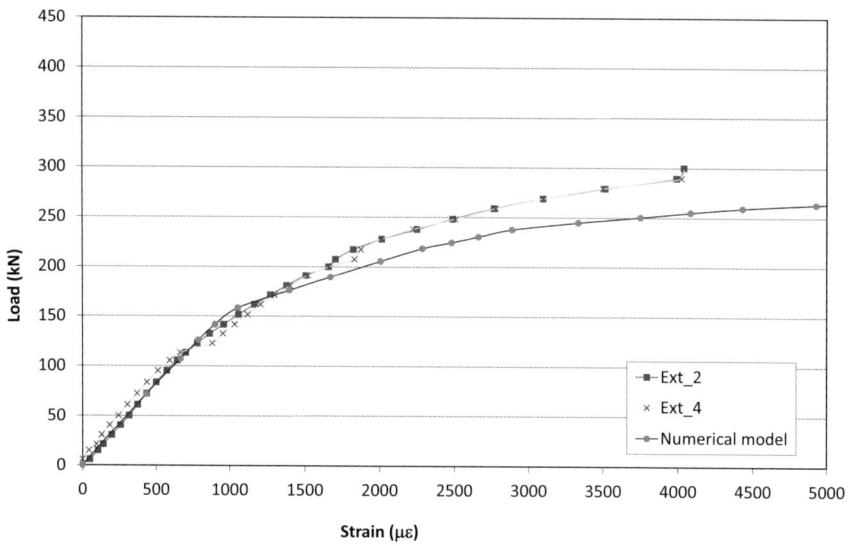

a) Strain gauges 2 and 4

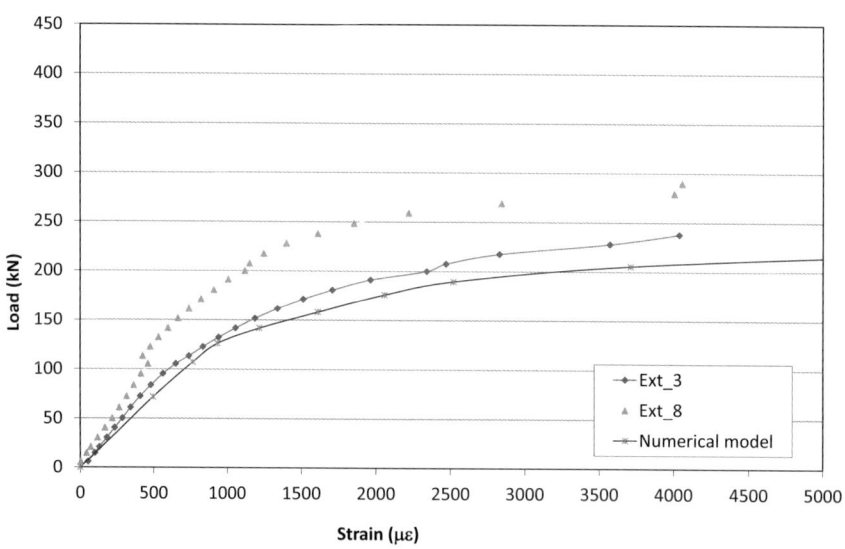

b) Strain gauges 3 and 8

Figure 24: Load *versus* strain (Experimental *versus* Numerical) - E5_INOX_50 [24]

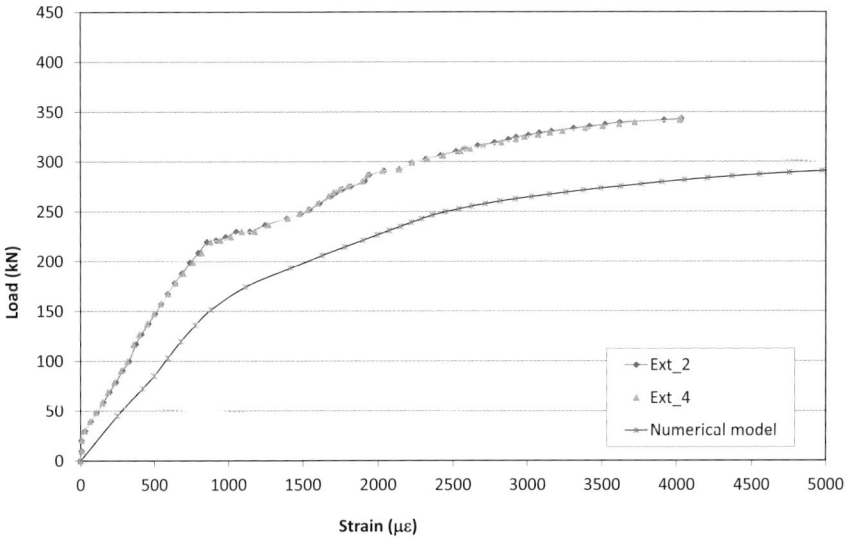

a) Strain gauges 2 and 4

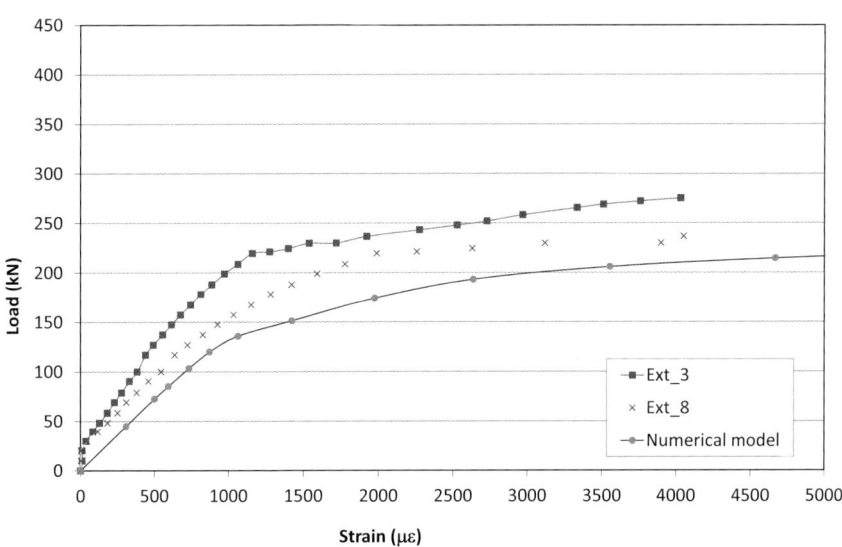

b) Strain gauges 3 and 8

Figure 25: Load *versus* strain (experimental *versus* numerical) - E7_INOX_30 [24]

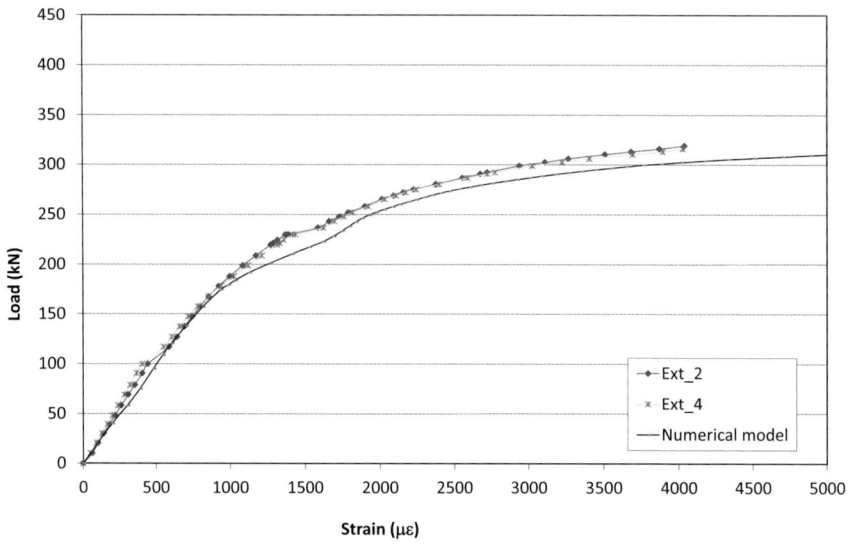

a) Strain gauges 3 and 8

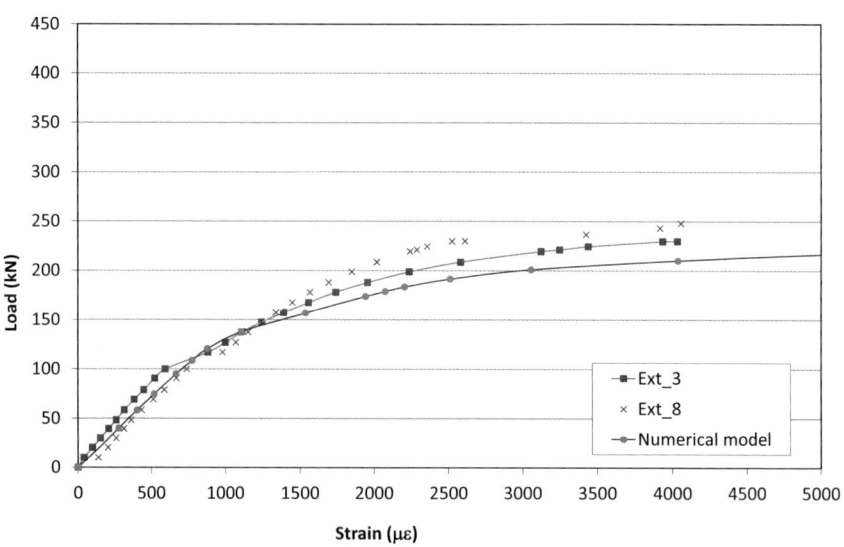

b) Strain gauges 3 and 8

Figure 26: Load *versus* strain (experimental *versus* numerical) - E9_INOX_23 [24]

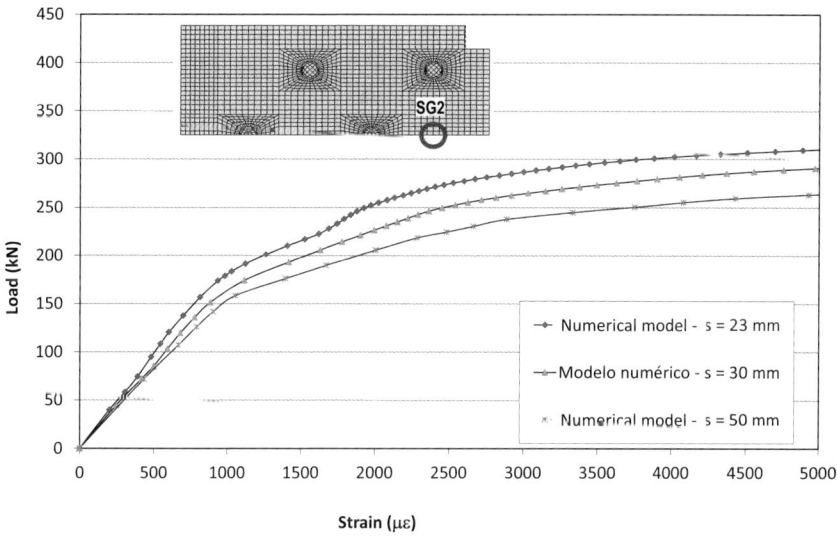

Figure 27: Load *versus* strain curves for all the numerical models [24]

5 Final considerations

This investigation presented an experimental and numerical programme to investigate the structural response of the carbon steel and stainless steel plates with staggered bolts under tension. Initially the experimental results were compared to theoretical results according to Eurocode 3 provisions [2, 3]. For carbon steel tests, a good agreement was reached between the design equation and the experiments, a fact that was not corroborated in the stainless steel tests where large differences were observed, mainly in terms of the ultimate load.

A possible explanation for these discrepancies could be related to the fact that the great majority of stainless steel structural design codes are still based on carbon steel analogies. At this point it is interesting to observe that the stainless steel codes need to be improved in order to correctly evaluate the stainless steel structural elements behaviour.

A finite element numerical model was also developed with the aid of the Ansys 11.0 program and considered material and geometrical nonlinearities through the Von Mises yield criterion and the updated Lagrangian formulation, respectively. The calibration of the numerical model was made against Kim and Kuwamura [22] experiments and numerical results, where the optimum mesh and element sizes were determined [24].

Table 15 presents a comparison between the numerical results and the Eurocode 3 provisions for the already mentioned stainless steel tests (Santos [23]). Differences of about 28% were found when the Eurocode 3 [2, 3] and the numerical models were compared.

Experim. tests	Experim. failure mode	Experim. ultimate load (*kN*)	Numerical failure mode	Numerical ultimate load (*kN*)	Difference Numerical x Experim. (%)	Difference Numerical x EC3 (%)
E5-INOX-S50	2F	480.0	2F	389	19.0	28.8
E7-INOX-S30	2F	459.0	2F / 3F	389	15.2	28.8
E9-INOX-S23	3F	436.0	3F	385	11.6	27.5

Table 15: Summary of the experimental tests

The numerical ultimate loads were less than their experimental counterparts for all the investigated specimens. This can be explained by the fact that the developed numerical models represent the joints in an idealized form, without imperfections or residual stresses. Another reason for these differences can be attributed to the fact that the stainless steel stress *versus* strain curve adopted in the finite element model was obtained through a series of coupons that are influenced by the rolling direction.

The problem related to the numerical and experimental assessment of stainless and carbon bolted tensioned members is certainly much more complicated and is influenced by several other design parameters. Further research in this area is currently being carried out, in order to consider imperfections, residual stresses and the coupons rolling directions.

On the other hand, differences varying from between 12% and 19% were found when the numerical and the experimental values were compared. These differences were also partly due to the natural conservatism present in most of the design standards Eurocode 3, part 1.4 [2]. This conservatism is largely due to the lack of experimental evidence regarding stainless steel structural response still present in the literature.

Acknowledgments

The authors would like to thank CAPES, CNPq and FAPERJ for the financial support for this research program. Thanks are also due to ACESITA and USIMINAS for donating the stainless and carbon steel plates used in the experiments.

References

[1] L. Gardner, N.R. Baddoo, "Fire testing and design of stainless steel structures", Journal of Constructional Steel Research, 62, 532-543, 2006.
[2] Eurocode 3, ENV 1993-1-4, 2003, "Design of steel structures – Part 1.4: General rules – Supplementary rules for stainless steel", CEN - European Committee for Standardisation 1996.

[3]   Eurocode 3, ENV 1993-1-1, 2003, "Design of steel structures - Structures – Part 1-1: General rules and rules for buildings", CEN, European Committee for Standardisation, Brussels, 2003.

[4]   B.A. Burgan, N.R. Baddoo, K.A. Gilsenan, "Structural design of stainless steel members - comparison between Eurocode 3, Part 1.4 and test results", Journal of Constructional Steel Research, 54(1), 51–73, 2000.

[5]   J. Kouhi, A. Talja, P. Salmi, T. Ala-Outinen, "Current R&D work on the use of stainless steel in construction in Finland", Journal of Constructional Steel Research, 54(1), 31–50, 2000.

[6]   G.J. Van Den Berg, "The effect of the non-linear stress–strain behaviour of stainless steels on member capacity", Journal of Constructional Steel Research, 54(1), 135–160, 2000.

[7]   L. Gardner, D.A. Nethercot, "Experiments on stainless steel hollow sections — Part 1: Material and cross-sectional behaviour", Journal of Constructional Steel Research, 60, 1291–1318.cap 3, 2004.

[8]   G. Graham, "Structural uses of stainless steel — buildings and civil engineering", Journal of Constructional Steel Research, 64, 1194-1198, 2008.

[9]   J. Bouchair, A. Averseng, A. Abidelah, "Analysis of the behaviour of stainless steel bolted connections", LaMI, Civil Engineering, Blaise Pascal University, rue des Meuniers, BP 206, 63174 Aubière cedex, France, 2008.

[10] M. D'Aniello, F. Portioli, L. Fiorino, R. Landolfo, "Experimental investigation on shear behaviour of riveted connections in steel structures", Engineering Structures, 33(2), 516-531, 2011.

[11] Eurocode 3, EN 1993-1-8, "Design of steel structures – Part 1.8: Design of joints", CEN, European Committee for Standardisation, Brussels, 2005.

[12] S.A. Oosterhof, R.G. Driver, "Effects of connection geometry on block shear failure of welded lap plate connections", Journal of Constructional Steel Research, 67(3), 525–532, 2011.

[13] D.J. Borello, M.D. Denavit, J.F. Hajjar, "Bolted steel slip-critical connections with fillers: I. Performance", Journal of Constructional Steel Research, 67(3), 379–388, 2011.

[14] M.D. Denavit, D.J. Borello, J.F. Hajjar, "Bolted steel slip-critical connections with fillers: II. Behavior", Journal of Constructional Steel Research, 67(3), 398–406, 2011.

[15] P. Moze, D. Beg, "Investigation of high strength steel connections with several bolts in double shear", Journal of Constructional Steel Research, 67(3), 333–347, 2011.

[16] P. Moze, D. Beg, "High strength steel tension splices with one or two bolts", Journal of Constructional Steel Research, 66(8-9), 1000-1010, 2010.

[17] H.J. Kim, J.A. Yura, "The effect of ultimate-to-yield ratio on the bearing strength of bolted connections", Journal of Constructional Steel Research, 49(3), 255–70, 1999.

[18] A. Aalberg, P.K. Larsen, "Bearing strength of bolted connections in high strength steel", in Mäkeläinen, et al., (Editors), Proceedings of Nordic steel construction conference 2001 - NSCC 2001, 859–866, 2001.

[19] A. Aalberg, P.K. Larsen, "The effect of steel strength and ductility on bearing failure of bolted connections", in Lamas, Silva, (Editors), Proceedings of the 3rd European conference on steel structures, Universidade de Coimbra, Portugal, 869–878, 2002.

[20] E.L. Salih, L. Gardner, D.A. Nethercot, "Bearing failure in stainless steel bolted connections", Engineering Structures, 33(2), 549–562, 2011.

[21] V.H. Cochrane, "Rules for Rivet Hole Deduction in Tension Members", Engineering News-Record, 80, 16 November 1922.

[22] T.S. Kim, H., Kuwamura, "Finite element modeling of bolted connections in thin-walled stainless steel plates under static shear", Thin-Walled Structures, 45(4), 407-421, 2007.

[23] J. de J. dos Santos, "Comportamento Estrutural de Elementos em Aço Inoxidável", MSc in Civil Engineering, State University of Rio de Janeiro, UERJ, Rio de Janeiro, Brazil, 2008. (in portuguese)

[24] A. T. da Silva, "Comportamento de Peças Tracionadas em Esruturas de Aço-Carbono e Aço Inoxidável", Graduate Project, Structural Engineering Department, State University of Rio de Janeiro, UERJ, Rio de Janeiro, Brazil, 2008. (in portuguese)

[25] O. Bursi, J.P. Jaspart, "Calibration of a Finite Element Model for Isolated Bolted End-Plate Steel Connections", Journal of Constructional Steel Researchers, 44(3), 225-262, 1997.

[26] ANSYS, Inc. Theory Reference (version 11.0), 2008.

Chapter 8

Numerical Evaluation of the Ultimate Bearing Capacity of Steel Structures

C.J. Gantes
School of Civil Engineering
National Technical University of Athens, Greece

Abstract

Numerical tools for understanding the behaviour, predicting all possible failure mechanisms, and evaluating the ultimate strength of steel structures by means of commercially available finite element software are critically reviewed. Failure dominated by either material yielding or instability is addressed, as well as interaction of failure modes. Steps include setting up an appropriate finite element model, obtaining critical buckling modes from linearized buckling analysis (LBA), and then using a linear combination of these modes as an imperfection pattern for a geometrically and material nonlinear imperfection analysis (GMNIA). Equilibrium paths accompanied by snapshots of deformation and stress distribution at characteristic points are highlighted as a powerful tool for evaluating the results of the GMNI analysis, identifying the dominant failure modes and thus proposing appropriate strengthening measures. Practical details of the implementation of the proposed strategy are discussed. Results for several case studies are used to demonstrate this methodology, including laced and battened built-up members, compression members with varying cross-section, and beams of long span roofs covering athletic or exhibition facilities.

Keywords: steel structures, geometric nonlinearity, material nonlinearity, ultimate limit state, strength, imperfections, advanced finite element analysis, equilibrium path.

1 Introduction

In recent years structural design codes have adopted limit state design (LSD), replacing the older concept of allowable stress design (ASD) [6, 24]. Limit state design requires the structure to satisfy two principal criteria: the ultimate limit state (ULS) and the serviceability limit state (SLS). A limit state is a set of performance criteria that must be met when the structure is subject to loads. To satisfy the ultimate limit state, the structure must not collapse when subjected to any of the pertinent design load combinations. To satisfy the serviceability limit state criteria, a structure must remain functional for its intended use subject to routine, everyday loading.

In the present chapter state-of-the-art numerical analysis tools are presented for predicting the collapse load of steel structures, in other words their strength in order to verify the ultimate limit state criteria. Collapse may be due to development at certain cross-sections of stresses exceeding the material capacity, a condition known as material nonlinearity, or due to buckling, associated with a sudden increase in deformations for a small increase in applied load, a condition also known as geometric nonlinearity, or due to a combination of both types of nonlinearity.

The prevailing approach proposed by modern codes for carrying out checks at the ultimate limit state consists of obtaining action effects by means of linear elastic analysis of the structure subjected to design loads, and comparing them for each member to resistances that account for both types of eventual nonlinearity. Usually this is accomplished by multiplying the cross-section resistance, assuming full exploitation of the material capacity of the cross-section, by appropriate reduction factors, representing the influence of geometric nonlinearity [48].

Local buckling effects are usually addressed by classifying cross-sections into appropriate slenderness categories, and accordingly using the plastic, elastic or a reduced, "effective" cross-section resistance. Global member buckling, such as flexural or lateral-torsional buckling, is addressed by the above mentioned reduction factors. These reduction factors are directly related to the member slenderness, depending on its length, boundary conditions and cross-sectional inertia characteristics, and incorporate the effects of initial imperfections and residual stresses. The reduction factors are obtained from so-called buckling curves, which have been calibrated for specific types of cross-sections by means of a wide range of experimental results.

The advantage of this approach for practical applications is that prediction of collapse, a highly nonlinear phenomenon, is accomplished by means of linear elastic analysis, which is simple and straightforward, can be carried out by means of widely available commercial software, is computationally inexpensive and does not require special engineering expertise in setting up the structural model, selecting the analysis parameters and interpreting the results. Furthermore, the results are quite reliable for ordinary structural systems, consisting of members with normal profile cross-sections as well as built-up cross-sections [31].

For certain common structural systems, particularly frames, more advanced procedures have been proposed in the literature. For example, in [34] simplified amplification factors were derived, representing material and geometric nonlinearity in frame instability. Such approximate methods, however, are restricted to specific cases and cannot be easily generalized to have a wider validity and applicability. Therefore, research efforts, promoted also by the rapid advances in computational structural analysis, have focused on the use of advanced numerical analysis procedures for obtaining the ultimate strength. A large part of this effort is directed towards frame structures (for example [17, 30, 37, 38, 40, 41, 52, 53]). Several other applications also have been addressed by investigators, such as tubular structures [16], transmission towers [5], space trusses [57], reticulated domes [33], storage pallet racks [7], out-of-plane effects [51], arches [22, 42], built-up columns [32] and offshore steel jacket structures [45]. Furthermore, thin-walled, plated and shell structures have also been investigated (for example [4, 23 ,28, 49]).

To a large extent, in the above mentioned work use is made of specialized, in-house software. On the other hand, modern design codes already allow prediction of collapse by means of more elaborate, nonlinear analysis with imperfections [24,25,26]. This approach may be more suitable than the use of reduction factors for members with unusual cross-sections and for uncommon structural systems, as such cases are not directly covered by buckling curves and interaction equations, and assumptions made in an effort to approximate their behaviour may lead to significant inaccuracies. This implies that practicing engineers applying these codes must have sufficient guidelines for setting up numerical models, selecting proper analysis methods and numerical algorithm parameters, and interpreting the results. Recent technical literature contributes in that direction ([3,18,27,35,39], among others).

Taking that direction, in the present work a systematic methodology is presented for prediction of collapse of steel structures by means of nonlinear numerical analysis, making use of commercially available finite element software. The strategy consists of setting up an appropriate finite element model, obtaining critical buckling modes from linearized buckling analysis (LBA), and then using a linear combination of these modes as the imperfection pattern for a geometrically and material nonlinear imperfection analysis (GMNIA). Equilibrium paths accompanied by snapshots of deformation and stress distribution at characteristic points are proposed as a powerful tool for evaluating the results of the GMNI analysis. Practical details of the implementation of the proposed strategy are discussed. The proposed methodology is demonstrated for a number of case studies of real structures. The aim of this chapter is to contribute towards making advanced numerical analysis methods accessible to practicing structural engineers, by providing guidelines for using such methods for structural design, and by demonstrating some of the capabilities afforded by such methods to the structural engineering community.

2 Pertinent finite element formulations

2.1 Nonlinear finite element formulation

To understand the basic principles of the presented methodology, first the fundamental nonlinear analysis equations in the context of the finite element method are summarized. Consider the motion of a general body in a stationary Cartesian coordinate system that experiences large displacements, large strains and a linear or nonlinear constitutive response. Supposing that the solutions for the static and kinematic variables are known from time 0 (initial configuration) up to the time t, the solution at the next required equilibrium position corresponding to time $t+\Delta t$ is sought. In the large displacement formulation, the equilibrium condition can be expressed, using the principle of virtual displacements, with the equation [11]:

$$\int_{^0V} {}^{t+\Delta t}_{0}S_{ij}\delta\, {}^{t+\Delta t}_{0}\varepsilon_{ij} d^0V = {}^{t+\Delta t}R \tag{1}$$

where ${}^{t+\Delta t}_{0}S_{ij}$ is the second Piola-Kirchhoff stress tensor

$${}^{t+\Delta t}_{0}S_{ij} = \frac{{}^0\rho}{{}^{t+\Delta t}\rho}\, {}^0_{t+\Delta t}x_{i,m}\, {}^{t+\Delta t}\tau_{mn}\, {}^0_{t+\Delta t}x_{j,n} \tag{2}$$

and ${}^{t+\Delta t}_{0}\varepsilon_{ij}$ is the Green-Lagrange strain tensor

$$\delta\, {}^{t+\Delta t}_{0}\varepsilon = \delta\frac{1}{2}\left({}^{t+\Delta t}_{0}u_{i,j} + {}^{t+\Delta t}_{0}u_{j,i} + {}^{t+\Delta t}_{0}u_{k,i}\, {}^{t+\Delta t}_{0}u_{k,j}\right) \tag{3}$$

while ${}^{t+\Delta t}R$ corresponds to the externally applied loads at time $t+\Delta t$, ${}^{t+\Delta t}\tau_{mn}$ are the Cartesian components of the Cauchy stress tensor in the deformed geometry at time $t+\Delta t$, ${}^{t+\Delta t}x_i$ denotes the Cartesian coordinates of material point at time $t+\Delta t$, ${}^{t+\Delta t}\rho$ is the mass density of the body at time $t+\Delta t$ and ${}^{t+\Delta t}_{0}u_i$ are the components of the displacement vector at the configuration at time $t+\Delta t$.

In the above quantities, the terminology used in [11] is adopted. Namely, in any quantity the left superscript indicates the configuration in which the quantity occurs and the left subscript indicates the configuration with respect to which the quantity is measured. In a total Lagrange (TL) formulation this subscript is equal to 0. Moreover, in this notation, a comma denotes differentiation with respect to the coordinate following.

The first step in linearizing the nonlinear equilibrium equation of Equation (1) is to decompose incrementally the stresses and the strains. The second Piola-Kirchhoff stresses ${}^{t+\Delta t}_{0}S_{ij}$ are decomposed into:

$${}^{t+\Delta t}_{0}S_{ij} = {}^{t}_{0}S_{ij} + {}_{0}S_{ij} \tag{4}$$

Similarly, the Green-Lagrange strains can be decomposed into:

$${}^{t+\Delta t}_{0}\varepsilon_{ij} = {}^{t}_{0}\varepsilon_{ij} + {}_{0}\varepsilon_{ij} \tag{5}$$

$${}_{0}\varepsilon_{ij} = {}_{0}e_{ij} + {}_{0}n_{ij} \tag{6}$$

$${}_{0}e_{ij} = \frac{1}{2}\left({}_{0}u_{i,j} + {}_{0}u_{j,i} + \underbrace{{}^{t}_{0}u_{k,i}\, {}_{0}u_{k,j} + {}_{0}u_{k,i}\, {}^{t}_{0}u_{k,j}}_{\text{Initial displacement effect}}\right) \tag{7}$$

$${}_{0}n_{ij} = \frac{1}{2}\, {}_{0}u_{k,i}\, {}_{0}u_{k,j} \tag{8}$$

where $_0S_{ij}$ are the incremental second Piola-Kirchhoff stress components and $_0\varepsilon_{ij}$ are the incremental Green-Lagrange strain components.

Next, the nonlinear equilibrium equations are formulated, taking into account the incremental decompositions of stresses and strains. Noting that $\delta^{t+\Delta t}_{0}\varepsilon_{ij} = \delta_0 \varepsilon_{ij}$, the equilibrium equation is:

$$\int_{^0V} {_0S_{ij}}\delta_0\varepsilon_{ij}d^0V + \int_{^0V} {_0^tS_{ij}}\delta_0 \eta_{ij}d^0V = {^{t+\Delta t}R} - \int_{^0V} {_0^tS_{ij}}\delta_0 e_{ij}d^0V \qquad (9)$$

The final step is the linearization of Equation (9). Using the approximations $_0S_{ij} = {_0C_{ijrs}} {_0e_{rs}}$ and $\delta_0 \varepsilon_{ij} = \delta_0 e_{ij}$ we obtain the following approximate equilibrium equation (Brendel and Ramm 1980):

$$\int_{^0V} {_0C_{ijrs}} {_0e_{rs}}\delta_0 e_{ij}d^0V + \int_{^0V} {_0^tS_{ij}}\delta_0 \eta_{ij}d^0V = {^{t+\Delta t}R} - \int_{^0V} {_0^tS_{ij}}\delta_0 e_{ij}d^0V \qquad (10)$$

where $_0C_{ijrs}$ is the incremental stress-strain tensor at time t, referred to the configuration at time 0.

Equation (10) is the well-known linearized equilibrium equation of the Total Lagrange (TL) formulation. The global stiffness matrix of the structure can be obtained with the usual assembly procedures, after a displacement field is introduced using a finite element discretization. According to [15], from the first integral of the left-hand side of Equation (10), two stiffness matrices arise, namely the linear elastic matrix $_0^tK_0$ and the initial displacement matrix $_0^tK_u$ (due to the terms of the initial displacement effect of Equation (7)). From the second integral of the left-hand side of Equation (10), we get the initial stress or geometric matrix $_0^tK_g$. Finally, the integral of the right-hand side of Equation (10) gives rise to the vector of the internal forces in configuration t, which are denoted as $_0^tF$.

Thus, in the finite element method context, Equation (10) can be written in matrix form as:

$$\left(\underbrace{_0^tK_0 + {_0^tK_u} + {_0^tK_g}}_{_0^tK} \right) \Delta U = {^{t+\Delta t}R} - {_0^tF} \qquad (11)$$

where $_0^tK$ is the tangent stiffness of the structure and ΔU is an increment to the current displacement vector.

2.2 Linearized buckling analysis

At collapse or buckling of the structure, the tangent stiffness is singular. The condition of instability of the structure then reads as:

$$det\left({}_0^t K\right) = det\left({}_0^t K_0 + {}_0^t K_u + {}_0^t K_g\right) = 0 \tag{12}$$

The following nonlinear buckling problems can be formulated [15,56]:

$$\left({}_0^t K_0 + {}_0^t K_u + {}^t\lambda\, {}_0^t K_g\right)\phi_i = 0 \tag{13}$$

$$\left({}_0^t K_0 + {}^t\lambda\left({}_0^t K_u + {}_0^t K_g\right)\right)\phi_i = 0 \tag{14}$$

which differ from each other in the way that the load parameter ${}^t\lambda$ of the buckling load ${}^t\lambda R$ is related to the elements of the tangent stiffness matrix. In the above equations ϕ_i denotes the ith buckling mode.

If the initial displacement matrix ${}_0^t K_u$ is omitted from the two nonlinear buckling problems, then the classical buckling problem can be obtained:

$$\left({}_0^t K_0 + {}^t\lambda\, {}_0^t K_g\right) f_i = 0 \tag{15}$$

3 Proposed numerical strategy

The engineers applying the proposed methodology should have good knowledge of theoretical and practical aspects of the finite element method for linear and nonlinear structural applications (as outlined, for example, in [8,9,10,11,13,21,36]). The proposed strategy for understanding the behaviour, predicting all possible failure mechanisms, evaluating the strength, and assessing the vulnerability of steel structures, consists of the following steps [27]:

Step 1. Setting up an appropriate finite element model

In this step special attention should be paid to the following:

(i) The finite element software, the specific elements used to model the structure, as well as the numerical solution algorithm employed for numerical analysis, should be verified, by means of comparisons to analytical solutions and/or numerical solutions from the literature.

(ii) The model should be able to predict all anticipated failure mechanisms. For example, modeling a column with beam elements is not appropriate for predicting failure where local or lateral buckling may be predominant. Shell or plate elements should be used instead, and the mesh should be fine enough to capture the curvature of the part of the cross-section where local buckling occurs.

(iii) Basic features of the model, such as element connectivity, boundary conditions and loads should be checked by means of simple, linear, static and modal analyses. The sum of support reactions can be used to check load application. Mode shapes and periods of vibration can be compared to expected values, in order to identify possible errors in mass or stiffness.

(iv) The mesh density should be verified by performing a convergence study. Element size should be successively divided in half, and representative quantities of the response should be plotted against mesh density until sufficient convergence is achieved. It should be noted that a finer mesh may be required for subsequent nonlinear analyses than the one obtained from the linear static convergence study. However, the latter one should be considered as a minimum.

Step 2. Carrying out linearized buckling analysis

Critical buckling loads obtained by means of linearized buckling analyses (LBA) are, in most cases, not safe predictions of strength. However, this type of analysis must always precede the subsequent, more exact, nonlinear analyses, for two main reasons:

(i) Critical buckling loads obtained by means of linearized buckling analyses are an initial indication, and in most cases an upper bound, of actual strength. Taking into account also the fact that this is a very fast and inexpensive type of analysis, it may be very useful as a tool for evaluating alternative structural scheme solutions during preliminary design, before resorting to the much more time-consuming and expensive nonlinear analyses.

(ii) Buckling modes obtained by means of linearized buckling analyses are commonly used as initial imperfections for geometrically and material nonlinear imperfection analyses (GMNIA). Such imperfections are necessary in order to trigger all possible failure mechanisms and to make sure that the critical failure mechanism is captured by the nonlinear analysis algorithm.

It should be noted that the choice of buckling modes as imperfection patterns is not unique, and has in many cases been found to lead to lower compliance with experimental results than other shapes of initial imperfections. For the case of cylindrical shell structures, a type of structures that is particularly sensitive to imperfections, in [46,47] it is demonstrated that different patterns can be decisive depending on the imperfection amplitude and that these amplitude-dependent patterns cannot be determined with certainty because of the considerable influence of material nonlinearity and because of the numerous postbuckling paths which cross each other. Imperfections affine to the collapse mode of the perfect shell were found to be more unfavourable than ideal buckling modes.

Furthermore, fabrication processes are responsible for imperfection patterns that can have significant consequences on the ultimate bearing capacity of steel structures. In [14] the effect of circumferential weld depressions on the buckling strength of cylindrical shells subjected to axial compression was studied, using experimentally measured initial imperfection. In [29] the response of large-diameter pipes manufactured by cold-forming plates through the UOE process, by pressing them into an O-shape between two semicircular dies, welding them closed and then circumferentially expanding them to obtain a highly circular shape, was investigated and it was found that this process degrades the mechanical properties of the pipe and results in a reduction in its collapse pressure upwards of 30%.

However, the choice of manufacturing-related imperfection patterns is usually dependent upon the structural type as well as the specific manufacturing method, cannot be easily generalized, and often lacks adequate data to be properly implemented. In case where such information is available, it is, in general, preferable to use these imperfection patterns, as they represent real strength more accurately. In the absence of such data, however, the use of buckling modes as imperfection shapes is considered to be the next best available choice.

One question arising when carrying out such an analysis concerns the number of modes that should be evaluated. Even though the first buckling mode is the one corresponding to the lowest critical buckling mode, it is, in general, not sufficient, for several reasons. A higher mode associated with an unstable post-buckling equilibrium path may be more critical than a lower one with a stable post-buckling equilibrium path, due to its sensitivity to imperfections. Even two modes with stable post-buckling equilibrium paths, but with nearly coinciding critical buckling loads, may interact in the presence of imperfections, and lead to unstable post-buckling behaviour (for example [12]). Therefore, several buckling modes should be obtained. There is no clear-cut criterion as to their number. One reasonable guideline is to obtain a sufficient number of modes, so that the critical buckling load of the last mode is at least double that of the first one [27].

Step 3. Carrying out nonlinear analysis with imperfections

A nonlinear analysis may include large displacement effects, usually called geometric nonlinearity, or material behaviour beyond the linear-elastic range, usually called material nonlinearity, or both. Geometric nonlinearity is usually critical for slender structures. Steel structures are in most cases slender, either in terms of small plate thickness with respect to width, a situation that may lead to local buckling, or in terms of long members with respect to their cross-sectional dimensions, which may lead to global buckling, of flexural, lateral or other type. Even in such cases of slender structures, where geometric nonlinearity is critical, the induced large displacements result in large strains and stresses, and thus to material behaviour beyond its linear-elastic range, in other words to material nonlinearity. Therefore, failure is a result of combined geometric and material nonlinearity, and the analysis should account for both.

Material nonlinearity is usually critical for stocky structures with low slenderness. Again, though, large displacements may be encountered when the material enters the inelastic range and its stiffness gets drastically reduced. Thus, failure may also be the result of combined geometric and material nonlinearity, and the analysis should account for both.

Furthermore, it is well known that the presence of initial imperfections, which may be associated with deviations from the perfect geometry, cross-sections, material, boundary conditions, or loads, is decisive for the nonlinear structural response, particularly for loads near failure. Therefore, imperfections must be included in the numerical model. Taking the above into account, a geometrically and material nonlinear imperfection analysis (GMNIA) is recommended for reliably understanding the behaviour, predicting all possible failure mechanisms, evaluating the strength, and assessing the vulnerability of steel structures. In many cases it is useful first to carry out a separate analyses accounting only for material nonlinearity (MNIA) and geometric nonlinearity (GNIA), which help us identify the prevailing failure mechanism of the structure in question. But the ultimate strength evaluation should always be based on GMNI analyses, accounting for both types of nonlinearity. Several decisions, related to the details of this type of analysis, arise and must be considered by the engineer:

(i) **Shape of initial imperfections:** As stated above, in the absence of experimental data on manufacturing-related imperfection patterns, it is widely recognized as appropriate to use a linear combination of the critical buckling modes obtained from linearized buckling analysis as the shape of initial imperfections. Regarding the number of critical buckling modes reference is made to the discussion in step 2 above. In the absence of any available information about the relative importance of individual modes, equal weights may be assumed for their linear combination.

(ii) **Size of initial imperfections:** With the exception of cases with steep unstable post-buckling equilibrium paths, for example frequently encountered in thin-walled shells, the size of initial imperfections is not very important for the type of response and the values of critical response quantities obtained from the analysis. For certain types of structures pertinent codes specify recommended magnitudes of imperfections, or manufacturing and erection tolerances that can be regarded as upper bounds of imperfection magnitudes [24]. In some cases, a parametric study for different sizes within the range of realistic construction tolerances may provide valuable insight into the effect of imperfection magnitude on ultimate strength.

(iii) **Details of numerical algorithm:** All numerical algorithms for nonlinear structural analysis rely on applying the loads in small steps, performing iterations of linear analyses within each step until a predefined convergence criterion is met, and using the final configuration of the structure at each step as a starting configuration for the next one. The two most frequently used algorithms are the Newton-Raphson (NR) method, updating the stiffness matrix at each iteration of each step, and the Modified Newton-Raphson (MNR) method, updating the stiffness matrix only at the beginning of each step [19,50]. MNR requires more iterations per step, and is computationally less expensive, but NR converges more successfully in problems where strong nonlinearity is encountered. Both methods are unable to track the structural response beyond the ultimate strength, the so-called limit point of the equilibrium path.

For following the post-limit-point part of the equilibrium path, which may provide useful information with respect to imperfection sensitivity as well as ductility, a category of numerical algorithms, known in the literature as arc-length methods, are available in many commercial finite element programs. Most such algorithms are equipped with an "automatic load incrementation" feature, which enables them to adjust the load step to the degree of nonlinearity, thus using larger steps at lower load levels, where the response is more linear, and gradually reducing the load step at higher loads and stronger nonlinearities [10, 20, 43, 44, 54, 55].

On the basis of the above, the choice of an arc-length algorithm with either MNR (as a first alternative) or NR (in cases of convergence difficulties) iterations is recommended for GMNI analyses for practical applications. If convergence difficulties are still encountered, the engineer may have to increase the default number of maximum iterations per step, or relax the convergence tolerances, but in the second case special attention is recommended, as erroneous convergence to configurations without physical meaning may be obtained.

(iv) **Evaluation of results:** It is recommended to evaluate the results of GMNI analyses by using a plot of the equilibrium path, accompanied by characteristic snapshots of the structure's deformation and stress distribution along this path. The equilibrium path is a plot of a characteristic response quantity (usually on the horizontal axis) with respect to the external action (usually on the vertical axis). Very often the external action is represented by a load multiplier, λ, which acts on the required design loads. Particular attention is needed for the selection of the characteristic response quantity of the horizontal axis, which should be representative of the overall structural response and not of some local phenomena, depending also on the critical failure mode. Multiple equilibrium paths for more than one response quantity may be useful in some cases.

Snapshots of the structure's deformation and stress distribution at characteristic points (for example, before and after change of slope or change of curvature, at the ultimate limit point, at the end of the curve) are very useful in appreciating the physical phenomena occurring with progressive external loading.

Important quantities of an equilibrium path include: (a) the slope of its initial, usually straight, part, representing the structure's initial stiffness; (b) the maximum value of the external action, representing the structure's strength; (c) the displacement corresponding to the maximum value of the external action, pointing out whether serviceability problems may be encountered before failure; (d) the maximum deformation at failure, as well as the area between the equilibrium path and the horizontal axis, particularly in the post-linear range, representing the structure's ductility; (e) the slope beyond the limit point, indicating higher imperfection sensitivity in cases of a steep decrease in the equilibrium path.

Comparison of representative numerical results with experimental measurements, if available, is absolutely encouraged as a means for calibrating and verifying the numerical results. The same holds for comparison to analytical results of similar cases, for which code specifications exist. Numerical analysis can then, with increased reliability, be extended to other cases of geometry, cross-sections, material, boundary conditions, or loads, thus performing "numerical experiments" that may help optimize the design.

Ultimate Bearing Capacity of Steel Structures 229

4 Case studies

The above methodology is next illustrated by means of a number of case studies, using the well-known and extensively tested finite element codes Abaqus [1] and Adina [2].

4.1 Built-up columns

Built-up columns are a preferred structural solution in the case of large spans and/or heavy loads. They usually consist of flanges offering bending rigidity, connected to each other either with lacing bars (laced built-up columns) or battens (battened built-up column). An example is the simply-supported laced built-up column shown in Figure 1(a), subjected to a concentrated axial load on top. The flanges and diagonal bars are modelled using shell and beam elements, respectively [32].

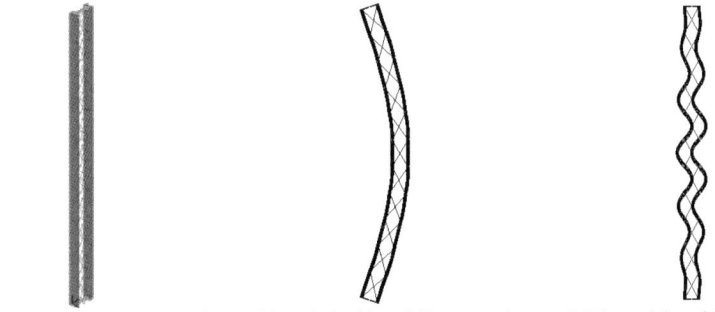

(a) Finite element model (b) Global buckling mode (c) Local buckling mode

Figure 1: Laced built-up column

By means of linearized buckling analysis, elastic critical loads corresponding to global and local buckling are determined (Figures 1(b) and (c)). The first one is related with buckling of the column as a whole, while the second relates to buckling of the panels between lacing bars as simply-supported members. These mode shapes are then used in the subsequent GMNI analysis for inserting global and local imperfections of appropriate, code-specified magnitudes.

The GMNIA results are presented by the equilibrium path of Figure 2. For reasons of comparison the values of global and local critical buckling loads as well as the equilibrium path ignoring material nonlinearity (GNI analysis) are shown also. Snapshots of stress and deformation at characteristic points A (near the limit point) and B (on the descending branch) of the GMNIA path are used for obtaining valuable information regarding the nature of failure (Figure 3). By comparing the GNIA equilibrium path to the critical buckling loads it is observed that the interaction between global and local buckling in the presence of imperfections significantly reduces the elastic capacity of the column. As seen from the GMNIA equilibrium path, further major reduction takes place due to material yielding, so that the ultimate load is several times lower than the corresponding elastic critical buckling loads.

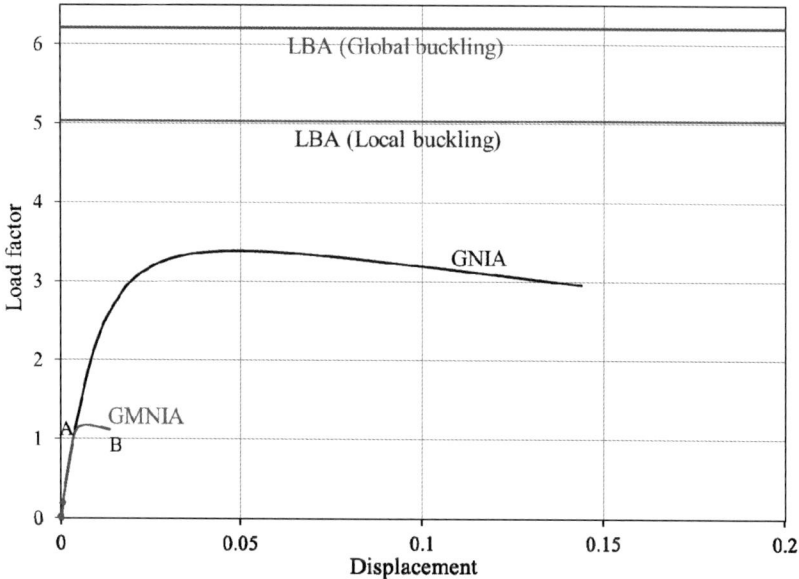

Figure 2: Equilibrium paths of laced built-up column

From the deformed configurations at points A and B it is confirmed that indeed both global and local buckling (at the more compressed side) take place. From the von Mises stress distribution it is also deduced that local material yielding is also encountered at the vicinity of the limit point, and yielding spreads on the descending branch of the equilibrium path.

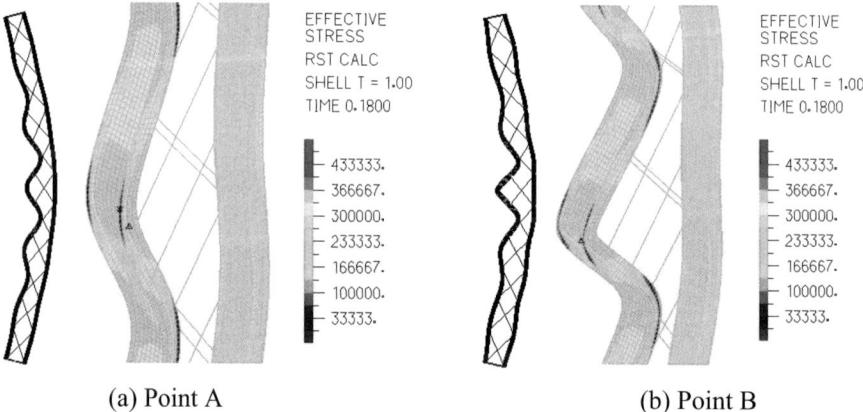

(a) Point A   (b) Point B

Figure 3: Deformation and von Mises stress distribution on characteristic points A and B along the GMNIA equilibrium path of laced built-up column

A second example is the battened built-up column shown in Figure 4(a), clamped at both ends and subjected to a concentrated axial load on top. Both flanges and diagonal bars are modeled using shell elements. Corresponding global and local buckling modes obtained from linearized buckling analysis are shown in Figures 4(b) and 4(c).

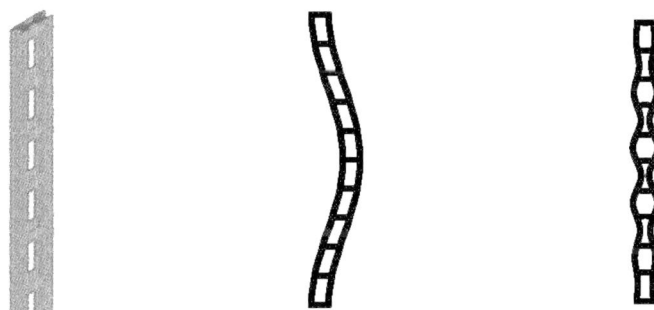

(a) Finite element model   (b) Global buckling mode   (c) Local buckling mode

Figure 4: Laced built-up column

The loads corresponding to these buckling modes are shown in Figure 5, along with GNIA and GMNIA equilibrium paths. In this case global buckling is by far more critical than local one, therefore the elastic equilibrium path exhibits no limit point but approaches asymptotically the value of the global critical load. Failure however is dominated by material yielding accompanied by moderate local buckling, as indicated by the sharp difference between GNIA and GMNIA equilibrium paths and by the snapshots of stress and deformation at characteristic points A (near the limit point) and B (on the descending branch) of the GMNIA path (Figure 6).

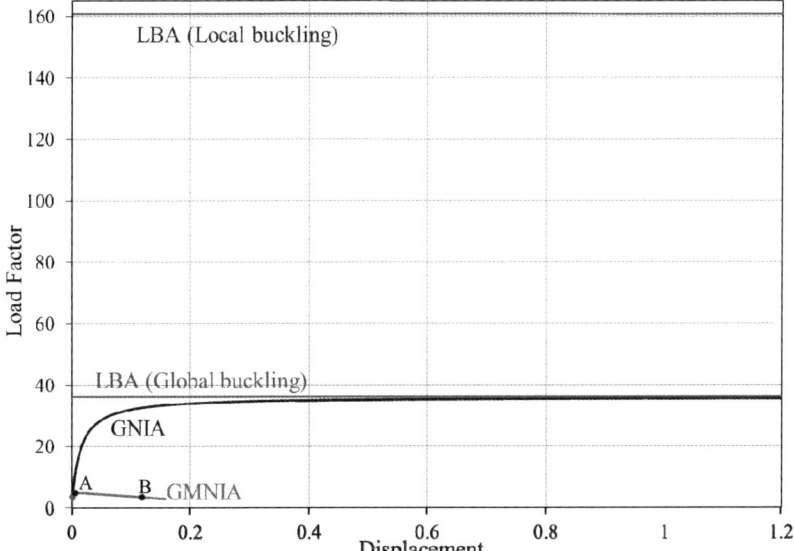

Figure 5: Equilibrium paths of battened built-up column

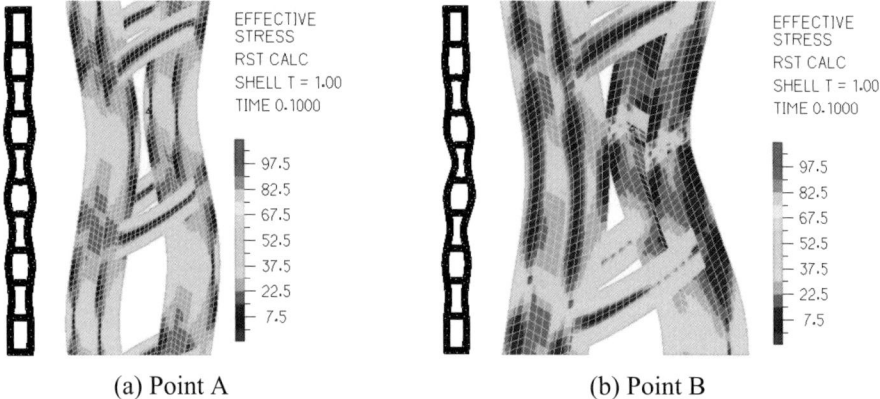

(a) Point A (b) Point B

Figure 6: Deformation and von Mises stress distribution on characteristic points A and B along the GMNIA equilibrium path of battened built-up column

4.2 Suspended roof over Aristotle's Lyceum

The suspended steel roof shown in Figure 7 will be constructed in Athens, Greece to provide cover for the archaeological site of Aristotle's Lyceum. Six parallel arches constitute the steel roof which is suspended from six pylons. For the design of the pylon, a compression member with hollow circular cross-section, varying over the height, and of the main arch, a member under compression and bending with varying box cross-section with protruding flanges, nonlinear numerical studies have been carried out using the finite element program ADINA.

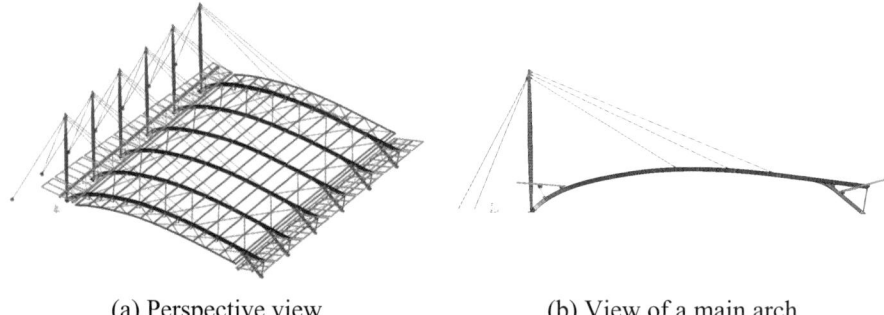

(a) Perspective view (b) View of a main arch

Figure 7: Suspended steel roof over Aristotle's Lyceum

In order to obtain the strength of a pylon with variable cross section against compression, the pylon has been modelled with shell finite elements, as shown in Figure 8(a). A design decision was taken to avoid local buckling as a potential failure mode, therefore class 1 or 2 cross-sections were used over the height. This was confirmed by the results of the linearized buckling analysis, which showed no local buckling modes among the first ten. Thus, only the first global flexural buckling mode, shown in Figure 8(b), was used as the initial imperfection.

Then, GNI and GMNI analyses were carried out and the corresponding equilibrium paths are shown in Figure 9. It is observed that the presence of imperfections reduces the elastic critical load with respect to the linear one. Failure is encountered on the linear ascending branch of the equilibrium path, and it is mainly due to material yielding, accompanied by moderate global flexural buckling. This is verified by the deformation and stress snapshots on three characteristic points before, on and beyond the critical point, shown in Figure 10. The formation of plastic hinges at two locations helps identify these cross-sections as weak, so that increased thickness at these areas would be the most appropriate way for improving strength.

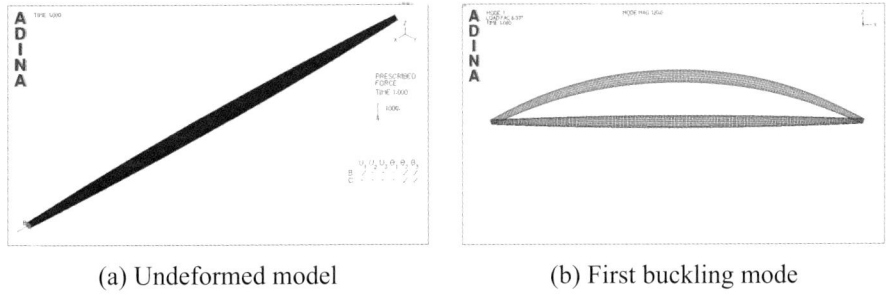

(a) Undeformed model   (b) First buckling mode

Figure 8: Finite element model of pylon of steel roof over Aristotle's Lyceum

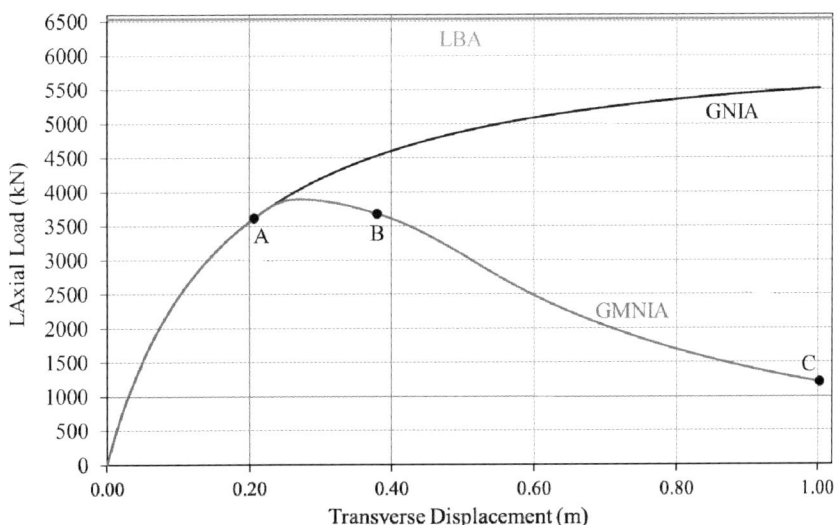

Figure 9: Equilibrium paths of pylon of steel roof over Aristotle's Lyceum

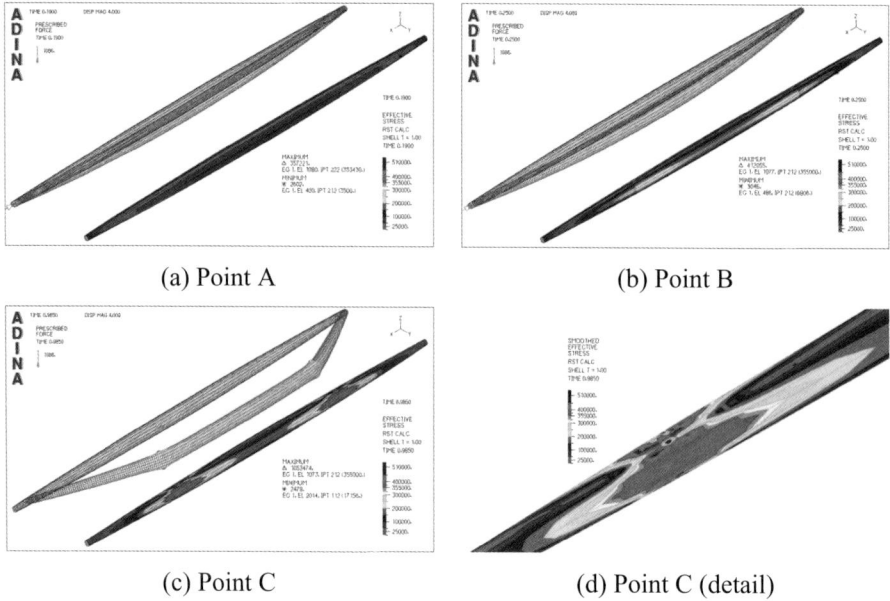

(a) Point A (b) Point B

(c) Point C (d) Point C (detail)

Figure 10: Deformation and von Mises stress distribution at characteristic points along the GMNIA equilibrium path of the pylon

In order to evaluate the behaviour of the main arches against lateral buckling, the use of code provisions is of questionable validity due to the unusual shape of the arches and of their cross-section. Therefore, a finite element model consisting of two main arches and their transverse and diagonal connecting members was analyzed (Figure 11(a)). The main arches were modelled with shell elements, bracing members with beam elements, and suspension cables by nonlinear, tension-only, prestressed truss elements. By considering two adjacent arches, the lateral support provided to the arches by their bracing system is taken into account. Cross-sections were class 1 and 2, so that local buckling was not critical.

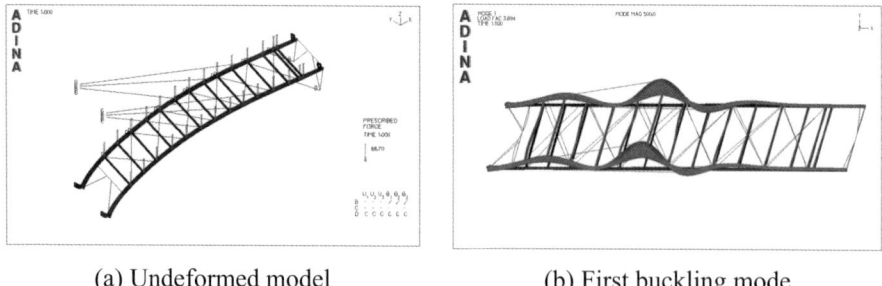

(a) Undeformed model (b) First buckling mode

Figure 11: Finite element model of two adjacent main arches of steel roof over Aristotle's Lyceum

Linearized buckling analysis was first performed in order to obtain the first buckling mode and the corresponding elastic critical buckling load (Figure 11(b)). This mode was then used as the initial imperfection for the subsequent GNI and GMNI analyses. The results are presented by the equilibrium paths of Figure 12, and the snapshots of stress and deformation at characteristic points A, B, C and D (Figures 13 and 14) along the GMNIA path. It is concluded that the box shaped cross-sections combined with the lateral support provided by the bracing system is sufficient for preventing lateral buckling, thus failure is primarily due to material yielding, with the formation of plastic hinges becoming evident from the deformation and stress pictures on the descending branch of the equilibrium path.

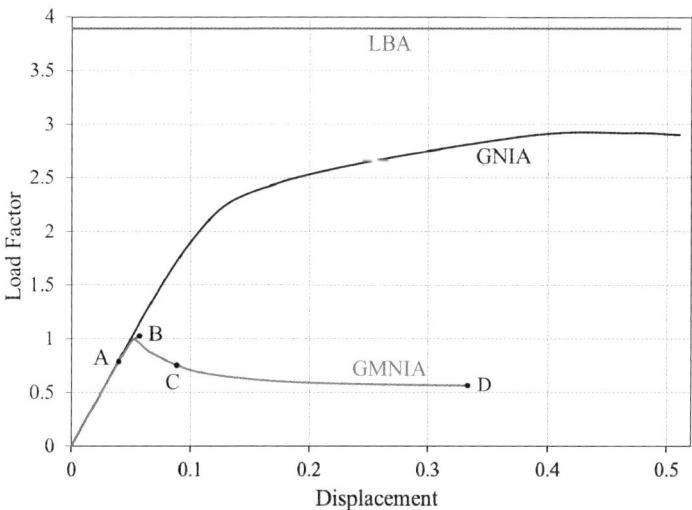

Figure 12: Equilibrium paths of main arches

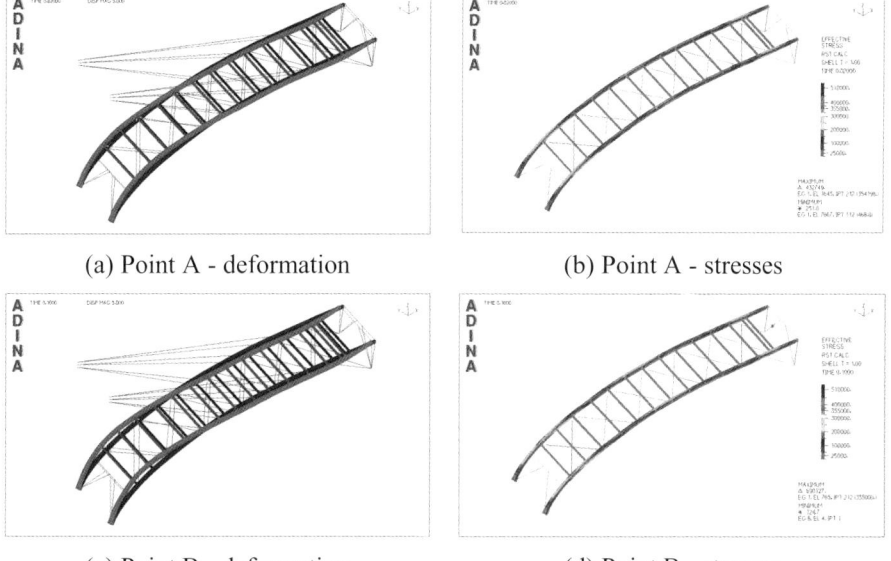

(a) Point A - deformation (b) Point A - stresses

(c) Point B - deformation (d) Point B - stresses

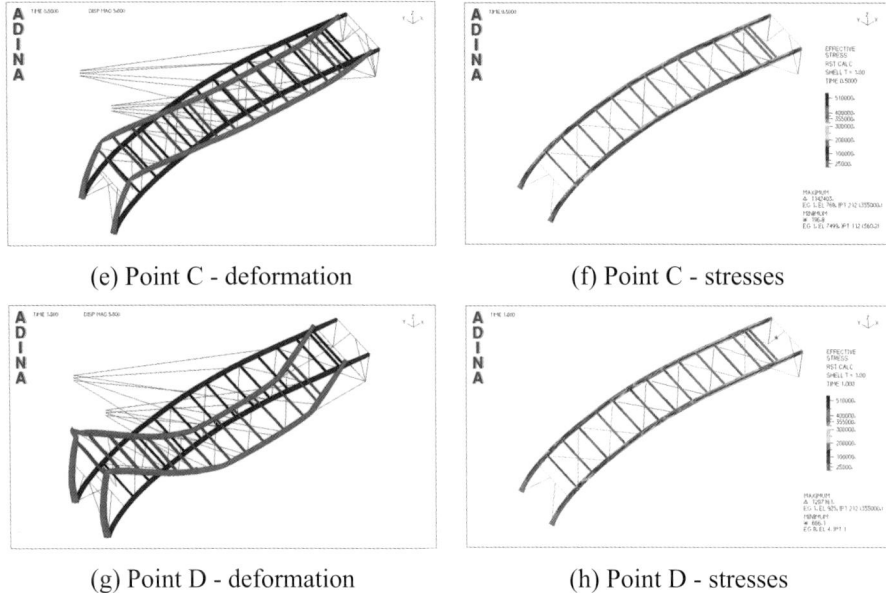

Figure 13: Deformation and von Mises stress distribution at characteristic points along the GMNIA equilibrium path of the main arches

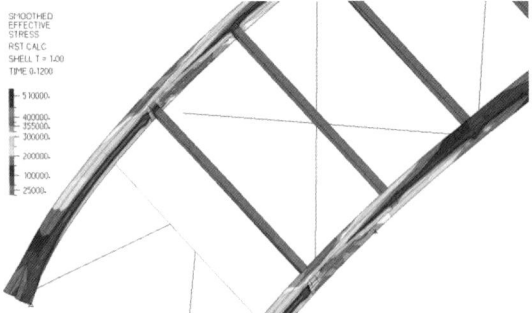

Figure 14: Von Mises stress distribution at characteristic point D along the GMNIA equilibrium path of the main arches (detail)

4.3 Roof of the Panathinaikos Football Stadium

The new football stadium for Panathinaikos F.C., shown in Figure 15, will be constructed in Votanikos, Athens, Greece. The roof has the shape of a cylindrical surface, and it consists of four structurally independent parts. The basic element of each part is a main truss-girder, simply supported on reinforced concrete pylons, arranged at the four corners of the stadium. Secondary beams are supported on these main girders and on the exterior peripheral reinforced concrete columns of the grandstands. These beams carry the roof cladding. Appropriately arranged cross-bracings contribute to the overall stability.

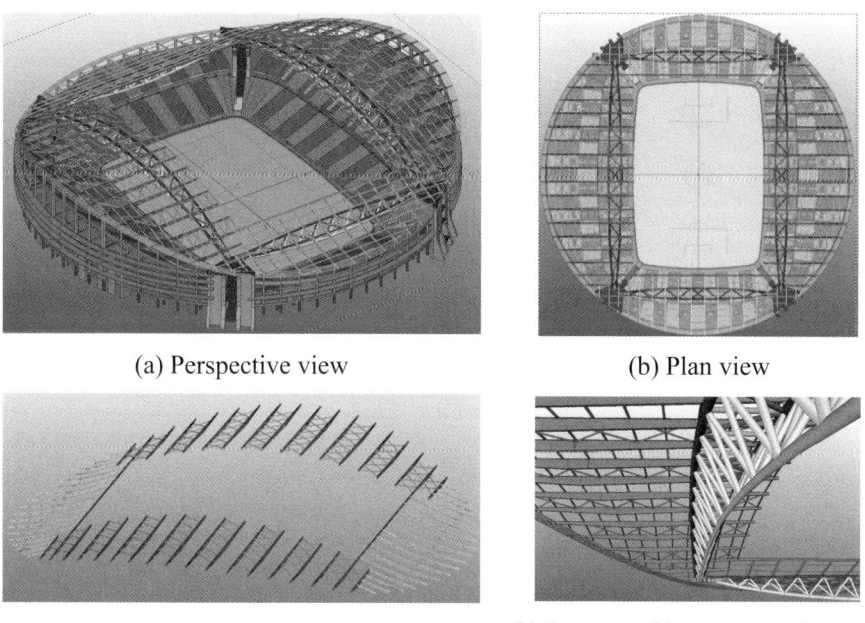

(a) Perspective view

(b) Plan view

(c) Beams and bracing

(b) Support of beams on main truss-girder

Figure 15: New football stadium of Panathinaikos F.C. in Votanikos

The beams have different lengths, depending on their location, so that they adjust to the overall roof geometry. They are made of narrow box cross-sections with protruding flanges that vary over their length. In order to evaluate their behaviour against lateral and local buckling, a finite element model was set up, consisting of two main beams, modelled with shell elements, and their cross-bracing modelled with beam elements (Figure 16). The first two buckling modes (Figure 17), corresponding to lateral and local buckling, respectively, were used as the initial imperfection for GMNI analysis. Failure was due to combined lateral buckling and material yielding, with a small contribution from local buckling, as indicated by the equilibrium path (Figure 18) and the corresponding deformation and stress distribution (Figure 19).

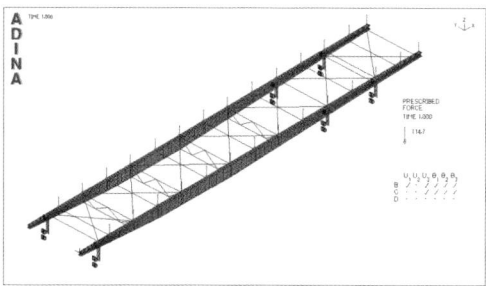

Figure 16: Finite element model of two beams and their cross-bracing in the steel roof of the Panathinaikos stadium

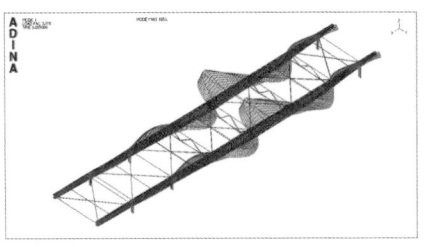 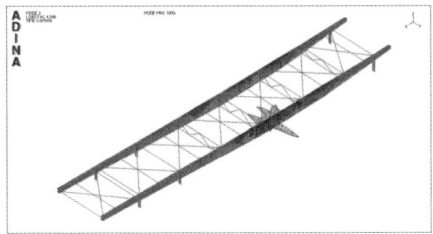

(a) First buckling mode (lateral)   (b) Second buckling mode (local)

Figure 17: Buckling modes of beams in the steel roof of the Panathinaikos stadium

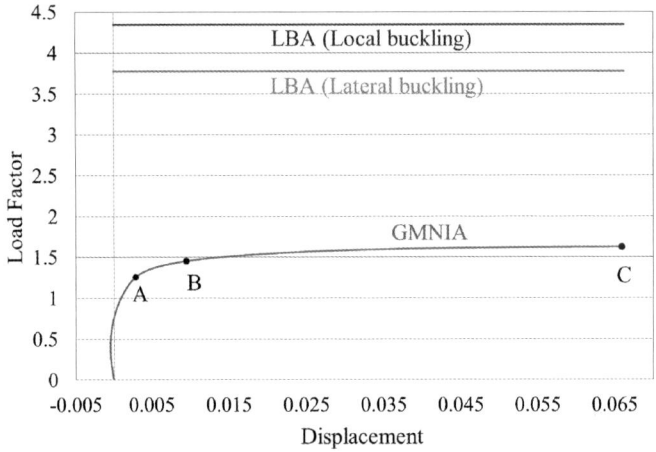

Figure 18: Equilibrium paths of the beams of the Panathinaikos stadium

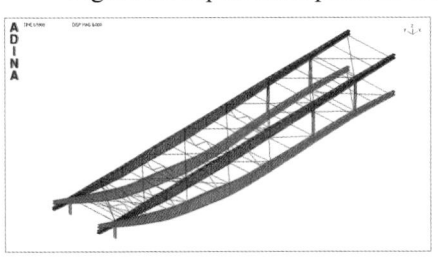 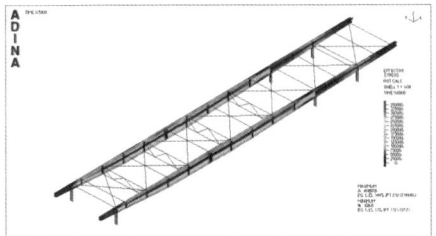

(a) Point A - deformation   (b) Point A - stresses

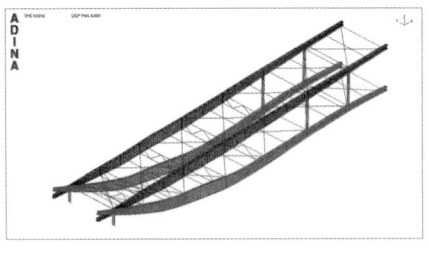 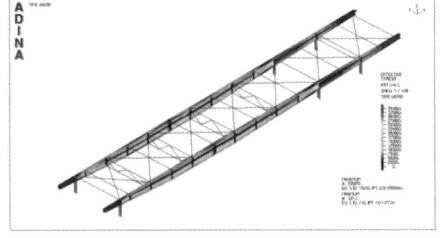

(c) Point B - deformation   (d) Point B - stresses

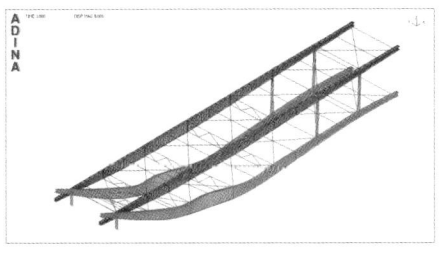 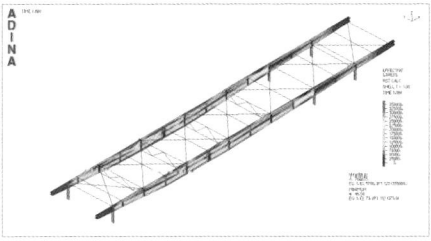

  (e) Point C - deformation   (f) Point C - stresses

Figure 19: Deformation and von Mises stress distribution at characteristic points along the GMNIA equilibrium path of the beams of the Panathinaikos stadium

5 Summary and conclusions

A methodology has been proposed for understanding the behaviour, predicting all possible failure mechanisms, and evaluating the ultimate strength of steel structures by means of linearized buckling and geometrically and material nonlinear imperfection analyses with commercially available finite element software. Failure dominated by either material yielding or instability is addressed, as well as interaction of failure modes. Equilibrium paths accompanied by snapshots of deformation and stress distribution at characteristic points are used for evaluating the results of the GMNI analysis, identifying the dominant failure modes and possibly proposing appropriate strengthening measures. The results of case studies involving laced and battened built-up members, compression members with varying cross-section, and beams of long span roofs covering athletic or exhibition facilities have been presented.

Acknowledgements

The author would like to thank his former and current graduate students Christoforos Dimopoulos, Konstantinos Kalochairetis and Konstantina Koulatsou for their contribution to the numerical case studies presented in this paper.

References

[1] ABAQUS/Standard and ABAQUS/Explicit - Version 6.8, "Abaqus Theory Manual", Dassault Systems.
[2] ADINA R&D (www.adina.com), "Theory and Modeling Guide" - Volume I: ADINA Solids & Structures, Report ARD 05-7, 2005.
[3] A. Agüero, F.J. Pallarés, "Proposal to evaluate the ultimate limit state of slender structures. Part 1: Technical aspects", Engineering Structures, 29(4), 483-497, 2007.

[4] F.G.A. Al-Bermaniand, S. Kitipornchai, "Elasto-plastic large deformation analysis of thin-walled structures", Engineering Structures, 12(1), 28-36, 1990.
[5] F.G.A. Al-Bermani, S. Kitipornchai, "Nonlinear analysis of transmission towers", Engineering Structures, 14(3), 139-151, 1992.
[6] American Institute of Steel Construction, "Load and Resistance Factor Design for Structural Steel Buildings", 1999.
[7] N. Baldassino, C. Bernuzzi, "Analysis and behaviour of steel storage pallet racks", Thin-Walled Structures, 37(4), 277-304, 2000.
[8] K.J. Bathe, A.P. Cimento, "Some practical procedures for the solution of nonlinear finite element equations", Computer Methods in Applied Mechanics and Engineering, 22(1), 59-85, 1980.
[9] K.J. Bathe, M.D. Snyder, A.P. Cimento, W.D. Rolph, "On some current procedures and difficulties in finite element analysis of elastic-plastic response", Computers & Structures, 12(4), 607-624, 1980.
[10] K.J. Bathe, E.N. Dvorkin, "On the automatic solution of nonlinear finite element equations", Computers & Structures, 17(5-6), 871-879, 1983.
[11] K.J. Bathe, "Finite element procedures", Englewood Cliffs: Prentice-Hall, 1995.
[12] Z.P. Bazant, L. Cedolin, "Stability of structures: elastic, inelastic, fracture, and damage theories", Oxford: Oxford University Press, 1991.
[13] T. Belytschko, W.K. Liu, B. Moran, "Nonlinear Finite Elements for Continua and Structures", John Wiley & Sons Ltd., 2000.
[14] P.A. Berry, J.M. Rotter, R.Q. Bridge, "Compression tests on cylinders with circumferential weld depressions", Journal of Engineering Mechanics, ASCE, 126(4), 405–413, 2000.
[15] B. Brendel, E. Ramm, "Linear and nonlinear stability analysis of cylindrical shells", Computers & Structures, 12, 549-558, 1980.
[16] S.L. Chan, "Inelastic post-buckling analysis of tubular beam-columns and frames", Engineering Structures, 11(1), 23–30, 1989.
[17] S.L. Chan, P.-T. Chui, "Non-linear static and cyclic analysis of semirigid steel frames", Elsevier Science, Amsterdam, 2000.
[18] W.F. Chen, S.E. Kim, "LRFD steel design using advanced analysis", Boca Raton, CRC Press, Florida, 1997.
[19] M.A. Crisfield, "A faster modified Newton-Raphson iteration", Computer Methods in Applied Mechanics and Engineering, 20(3), 267-278, 1979.
[20] M.A. Crisfield, "A fast incremental/iterative solution procedure that handles snap-through", Computers & Structures, 13(1-3), 55-62, 1981.
[21] M.A. Crisfield, G. Jelenic, Y. Mi, H.-G. Zhong, Z. Fan, "Some aspects of the non-linear finite element method", Finite Elements in Analysis and Design, 27(1), 19-40, 1997.
[22] C.A. Dimopoulos, C.J. Gantes, "Nonlinear In-Plane Behavior of Circular Steel Arches with Hollow Circular Cross-Section", Journal of Constructional Steel Research, 64(12), 1436-1445, 2008.
[23] M. Elgaaly, "Post-buckling behavior of thin steel plates using computational models", Advances in Engineering Software, 31(8-9), 511-517, 2000.

[24] European Committee for Standardisation, "Eurocode 3: Design of steel structures, Part 1-1: General rules and rules for buildings", 2004.
[25] European Committee for Standardisation, "Eurocode 3: Design of steel structures, Part 1.5: Plated structural elements", 2004.
[26] European Committee for Standardization, "Eurocode 3 – Design of Steel Structures – Part 1.6: Strength and Stability of Shell Structures", 2006.
[27] C.J. Gantes, K.A. Fragkopoulos, "Strategy for Numerical Verification of Steel Structures at the Ultimate Limit State", Structure and Infrastructure Engineering, 6(1-2), 225-255, Feb. 2010.
[28] M. Gettel, W. Schneider, "Buckling strength verification of cantilevered cylindrical shells subjected to transverse load using Eurocode 3", Journal of Constructional Steel Research, 63(11), 1467-1478, 2007.
[29] M.D. Herynk, S. Kyriakides, A. Onoufriou, H.D. Yun, "Effects of the UOE/UOC pipe manufacturing processes on pipe collapse pressure", International Journal of Mechanical Sciences, 49, 533–553, 2007.
[30] S.H. Hsieh, G.G. Deierlein, "Nonlinear analysis of three-dimensional steel frames with semi-rigid connections", Computers & Structures, 41(5), 995-1009, 1991.
[31] B. Johansson, R. Maquoi, G. Sedlacek, "New design rules for plated structures in Eurocode 3", Journal of Constructional Steel Research, 57(3), 279-311, 2001.
[32] K.E. Kalochairetis, C.J. Gantes, "Numerical and Analytical Investigation of Collapse Loads of Laced Built-up Columns", Computers & Structures, 89(11-12), 1166-1176, June 2011.
[33] S. Kato, I. Mutoh, M. Shomura, "Collapse of semi-rigidly jointed reticulated domes with initial geometric imperfections", Journal of Constructional Steel Research, 48(2-3), 145-167, 1998.
[34] A.R. Kemp, "Simplified amplification factors representing material and geometric inelasticity in frame instability", Engineering Structures, 22(12), 1609-1619, 2000.
[35] S.-E. Kim, M.-H. Park, S.-H. Choi, "Direct design of three-dimensional frames using practical advanced analysis", Engineering Structures, 23(11), 1491-1502, 2001.
[36] M. Kojic, K.J. Bathe, "Inelastic analysis of solids and structures (Series in Computational Fluid and Solid Mechanics)", Springer Verlag, Berlin, 2004.
[37] J.Y.R. Liew, D.W. White, W.F. Chen, "Limit states design of semi-rigid frames using advanced analysis, Part 2: Analysis and design", Journal of Constructional Steel Research, 26(1), 29-57, 1993.
[38] J.Y.R. Liew, W.F. Chen, H. Chen, "Advanced inelastic analysis of frame structures", Journal of Constructional Steel Research, 55(1-3), 245-265, 2000.
[39] J.K. Paik, A.K. Thayamballi, "Ultimate Limit State Design of Steel-Plated Structures", Wiley, New Jersey, 2003.
[40] Y.-L. Pi, N.S. Trahair, "Nonlinear inelastic analysis of steel beam columns – theory", Journal of Structural Engineering, ASCE, 120(7), 2041–2061, 1994.

[41] Y.-L. Pi, N.S. Trahair, "Nonlinear inelastic analysis of steel beam columns – applications", Journal of Structural Engineering, ASCE, 120(7), 2062–2085, 1994.
[42] Y.-L. Pi, M.A. Bradford, "In-plane strength and design of fixed steel I-section arches", Engineering Structures, 26(3), 291-301, 2004.
[43] E. Ramm, "Strategies for tracing nonlinear responses near limit points", in W. Wunderlich, E. Stein, K.J. Bathe, (Editors), Nonlinear Finite Element Analysis in Structural Mechanics, 63-89, Springer-Verlag, New York, 1981.
[44] E. Riks, "An incremental approach to the solution of snapping and buckling problems", Int. Journal of Solids and Structures, 15(7), 529-551, 1979.
[45] P.F.N. Rodrigues, B.P. Jacob, "Collapse analysis of steel jacket structures for offshore oil exploitation", Journal of Constructional Steel Research, 61(8), 1147-1171, 2005.
[46] W. Schneider, A. Brede, "Consistent equivalent geometric imperfections for the numerical buckling strength verification of cylindrical shells under uniform external pressure", Thin-Walled Structures, 43(2), 175-188, 2005.
[47] W. Schneider, I. Timmel, K. Höhn, "The conception of quasi-collapse-affine imperfections: A new approach to unfavourable imperfections of thin-walled shell structures", Thin-Walled Structures, 43(8), 1202-1224, 2005.
[48] G. Sedlacek, C. Müller, "The European standard family and its basis", Journal of Constructional Steel Research, 62(11), 1047-1059, 2006.
[49] N.E. Shanmugam, J.Y.R. Liew, S.L. Lee, "Ultimate strength design of biaxially loaded steel box beam-columns", Journal of Constructional Steel Research, 26(2-3), 99-123, 1993.
[50] J.A. Stricklin, W.E. Haisler, von W.A. Riesemann, "Evaluation of solution procedures for material and/or geometrically nonlinear structural analysis", AIAA Journal, 11(3), 292-299, 1973.
[51] N.S. Trahair, S.-L. Chan, "Out-of-plane advanced analysis of steel structures", Engineering Structures, 25(13), 1627-1637, 2003.
[52] D.W. White, "Plastic-hinge methods for advanced analysis of steel frames", Journal of Constructional Steel Research, 24(2), 121-152, 1993.
[53] D.W. White, J.F. Hajjar, "Stability of steel frames: the cases for simple elastic and rigorous inelastic analysis/design procedures", Engineering Structures, 22(5), 155-167, 2000.
[54] P. Wriggers, W. Wagner, C. Miehe, "A quadratically convergent procedure for the calculation of stability points in finite element analysis", Computation Methods in Applied Mechanics and Engineering, 70, 329-347, 1987.
[55] P. Wriggers, J.C. Simo, "A general procedure for the direct computation of turning and bifurcation points", International Journal for Numerical Journal in Engineering, 30, 155-176, 1990.
[56] H. Wriggers, E. Onate, P. Wriggers, "Direct Computation of Instability Points with Inequality Constraints Using the Finite Element Method", International Center for Numerical Methods in Engineering, 2001.
[57] Y.B. Yang, C.T. Yang, T.P. Chang, P.K. Chang, "Effects of member buckling and yielding on ultimate strengths of space trusses", Engineering Structures, 19(2), 179-191, 1997.

Chapter 9

©Saxe-Coburg Publications, 2011.
Civil and Structural Engineering Computational Technology
B.H.V. Topping and Y. Tsompanakis, (Editors)
Saxe-Coburg Publications, Stirlingshire, Scotland, 243-277.

Recent Developments in the Analysis of Stiffened Plates

E.J. Sapountzakis
Institute of Structural Analysis and Antiseismic Research
School of Civil Engineering
National Technical University of Athens, Greece

Abstract

The small and large deflection analysis of plates stiffened by arbitrarily placed parallel beams subjected to an arbitrary loading is reviewed. The most general case is treated by employing a model according to which the stiffening beams are isolated from the plate by sections in the lower outer surface of the plate, making the hypothesis that the plate and the beams can slip in all directions of the connection without separation (*i.e.* uplift neglected). The arising tractions in all directions at the fictitious interfaces are integrated with respect to each half of the interface width resulting in two interface lines, along which the loading of the beams as well as the additional loading of the plate is defined. The unknown distribution of the aforementioned integrated tractions is established by applying continuity conditions in all directions at the two interface lines taking into account their relation with the interface slip through the shear connector stiffness. The adopted model permits the evaluation of the shear forces at the interfaces in both directions, the knowledge of which is very important in the design of prefabricated ribbed plates. In order to illustrate the application and potential of the proposed model numerical examples are presented and discussed. Finally, the paper closes with some concluding remarks and future perspectives.

Keywords: elastic stiffened plate, reinforced plate with beams, bending, nonuniform torsion, warping, ribbed plate, boundary element method, slab-and-beam structure, nonlinear analysis.

1 Introduction

Structural plate systems stiffened by beams are widely used in buildings, bridges, ships, aircraft and machines. Stiffening of the plate is used to increase its load carrying capacity and to prevent buckling especially in the case of in-plane loading. Moreover, composite reinforced concrete slabs stiffened by steel or prestressed concrete beams have been widespread in recent years due to the economic and structural advantages of such systems. However, these structures are prone to failure of the bond between the plate and the beams. It is the behavior of this bond that gives composite construction its unique peculiarities, while interface slip can cause significant redistribution of strain and stress. Moreover bearing in mind the importance of weight saving in engineering structures, the study of nonlinear effects on the analysis (large deflection analysis) of stiffened plates becomes essential. This non-linearity results from retaining the squares of the slopes in the strain–displacement relations (intermediate non-linear theory), thus avoiding the inaccuracies arising from a linearized second – order analysis.

The small deflection analysis of stiffened plates under static loading has received a considerable amount of literature in the past few decades. These structural systems were initially approximated by smearing – out the stiffness properties of the beams to get an equivalent orthotropic homogeneous slab of constant thickness [1-4]. This approximation may be applicable only when the stiffened plate satisfies two limitations. The first one is that ratios of spacing between two consecutive stiffeners to slab boundary dimensions are small enough to ensure approximate homogeneity of stiffness. The second limitation is that the ratio of stiffener rigidity to the slab rigidity must not become so large that the beam action is predominant.

Subsequently, in more refined approximations the adopted models for the analysis of the plate - beams system isolated the beams from the plate and employed numerical methods for the solution of the plate and the beams such as a semianalytical method [5], a methodology based on energy principles [6-7], the differential quadrature method [8], the finite strip or the finite element method (FEM) [9-20], the boundary element method (BEM) [21-33] or a combination of these methods [34-35]. Moreover, extensive literature exists on composite concrete slabs stiffened by steel beams with deformable connection, employing the aforementioned simplified models [36-43]. In all these approximations the solution of the bending problem of stiffened plates is not general since either the analysis of the plate and the beams is performed on the undeformed shape ignoring second-order effects or the shear longitudinal or transverse forces at the interfaces have been neglected or the torsional and warping behavior of the stiffening beams has been neglected excluding in this way the placement of an eccentric stiffener. Later, Sapountzakis and Mokos [44] refined the structural model proposed by Sapountzakis and Katsikadelis [21] by presenting a more general solution for the analysis of plates stiffened by parallel beams taking into account tractions in all directions at the fictitious plate–beam interfaces in this way enabling the analysis to include eccentric beams. In this latter analysis the distribution of the interface transverse shear force is assumed to be constant along the width of the beam flange. Finally, a general solution for the analysis of plates stiffened by arbitrarily placed parallel beams of arbitrary doubly symmetric cross section subjected to an arbitrary loading has been presented in [45-48] by improving once again the structural model of Sapountzakis and Mokos in [44], so that a nonuniform distribution of the interface transverse shear force and the nonuniform torsional response of the beams are taken into account.

Contrary to the many aforementioned research efforts, the large deflection analysis of stiffened plates has generated a limited amount of literature. The stiffened plate again was initially approximated by an equivalent orthotropic homogeneous slab of constant thickness and simple boundary conditions and solved employing the Galerkin method [49] or using an approximate analytical procedure for the modal equation [50]. Subsequently, in more refined approximations the adopted models for the analysis of the plate - beams system employed numerical methods for the solution of the plate and the beams such as the finite strip or the finite element method [51-59], the dynamic relaxation method [60-63], Fourier series for the displacement expressions [64] and the boundary element method [65]. In these research efforts the restrictions mentioned also apply for the small deflection analysis, such as the ignorance of tractions in a direction at the fictitious plate – beams interfaces, the nonuniform distribution of the interface transverse shear force, the nonuniform torsional response of the beams *etc.* Finally, a general solution for the large deflection analysis of plates stiffened by arbitrarily placed parallel beams of arbitrary doubly symmetric cross section subjected to an arbitrary loading is presented in [66] based on the structural model of Sapountzakis and Mokos, proposed in [45].

In the following sections of this paper, the application of this latter model in the large deflection analysis of stiffened plates is presented, while the small deflection analysis is treated as a special case. According to this model, the stiffening beams are isolated from the plate by sections in the lower outer surface of the plate, making the hypothesis that the plate and the beams can slip in all directions of the connection without separation (*i.e.* uplift neglected) and taking into account the arising tractions in all directions at the fictitious interfaces. These tractions are integrated with respect to each half of the interface width resulting in two interface lines, along which the loading of the beams as well as the additional loading of the plate is defined. The utilization of two interface lines for each beam enables the nonuniform torsional response of the beams to be taken into account as the angle of twist is indirectly equated with the corresponding plate slope. The unknown distribution of the aforementioned integrated tractions is established by applying continuity conditions in all directions at the two interface lines taking into account their relation with the interface slip through the shear connector stiffness. Any distribution of connectors in each direction of the interfaces can be handled. Six boundary value problems are formulated and solved using the Analog Equation Method (AEM) [67], a boundary element based method. The solution of the aforementioned plate and beam problems, which are nonlinearly coupled, is achieved using iterative numerical methods. The model adopted permits the evaluation of the shear forces at the interfaces in both directions, knowledge of which is very important in the design of prefabricated ribbed plates. Numerical examples of great practical interest demonstrate the improved accuracy of this model which better describes the actual response of the plate - beams system.

2 Statement of the problem

Consider a thin plate of homogeneous, isotropic and linearly elastic material with modulus of elasticity E, shear modulus G and Poisson's ratio v, having constant thickness h_p and occupying the two-dimensional multiple connected region Ω of

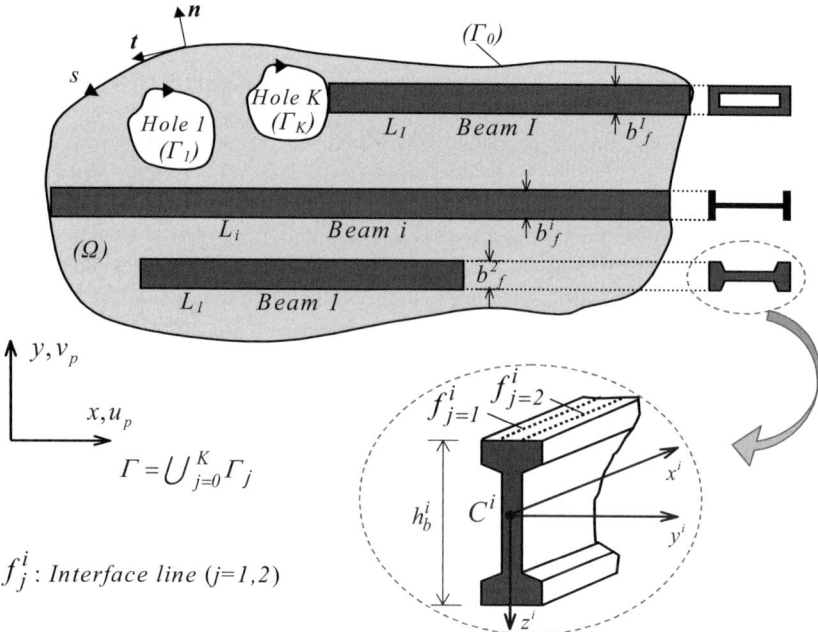

Figure 1: Two dimensional region Ω occupied by the plate

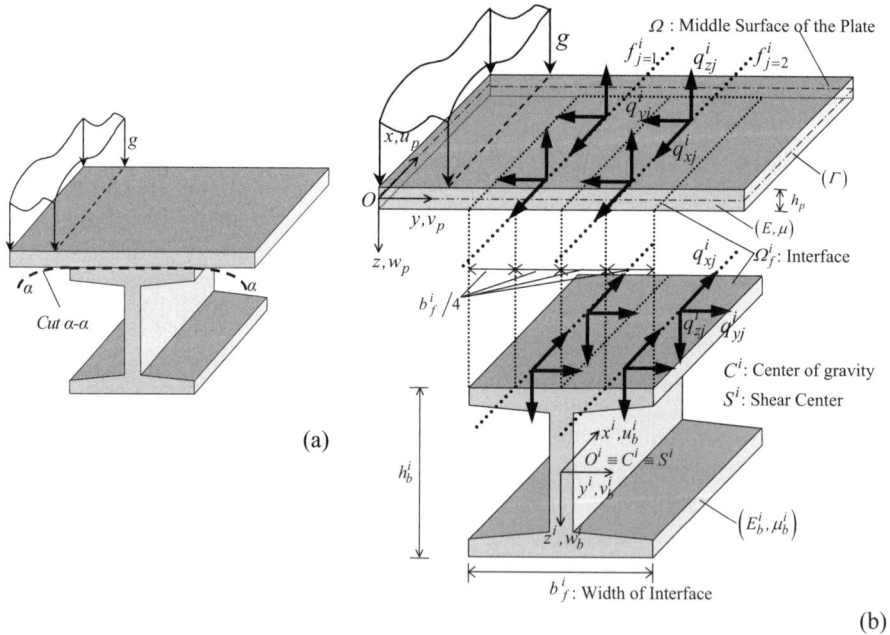

Figure 2: Thin elastic plate stiffened by beams (a) and isolation of the beams from the plate (b)

the x,y plane bounded by the Γ_j ($j=0,1,2,...,K$) boundary curves, which are piecewise smooth, i.e. they may have a finite number of corners, as shown in Figure 1. The plate is stiffened by a set of $i=1,2,...,I$ arbitrarily placed parallel beams of arbitrary doubly symmetric cross section of area A_b^i, which may have either internal or boundary point supports. The material of the beams is considered to be homogeneous, isotropic and linearly elastic with modulus of elasticity E_b^i, shear modulus G_b^i and Poisson's ratio v_b^i. For the sake of convenience the x axis is taken parallel to the beams. The stiffened plate is subjected to the lateral load $g = g(\mathbf{x})$, $\mathbf{x}:\{x,y\}$. Due to this loading the plate and the beams can slip in all directions of the connection without separation (i.e. uplift is neglected). For the analysis of the aforementioned problem a global coordinate system Oxy for the analysis of the plate and local coordinate ones $O^i x^i y^i$ corresponding to the centroid axes of each beam are employed.

The solution of the problem at hand is approached employing the structural model proposed by Sapountzakis and Mokos in [45]. According to this model, the stiffening beams are isolated from the plate by sections in its lower outer surface, taking into account the arising tractions at the fictitious interfaces (Figure2). Integration of these tractions along each half of the width of the i-th beam results in line forces per unit length in all directions in two interface lines, which are denoted by q_{xj}^i, q_{yj}^i and q_{zj}^i ($j=1,2$) encountering in this way the nonuniform distribution of the interface transverse shear forces q_y^i, which in previous models [44] was ignored. The aforementioned integrated tractions result in the loading of the i-th beam as well as the additional loading of the plate. Their distribution is unknown and can be established by imposing displacement continuity conditions in all directions along the two interface lines, enabling in this way the nonuniform torsional response of the beams to be taken into account, which in previous models [44] was also ignored.

The additional loading arising at the middle surface of the plate and the loading along the centroid axis (coinciding with the shear center axis) of each beam can be summarized as follows:

(a) In the plate (at the traces of the two interface lines j=1,2 of the i-th plate-beam interface)
  (i) A lateral line load q_{zj}^i.
  (ii) A lateral line load $\partial m_{pyj}^i / \partial x$ due to the eccentricity of the component q_{xj}^i from the middle surface of the plate. $m_{pyj}^i = q_{xj}^i h_p / 2$ is the bending moment.

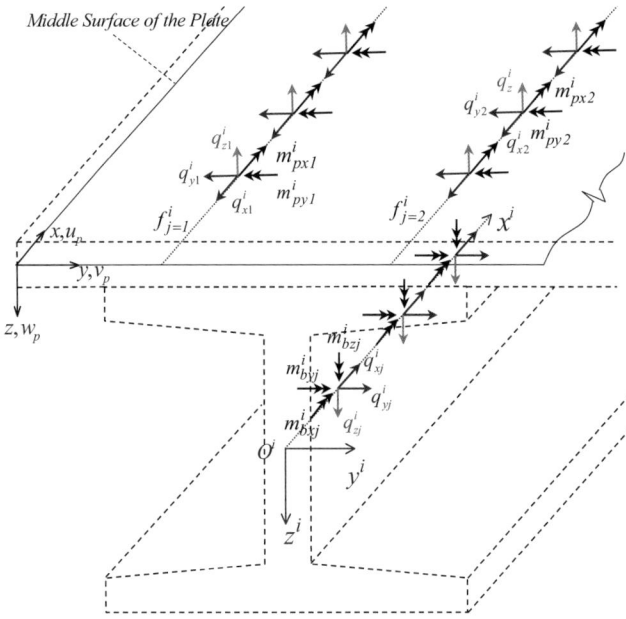

Figure 3: Structural model and directions of the additional loading of the plate and the i-th beam.

(iii) A lateral line load $\partial m^i_{pxj}/\partial y$ due to the eccentricity of the component q^i_{yj} from the middle surface of the plate. $m^i_{pxj} = -q^i_{yj} h_p/2$ is the bending moment.

(iv) An inplane line body force q^i_{xj} at the middle surface of the plate.

(v) An inplane line body force q^i_{yj} at the middle surface of the plate.

(b) In each (i-th) beam ($O^i x^i y^i z^i$ system of axes)

(i) A perpendicularly distributed line load q^i_{zj} along the beam centroid axis $O^i x^i$.

(ii) A transversely distributed line load q^i_{yj} along the beam centroid axis $O^i x^i$.

(iii) An axially distributed line load q^i_{xj} along the beam centroid axis $O^i x^i$.

(iv) A distributed bending moment $m^i_{byj} = q^i_{xj} e^i_{zj}$ about $O^i y^i$ local beam centroid axis due to the eccentricities e^i_{zj} of the components q^i_{xj} from the beam centroid axis. $e^i_{z1} = e^i_{z2} = -h^i_b/2$ are the eccentricities.

(v) A distributed bending moment $m_{bzj}^i = -q_{xj}^i e_{yj}^i$ about $O^i z^i$ local beam centroid axis due to the eccentricities e_{yj}^i of the components q_{xj}^i from the beam centroid axis. $e_{y1}^i = -b_f^i/4$, $e_{y2}^i = b_f^i/4$ are the eccentricities.

(vi) A distributed twisting moment $m_{bxj}^i = q_{zj}^i e_{yj}^i - q_{yj}^i e_{zj}^i$ about $O^i x^i$ local beam shear center axis due to the eccentricities e_{zj}^i, e_{yj}^i of the components q_{yj}^i, q_{zj}^i from the beam shear center axis, respectively. $e_{z1}^i = e_{z2}^i = -h_b^i/2$ and $e_{y1}^i = -b_f^i/4$, $e_{y2}^i = b_f^i/4$ are the eccentricities.

The structural models and the aforementioned additional loading of the plate and the beams are shown in Figure 3.

On the basis of the above considerations the response of the plate and of the beams may be described by the following boundary value problems.

For the plate

Taking into account geometrical nonlinear effects, the Von Kármán plate theory assumption is adopted, according to which the deflection of the plate is no longer small compared to the plate thickness, while it remains small in comparison with the rest dimensions of the plate. Within the context of this assumption, the displacement field of an arbitrary point of the plate, as implied by the Kirchhoff hypothesis, is given as

$$\bar{u}_p(x,y,z) = u_p(x,y) - z\frac{\partial w_p}{\partial x} \quad (1a)$$

$$\bar{v}_p(x,y,z) = v_p(x,y) - z\frac{\partial w_p}{\partial y} \quad (1b)$$

$$\bar{w}_p(x,y,z) = w_p(x,y) \quad (1c)$$

where $\bar{u}_p$, $\bar{v}_p$, $\bar{w}_p$ are the inplane and transverse displacement components of an arbitrary point of the plate and $u_p = u_p(x,y)$, $v_p = v_p(x,y)$ and $w_p = w_p(x,y)$ are the corresponding components of a point at its middle surface. Employing the strain-displacement relations of the three-dimensional elasticity for moderate large displacements [68-69], the strain components can be written as

$$\varepsilon_{xx} = \frac{\partial \bar{u}_p}{\partial x} + \frac{1}{2}\left(\frac{\partial \bar{w}_p}{\partial x}\right)^2 \quad \varepsilon_{yy} = \frac{\partial \bar{v}_p}{\partial y} + \frac{1}{2}\left(\frac{\partial \bar{w}_p}{\partial y}\right)^2 \quad \gamma_{xy} = \frac{\partial \bar{u}_p}{\partial y} + \frac{\partial \bar{v}_p}{\partial x} + \frac{\partial \bar{w}_p}{\partial x}\frac{\partial \bar{w}_p}{\partial y}$$

$$(2a,b,c)$$

$$\varepsilon_{zz} = \gamma_{xz} = \gamma_{yz} = 0 \quad (2d)$$

Substituting Equations (1) and (2) to the stress-strain relations defined by the Hooke's law, the non-vanishing components of the second Piola – Kirchhoff stress tensor are obtained as

$$S_{xx} = \frac{E}{(1-v^2)} \left\{ \left[\frac{\partial u_p}{\partial x} - z\frac{\partial^2 w_p}{\partial x^2} + \frac{1}{2}\left(\frac{\partial w_p}{\partial x}\right)^2 \right] + v\left[\frac{\partial v_p}{\partial y} - z\frac{\partial^2 w_p}{\partial y^2} + \frac{1}{2}\left(\frac{\partial w_p}{\partial y}\right)^2 \right] \right\} \quad (3a)$$

$$S_{yy} = \frac{E}{(1-v^2)} \left\{ v\left[\frac{\partial u_p}{\partial x} - z\frac{\partial^2 w_p}{\partial x^2} + \frac{1}{2}\left(\frac{\partial w_p}{\partial x}\right)^2 \right] + \left[\frac{\partial v_p}{\partial y} - z\frac{\partial^2 w_p}{\partial y^2} + \frac{1}{2}\left(\frac{\partial w_p}{\partial y}\right)^2 \right] \right\} \quad (3b)$$

$$S_{xy} = \frac{E}{2(1+v)} \left(\frac{\partial u_p}{\partial y} + \frac{\partial v_p}{\partial x} - 2z\frac{\partial^2 w_p}{\partial x \partial y} + \frac{\partial w_p}{\partial x}\frac{\partial w_p}{\partial y} \right) \quad (3c)$$

Subsequently, integrating the stress components over the plate thickness, the stress resultants acting on the plate are written as

$$N_{px} = C\left[\frac{\partial u_p}{\partial x} + v\frac{\partial v_p}{\partial x} + \frac{1}{2}\left(\frac{\partial w_p}{\partial x}\right)^2 + \frac{1}{2}v\left(\frac{\partial w_p}{\partial y}\right)^2 \right] \quad (4a)$$

$$N_{py} = C\left[\frac{\partial v_p}{\partial x} + v\frac{\partial u_p}{\partial x} + \frac{1}{2}\left(\frac{\partial w_p}{\partial y}\right)^2 + \frac{1}{2}v\left(\frac{\partial w_p}{\partial x}\right)^2 \right] \quad (4b)$$

$$N_{pxy} = C\frac{1-v}{2}\left(\frac{\partial u_p}{\partial y} + \frac{\partial v_p}{\partial x} + \frac{\partial w_p}{\partial y}\frac{\partial w_p}{\partial x} \right) \quad (4c)$$

$$M_{px} = -D\left(\frac{\partial^2 w}{\partial x^2} + v\frac{\partial^2 w}{\partial y^2}\right) \quad M_{py} = -D\left(\frac{\partial^2 w}{\partial y^2} + v\frac{\partial^2 w}{\partial x^2}\right) \quad M_{pxy} = -D(1-v)\frac{\partial^2 w}{\partial x \partial y}$$

(4d, e, f)

where $C = Eh_p/(1-v^2)$, $D = Eh_p^3/12(1-v^2)$ are the membrane and the bending rigidities of the plate, respectively. Applying the principle of virtual work, the system of partial differential equations of equilibrium of the plate in terms of the stress resultants, is obtained as

$$\frac{\partial N_{px}}{\partial x} + \frac{\partial N_{pxy}}{\partial y} = \sum_{i=1}^{I}\left(\sum_{j=1}^{2} q_{xj}^i \delta(y-y_j)\right) \quad (5a)$$

$$\frac{\partial N_{py}}{\partial y} + \frac{\partial N_{pxy}}{\partial x} = \sum_{i=1}^{I}\left(\sum_{j=1}^{2} q_{yj}^i \delta(y-y_j)\right) \quad (5b)$$

$$\frac{\partial^2 M_{px}}{\partial x^2} + 2\frac{\partial^2 M_{pxy}}{\partial x \partial y} + \frac{\partial^2 M_{py}}{\partial y^2} + N_{px}\frac{\partial^2 w_p}{\partial x^2} + 2N_{pxy}\frac{\partial^2 w_p}{\partial x \partial y} + N_{py}\frac{\partial^2 w_p}{\partial y^2} =$$

$$= -g + \sum_{i=1}^{I}\left[\sum_{j=1}^{2}\left(q_{zj}^i - \frac{\partial m_{pxj}^i}{\partial y} + \frac{\partial m_{pyj}^i}{\partial x} - q_{xj}^i \frac{\partial w_{pj}^i}{\partial x} - q_{yj}^i \frac{\partial w_{pj}^i}{\partial y}\right)\delta(y-y_j)\right] \quad (5c)$$

where $\delta(y-y_i)$ is the Dirac's delta function in the y direction. Employing relations (4), the governing differential Equations (5) in the domain Ω can be expressed in terms of the displacement components as

$$\nabla^2 u_p + \frac{1+\nu}{1-\nu}\frac{\partial}{\partial x}\left(\frac{\partial u_p}{\partial x} + \frac{\partial v_p}{\partial y}\right) + \left(\frac{2}{1-\nu}\frac{\partial^2 w_p}{\partial x^2} + \frac{\partial^2 w_p}{\partial y^2}\right)\frac{\partial w_p}{\partial x} + \frac{1+\nu}{1-\nu}\frac{\partial^2 w_p}{\partial x \partial y}\frac{\partial w_p}{\partial y} =$$

$$= \frac{1}{Gh_p}\sum_{i=1}^{I}\left(\sum_{j=1}^{2} q_{xj}^i \delta(y-y_j)\right) \quad (6a)$$

$$\nabla^2 v_p + \frac{1+\nu}{1-\nu}\frac{\partial}{\partial y}\left(\frac{\partial u_p}{\partial x} + \frac{\partial v_p}{\partial y}\right) + \left(\frac{2}{1-\nu}\frac{\partial^2 w_p}{\partial y^2} + \frac{\partial^2 w_p}{\partial x^2}\right)\frac{\partial w_p}{\partial y} + \frac{1+\nu}{1-\nu}\frac{\partial^2 w_p}{\partial x \partial y}\frac{\partial w_p}{\partial x} =$$

$$= \frac{1}{Gh_p}\sum_{i=1}^{I}\left(\sum_{j=1}^{2} q_{yj}^i \delta(y-y_j)\right) \quad (6b)$$

$$D\nabla^4 w_p - C\left\{\left[\left(\frac{\partial u_p}{\partial x} + \frac{1}{2}\left(\frac{\partial w_p}{\partial x}\right)^2\right) + \nu\left(\frac{\partial v_p}{\partial y} + \frac{1}{2}\left(\frac{\partial w_p}{\partial y}\right)^2\right)\right]\frac{\partial^2 w_p}{\partial x^2} + (1-\nu)\cdot\right.$$

$$\cdot\left(\frac{\partial u_p}{\partial y} + \frac{\partial v_p}{\partial x} + \frac{\partial w_p}{\partial x}\frac{\partial w_p}{\partial y}\right)\frac{\partial^2 w_p}{\partial x \partial y} + \left[\left(\frac{\partial v_p}{\partial y} + \frac{1}{2}\left(\frac{\partial w_p}{\partial y}\right)^2\right) + \nu\left(\frac{\partial u_p}{\partial x} + \frac{1}{2}\left(\frac{\partial w_p}{\partial x}\right)^2\right)\right]\cdot$$

$$\left.\cdot\frac{\partial^2 w_p}{\partial y^2}\right\} = g - \sum_{i=1}^{I}\left[\sum_{j=1}^{2}\left(q_{zj}^i - \frac{\partial m_{pxj}^i}{\partial y} + \frac{\partial m_{pyj}^i}{\partial x} - q_{xj}^i\frac{\partial w_{pj}^i}{\partial x} - q_{yj}^i\frac{\partial w_{pj}^i}{\partial y}\right)\delta(y-y_j)\right]$$

$$(6c)$$

The corresponding boundary conditions can be written as

$$\alpha_{p1}u_{pn} + \alpha_{p2}N_{pn} = \alpha_{p3} \qquad \beta_{p1}u_{pt} + \beta_{p2}N_{pt} = \beta_{p3} \quad (7a,b)$$

$$\gamma_{p1}w_p + \gamma_{p2}R_{pn} = \gamma_{p3} \qquad \delta_{p1}\frac{\partial w_p}{\partial n} + \delta_{p2}M_{pn} = \delta_{p3} \quad (7c,d)$$

$$\varepsilon_{1k}w_p + \varepsilon_{2k}\|Tw_p\|_k = \varepsilon_{3k}, \qquad \varepsilon_{2k} \neq 0 \quad (7e)$$

where α_{pl}, β_{pl}, γ_{pl}, δ_{pl} $(l=1,2,3)$ are functions specified at the boundary Γ

and ε_{lk} ($l=1,2,3$) are functions specified at the k corners of the plate; u_{pn}, u_{pt} and N_{pn}, N_{pt} are the boundary membrane displacements and forces in the normal and tangential directions to the boundary, respectively. R_{pn} and M_{pn} are the effective reaction along the boundary and the bending moment normal to it, respectively, which, employing intrinsic coordinates (*i.e.* the distance along the outward normal n to the boundary and the arc length s) are written as

$$R_{pn} = -D\left[\frac{\partial}{\partial n}\nabla^2 w_p - (v-1)\frac{\partial}{\partial s}\left(\frac{\partial^2 w_p}{\partial s \partial n} - \kappa\frac{\partial w_p}{\partial s}\right)\right] + N_{pn}\frac{\partial w_p}{\partial n} + N_{pt}\frac{\partial w_p}{\partial s} \quad (8a)$$

$$M_{pn} = -D\left[\nabla^2 w_p + (v-1)\left(\frac{\partial^2 w_p}{\partial s^2} + \kappa\frac{\partial w_p}{\partial n}\right)\right] \quad (8b)$$

in which $\kappa(s)$ is the curvature of the boundary. Finally, $\|Tw_p\|_k$ is the discontinuity jump of the twisting moment Tw_p at the corner k of the plate, while Tw_p is given along the boundary as

$$Tw_p = D(v-1)\left(\frac{\partial^2 w_p}{\partial s \partial n} - \kappa\frac{\partial w_p}{\partial s}\right) \quad (9)$$

In the special case of a small deflection analysis, the nonlinear terms arising from the squares or the products of derivatives are ignored and the governing differential Equations (6) are simplified to

$$\nabla^2 u_p + \frac{1+v}{1-v}\frac{\partial}{\partial x}\left[\frac{\partial u_p}{\partial x} + \frac{\partial v_p}{\partial y}\right] = \frac{1}{Gh_p}\sum_{i=1}^{I}\left(\sum_{j=1}^{2} q_{xj}^i \delta_j^i (y-y_i)\right) \quad (10a)$$

$$\nabla^2 v_p + \frac{1+v}{1-v}\frac{\partial}{\partial y}\left[\frac{\partial u_p}{\partial x} + \frac{\partial v_p}{\partial y}\right] = \frac{1}{Gh_p}\sum_{i=1}^{I}\left(\sum_{j=1}^{2} q_{yj}^i \delta_j^i (y-y_i)\right) \quad (10b)$$

$$D\nabla^4 w_p - \left(N_x\frac{\partial^2 w_p}{\partial x^2} + 2N_{xy}\frac{\partial^2 w_p}{\partial x \partial y} + N_y\frac{\partial^2 w_p}{\partial y^2}\right) =$$

$$g - \sum_{i=1}^{I}\sum_{j=1}^{2}\left(q_{zj}^i - \frac{\partial m_{pxj}^i}{\partial y} + \frac{\partial m_{pyj}^i}{\partial x} - q_{xj}^i\frac{\partial w_{pj}^i}{\partial x} - q_{yj}^i\frac{\partial w_{pj}^i}{\partial y}\right)\delta_j^i(y-y_j) \quad (10c)$$

where $N_x = N_x(x,y)$, $N_y = N_y(x,y)$, $N_{xy} = N_{xy}(x,y)$ are the linearized membrane forces per unit length of the plate cross section given as

$$N_x = C\left(\frac{\partial u_p}{\partial x} + v\frac{\partial v_p}{\partial y}\right) \quad N_y = C\left(v\frac{\partial u_p}{\partial x} + \frac{\partial v_p}{\partial y}\right) \quad N_{xy} = C\frac{1-v}{2}\left(\frac{\partial u_p}{\partial y} + \frac{\partial v_p}{\partial x}\right)$$
(11a,b,c)

The boundary conditions (7a-7d) are the most general boundary conditions for the plate problem including also the elastic support, while the corner condition (7e) holds for free or transversely elastically restrained edges k. It is apparent that all types of the conventional boundary conditions can be derived from these equations by specifying appropriately the functions a_{pl}, β_{pl}, γ_{pl} and δ_{pl} ($l = 1,2,3$) (e.g. for a clamped edge it is $a_{p1} = \beta_{p1} = \gamma_{p1} = \delta_{p1} = 1$, $a_{p2} = a_{p3} = \beta_{p2} = \beta_{p3} = \gamma_{p2} = \gamma_{p3} = \delta_{p2} = \delta_{p3} = 0$).

For each (i-th) beam

Each beam undergoes transverse deflection with respect to z^i and y^i axes, as well as axial deformation and a nonuniform angle of twist along x^i axis. The displacement field of an arbitrary point of a cross section (taking into account moderate large displacements and rotations) can be derived with respect to those of its centroid as

$$\bar{u}_b^i(x^i,y^i,z^i) = u_b^i(x^i) - y^i\theta_{bz}^i(x^i) + z^i\theta_{by}^i(x^i) + \frac{d\theta_{bx}^i}{dx}\varphi_S^P(y^i,z^i) \quad (12a)$$

$$\bar{v}_b^i(x^i,y^i,z^i) = v_b^i(x^i) - z^i\sin(\theta_{bx}^i(x^i)) - y^i\left[1-\cos(\theta_{bx}^i(x^i))\right] \quad (12b)$$

$$\bar{w}_b^i(x^i,y^i,z^i) = w_b^i(x^i) + y^i\sin(\theta_{bx}^i(x^i)) - z^i\left[1-\cos(\theta_{bx}^i(x^i))\right] \quad (12c)$$

$$\theta_{by}^i(x^i) = \frac{dv_b^i}{dx^i}\sin(\theta_{bx}^i(x^i)) - \frac{dw_b^i}{dx^i}\cos(\theta_{bx}^i(x^i)) \quad (12d)$$

$$\theta_{bz}^i(x^i) = \frac{dv_b^i}{dx^i}\cos(\theta_{bx}^i(x^i)) + \frac{dw_b^i}{dx^i}\sin(\theta_{bx}^i(x^i)) \quad (12e)$$

where $\bar{u}_b^i$, $\bar{v}_b^i$, $\bar{w}_b^i$ are the axial and transverse displacement components with respect to the $O^i y^i z^i$ system of axes; $u_b^i = u_b^i(x^i)$, $v_b^i = v_b^i(x^i)$ and $w_b^i = w_b^i(x^i)$ are the corresponding components of the centroid O^i; $\theta_{by}^i = \theta_{by}^i(x^i)$, $\theta_{bz}^i = \theta_{bz}^i(x^i)$ are the angles of rotation of the cross section due to bending, with respect to its centroid; $d\theta_{bx}^i/dx$ denotes the rate of change of the angle of twist $\theta_{bx}^i(x^i)$ regarded as the torsional curvature and φ_S^P is the primary warping function with respect to the cross section's shear center (coinciding with its centroid) [70]. Employing again the strain-displacement relations of the three-dimensional elasticity for moderate large displacements [68-69], the strain components are given as

$$\varepsilon_{xx} = \frac{\partial \overline{u}_b^i}{\partial x^i} + \frac{1}{2}\left[\left(\frac{\partial \overline{v}_b^i}{\partial x^i}\right)^2 + \left(\frac{\partial \overline{w}_b^i}{\partial x^i}\right)^2\right] \quad (13a)$$

$$\gamma_{xz} = \frac{\partial \overline{w}_b^i}{\partial x^i} + \frac{\partial \overline{u}_b^i}{\partial z^i} + \left(\frac{\partial \overline{v}_b^i}{\partial x^i}\frac{\partial \overline{v}_b^i}{\partial z^i} + \frac{\partial \overline{w}_b^i}{\partial x^i}\frac{\partial \overline{w}_b^i}{\partial z^i}\right) \quad (13b)$$

$$\gamma_{xy} = \frac{\partial \overline{v}_b^i}{\partial x^i} + \frac{\partial \overline{u}_b^i}{\partial y^i} + \left(\frac{\partial \overline{v}_b^i}{\partial x^i}\frac{\partial \overline{v}_b^i}{\partial y^i} + \frac{\partial \overline{w}_b^i}{\partial x^i}\frac{\partial \overline{w}_b^i}{\partial y^i}\right) \quad (13c)$$

$$\varepsilon_{yy} = \varepsilon_{zz} = \gamma_{yz} = 0 \quad (13d)$$

Employing the Hooke's stress-strain law and integrating the arising stress components over the beam's cross section, the stress resultants of the beam are derived as

$$N_b^i = E_b^i A_b^i \left[\frac{du_b^i}{dx^i} + \frac{1}{2}\left(\left(\frac{dv_b^i}{dx^i}\right)^2 + \left(\frac{dw_b^i}{dx^i}\right)^2 + \frac{I_p^i}{A}\left(\frac{d\theta_{bx}^i}{dx^i}\right)^2\right)\right] \quad (14a)$$

$$M_{by}^i = -E_b^i I_y^i \left(\frac{d^2 w_b^i}{dx^{i2}}\cos\theta_{bx}^i - \frac{d^2 v_b^i}{dx^{i2}}\sin\theta_{bx}^i\right) \quad (14b)$$

$$M_{bz}^i = E_b^i I_z^i \left(\frac{d^2 v_b^i}{dx^{i2}}\cos\theta_{bx}^i + \frac{d^2 w_b^i}{dx^{i2}}\sin\theta_{bx}^i\right) \quad (14c)$$

$$M_{bt}^{Pi} = G_b^i I_t^i \frac{d\theta_{bx}^i}{dx^i} \qquad M_{bw}^i = -E_b^i C_S^i \frac{d^2 \theta_{bx}^i}{dx^{i2}} \quad (14d,e)$$

$$M_{bR}^i = N_b^i \frac{I_p^i}{A_b^i} + \frac{1}{2}E_b^i \left(I_R^i - \frac{I_p^{i\,2}}{A_b^i}\right)\left(\frac{d\theta_{bx}^i}{dx^i}\right)^2 \quad (14f)$$

where M_{bt}^{Pi} is the primary twisting moment [70] resulting from the primary shear stress distribution; M_{bw}^i is the warping moment due to torsional curvature and M_{bR}^i is a higher order stress resultant [71]. Furthermore I_y^i, I_z^i are the principal moments of inertia; I_p^i is the polar moment of inertia, while I_t^i and C_S^i are the torsion and warping constants of the i-th beam with respect to the cross section's shear center (coinciding with its centroid), respectively, given as

$$I_t^i = \int_\Omega \left(y^{i2} + z^{i2} + y^i\frac{\partial \varphi_S^P}{\partial z} - z^i\frac{\partial \varphi_S^P}{\partial y}\right)d\Omega \qquad C_S^i = \int_\Omega \left(\varphi_S^P\right)^2 d\Omega \quad (15a,b)$$

Moreover, I_R^i is the fourth moment of inertia with respect to the cross section's shear center (coinciding with its centroid), defined as

$$I_R^i = \int_\Omega \left(y^{i2} + z^{i2}\right)^2 d\Omega \tag{16}$$

The principle of virtual work is once again employed and the equilibrium equations concerning moderate large deflections and twisting rotations of beams of a doubly symmetric cross section are written as

$$-\frac{dN_b^i}{dx^i} = \sum_{j=1}^2 q_{xj}^i \tag{17a}$$

$$-\frac{d}{dx^i}\left(N_b^i \frac{dv_b^i}{dx^i}\right) + \frac{d^2}{dx^{i2}}\left(M_{bz}^i \cos\theta_{bx}^i\right) + \frac{d^2}{dx^{i2}}\left(M_{by}^i \sin\theta_{bx}^i\right) = \sum_{j=1}^2\left(q_{yj}^i - q_{xj}^i \frac{dv_b^i}{dx^i} - \frac{dm_{bzj}^i}{dx^i}\right) \tag{17b}$$

$$-\frac{d}{dx^i}\left(N_b^i \frac{dw_b^i}{dx^i}\right) - \frac{d^2}{dx^{i2}}\left(M_{by}^i \cos\theta_{bx}^i\right) + \frac{d^2}{dx^{i2}}\left(M_{bz}^i \sin\theta_{bx}^i\right) = \sum_{j=1}^2\left(q_{zj}^i - q_{xj}^i \frac{dw_b^i}{dx^i} + \frac{dm_{byj}^i}{dx^i}\right) \tag{17c}$$

$$M_{by}^i\left(\frac{d^2 w_b^i}{dx^{i2}}\sin\theta_{bx}^i + \frac{d^2 v_b^i}{dx^{i2}}\cos\theta_{bx}^i\right) + M_{bz}^i\left(\frac{d^2 w_b^i}{dx^{i2}}\cos\theta_{bx}^i - \frac{d^2 v_b^i}{dx^{i2}}\sin\theta_{bx}^i\right) - \frac{d}{dx^i}M_{bt}^{Pi} -$$

$$-\frac{d^2}{dx^{i2}}M_{bw}^i - \frac{d}{dx^i}\left(M_{bR}^i \frac{d\theta_{bx}^i}{dx^i}\right) = \sum_{j=1}^2\left(q_{zj}^i y_{q_z j} \cos\theta_{bx}^i - q_{zj}^i z_{q_z j}\sin\theta_{bx}^i - \right.$$

$$\left. -q_{yj}^i z_{q_y j}\cos\theta_{bx}^i - q_{yj}^i y_{q_y j}\sin\theta_{bx}^i - q_{xj}^i \frac{I_p^i}{A_b^i}\frac{d\theta_{bx}^i}{dx^i}\right) \tag{17d}$$

Substituting the expressions of the stress resultants of Equations (14) in Equations (17), the governing differential equations are written in terms of the displacement components as

$$-E_b^i A_b^i\left(\frac{d^2 u_b^i}{dx^{i2}} + \frac{dw_b^i}{dx^i}\frac{d^2 w_b^i}{dx^{i2}} + \frac{dv_b^i}{dx^i}\frac{d^2 v_b^i}{dx^{i2}} + \frac{I_p^i}{A}\frac{d\theta_{bx}^i}{dx^i}\frac{d^2 \theta_{bx}^i}{dx^{i2}}\right) = \sum_{j=1}^2 q_{xj}^i \tag{18a}$$

$$E_b^i I_z^i \frac{d^4 v_b^i}{dx^{i4}} - N_b^i \frac{d^2 v_b^i}{dx^{i2}} + \left(E_b^i I_z^i - E_b^i I_y^i\right) \cdot$$

$$\cdot\left[\frac{d^4 w_b^i}{dx^{i4}}\theta_{xb}^i + 2\frac{d^3 w_b^i}{dx^{i3}}\frac{d\theta_{bx}^i}{dx^i} + \frac{d^2 w_b^i}{dx^{i2}}\frac{d^2\theta_{bx}^i}{dx^{i2}} - \frac{d^4 v_b^i}{dx^{i4}}\left(\theta_{bx}^i\right)^2 - 4\frac{d^3 v_b^i}{dx^{i3}}\frac{d\theta_{bx}^i}{dx^i}\theta_{bx}^i -\right.$$

$$\left. -2\frac{d^2 v_b^i}{dx^{i2}}\frac{d^2\theta_{bx}^i}{dx^{i2}}\theta_{bx}^i - 2\frac{d^2 v_b^i}{dx^{i2}}\left(\frac{d\theta_{bx}^i}{dx^i}\right)^2\right] = \sum_{j=1}^2\left(q_{yj}^i - q_{xj}^i\frac{dv_b^i}{dx^i} - \frac{dm_{bzj}^i}{dx^i}\right) \tag{18b}$$

$$E_b^i I_y^i \frac{d^4 w_b^i}{dx^{i4}} - N_b^i \frac{d^2 w_b^i}{dx^{i2}} + \left(E_b^i I_z^i - E_b^i I_y^i\right) \cdot$$

$$\cdot \left[\frac{d^4 v_b^i}{dx^{i4}} \theta_{bx}^i + 2\frac{d^3 v_b^i}{dx^{i3}} \frac{d\theta_{bx}^i}{dx^i} + \frac{d^2 v_b^i}{dx^{i2}} \frac{d^2 \theta_{bx}^i}{dx^{i2}} + \frac{d^4 w_b^i}{dx^{i4}} \left(\theta_{bx}^i\right)^2 + 4\frac{d^3 w_b^i}{dx^{i3}} \frac{d\theta_{bx}^i}{dx^i} \theta_{bx}^i + \right.$$

$$\left. +2\frac{d^2 w_b^i}{dx^{i2}} \frac{d^2 \theta_{bx}^i}{dx^{i2}} \theta_{bx}^i + 2\frac{d^2 w_b^i}{dx^{i2}} \left(\frac{d\theta_{bx}^i}{dx^i}\right)^2 \right] = \sum_{j=1}^{2}\left(q_{zj}^i - q_{xj}^i \frac{dw_b^i}{dx^i} + \frac{dm_{byj}^i}{dx^i}\right) \quad (18\text{c})$$

$$E_b^i C_S^i \frac{d^4 \theta_{bx}^i}{dx^{i4}} - G_b^i I_t^i \frac{d^2 \theta_{bx}^i}{dx^{i2}} - \frac{3}{2}E_b^i\left(I_R^i - \frac{I_p^{i\,2}}{A_b^i}\right)\left(\frac{d\theta_{bx}^i}{dx^i}\right)^2 \frac{d^2 \theta_{bx}^i}{dx^{i2}} - N_b^i \frac{I_p^i}{A_b^i}\frac{d^2 \theta_{bx}^i}{dx^{i2}} +$$

$$+\left(E_b^i I_z^i - E_b^i I_y^i\right)\left[\frac{d^2 v_b^i}{dx^{i2}} \frac{d^2 w_b^i}{dx^{i2}} - \left(\frac{d^2 v_b^i}{dx^{i2}}\right)^2 \theta_{bx}^i + \left(\frac{d^2 w_b^i}{dx^{i2}}\right)^2 \theta_{bx}^i\right] = \sum_{j=1}^{2}\left[m_{bxj}^i + \right.$$

$$\left. +\left(\frac{1}{2}\left(\theta_{bx}^i\right)^2 z_{q_y j} - \theta_{bx}^i y_{q_y j}\right)q_{yj}^i + \left(-\frac{1}{2}\left(\theta_{bx}^i\right)^2 y_{q_z j} - \theta_{bx}^i z_{q_z j}\right)q_{zj}^i - q_{xj}^i \frac{I_p^i}{A_b^i}\frac{d\theta_{bx}^i}{dx^i}\right]$$

(18d)

Considering the angle of twisting rotation θ_{bx}^i and the transverse displacement v_b^i along y^i direction of the i-th beam to have relatively small values, all the nonlinear terms of v_b^i and θ_{bx}^i can be neglected and Equations (18) are simplified to

$$-E_b^i A_b^i \left(\frac{d^2 u_b^i}{dx^{i2}} + \frac{dw_b^i}{dx^i}\frac{d^2 w_b^i}{dx^{i2}}\right) = \sum_{j=1}^{2} q_{xj}^i \quad (19\text{a})$$

$$E_b^i I_z^i \frac{d^4 v_b^i}{dx^{i4}} - N_b^i \frac{d^2 v_b^i}{dx^{i2}} + \left(E_b^i I_z^i - E_b^i I_y^i\right) \cdot$$

$$\cdot \left(\frac{d^4 w_b^i}{dx^{i4}} \theta_{bx}^i + 2\frac{d^3 w_b^i}{dx^{i3}}\frac{d\theta_{bx}^i}{dx^i} + \frac{d^2 w_b^i}{dx^{i2}}\frac{d^2 \theta_{bx}^i}{dx^{i2}}\right) = \sum_{j=1}^{2}\left(q_{yj}^i - q_{xj}^i \frac{dv_b^i}{dx^i} - \frac{dm_{bzj}^i}{dx^i}\right) \quad (19\text{b})$$

$$E_b^i I_y^i \frac{d^4 w_b^i}{dx^{i4}} - N_b^i \frac{d^2 w_b^i}{dx^{i2}} = \sum_{j=1}^{2}\left(q_{zj}^i - q_{xj}^i \frac{dw_b^i}{dx^i} + \frac{dm_{byj}^i}{dx^i}\right) \quad (19\text{c})$$

$$E_b^i C_S^i \frac{d^4 \theta_{bx}^i}{dx^{i4}} - G_b^i I_t^i \frac{d^2 \theta_{bx}^i}{dx^{i2}} - N_b^i \frac{I_p^i}{A_b^i}\frac{d^2 \theta_{bx}^i}{dx^{i2}} + \left(E_b^i I_z^i - E_b^i I_y^i\right)\left[\frac{d^2 v_b^i}{dx^{i2}}\frac{d^2 w_b^i}{dx^{i2}} + \left(\frac{d^2 w_b^i}{dx^{i2}}\right)^2 \theta_{bx}^i\right] =$$

$$= \sum_{j=1}^{2}\left(m_{bxj}^i - q_{xj}^i \frac{I_p^i}{A_b^i}\frac{d\theta_{bx}^i}{dx^i}\right) \quad (19\text{d})$$

while the expression of the axial stress resultant N_b^i of Equations (14a) is written as

$$N_b^i - E_b^i A_b^i \left[\frac{du_b^i}{dx^i} + \frac{1}{2}\left(\frac{dw_b^i}{dx^i}\right)^2 \right] \qquad (20)$$

Moreover, the corresponding boundary conditions of the i-th beam at its ends $x = 0,l$ are given as

$$\alpha_{b1}^i u_b^i + \alpha_{b2}^i N_b^i = \alpha_{b3}^i \qquad (21)$$

$$\beta_{b1}^i v_b^i + \beta_{b2}^i R_{by}^i = \beta_{b3}^i \qquad \bar{\beta}_{b1}^i \theta_{bz}^i + \bar{\beta}_{b2}^i M_{bz}^i = \bar{\beta}_{b3}^i \qquad (22a,b)$$

$$\gamma_{b1}^i w_b^i + \gamma_{b2}^i R_{bz}^i = \gamma_{b3}^i \qquad \bar{\gamma}_{b1}^i \theta_{by}^i + \bar{\gamma}_{b2}^i M_{by}^i = \bar{\gamma}_{b3}^i \qquad (23a,b)$$

$$\delta_{b1}^i \theta_{bx}^i + \delta_{b2}^i M_{bt}^i = \delta_{b3}^i \qquad \bar{\delta}_{b1}^i \frac{d\theta_{bx}^i}{dx} + \bar{\delta}_{b2}^i M_{bw}^i = \bar{\delta}_{b3}^i \qquad (24a,b)$$

where the angles of rotation of the cross section due to bending θ_{by}^i, θ_{bz}^i given from Equations (12d), (12e) are also simplified to

$$\theta_{by}^i = -\frac{dw_b^i}{dx^i} \qquad \theta_{bz}^i = \frac{dv_b^i}{dx^i} + \frac{dw_b^i}{dx^i} \theta_{bx}^i \qquad (25a,b)$$

R_{by}^i, R_{bz}^i and M_{bz}^i, M_{by}^i are the reactions and bending moments with respect to y^i, z^i axes, respectively, which after applying the aforementioned simplifications are given as

$$R_{by}^i = N_b^i \frac{dv_b^i}{dx^i} + E_b^i I_z^i \left(-\frac{d^2 w_b^i}{dx^{i2}} \frac{d\theta_{bx}^i}{dx^i} - \frac{d^3 w_b^i}{dx^{i3}} \theta_{bx}^i - \frac{d^3 v_b^i}{dx^{i3}} \right) + E_b^i I_y^i \left(\frac{d^3 w_b^i}{dx^{i3}} \theta_{bx}^i + \frac{d^2 w_b^i}{dx^{i2}} \frac{d\theta_{bx}^i}{dx^i} \right) \qquad (26a)$$

$$R_{bz}^i = N_b^i \frac{dw_b^i}{dx^i} - E_b^i I_y^i \frac{d^3 w_b^i}{dx^{i3}} \qquad (26b)$$

$$M_{bz}^i = E_b^i I_z^i \left(\frac{d^2 w_b^i}{dx^{i2}} \theta_{bx}^i + \frac{d^2 v_b^i}{dx^{i2}} \right) - E_b^i I_y^i \frac{d^2 w_b^i}{dx^{i2}} \theta_{bx}^i \qquad (26c)$$

$$M_{by}^i = -E_b^i I_Y^i \frac{d^2 w_b^i}{dx^{i2}} \qquad (26d)$$

M_{bt}^i, M_{bw}^i are the torsional and warping moments at the boundaries of the beam, respectively, given as

$$M_{bt}^i = G_b^i I_t^i \frac{d\theta_{bx}^i}{dx^i} - E_b^i C_S^i \frac{d^3\theta_{bx}^i}{dx^{i3}} + N_b^i \frac{I_p^i}{A_b^i} \frac{d\theta_{bx}^i}{dx^i} \tag{27a}$$

$$M_{bw}^i = -E_b^i C_S^i \frac{d^2\theta_{bx}^i}{dx^2} \tag{27b}$$

Finally, α_{bk}^i, β_{bk}^i, $\bar{\beta}_{bk}^i$, γ_{bk}^i, $\bar{\gamma}_{bk}^i$, δ_{bk}^i, $\bar{\delta}_{bk}^i$ ($k = 1,2,3$) are functions specified at the boundaries of the i-th beam ($x = 0,l$). The boundary conditions (21-24) are the most general boundary conditions for the beam problem including also the elastic support. It is apparent that all types of the conventional boundary conditions (clamped, simply supported, free or guided edge) can be derived from these equations by specifying appropriately the aforementioned coefficients.

In the special case of a small deflection analysis, the nonlinear terms arising from the squares or the products of derivatives are ignored and the governing differential Equations (19) are simplified to

$$-E_b^i A_b^i \frac{d^2 u_b^i}{dx^{i2}} = \sum_{j=1}^{2} q_{xj}^i \tag{28a}$$

$$E_b^i I_z^i \frac{d^4 v_b^i}{dx^{i4}} - N_b^i \frac{d^2 v_b^i}{dx^{i2}} = \sum_{j=1}^{2} \left(q_{yj}^i - q_{xj}^i \frac{dv_b^i}{dx^i} - \frac{dm_{bzj}^i}{dx^i} \right) \tag{28b}$$

$$E_b^i I_y^i \frac{d^4 w_b^i}{dx^{i4}} - N_b^i \frac{d^2 w_b^i}{dx^{i2}} = \sum_{j=1}^{2} \left(q_{zj}^i - q_{xj}^i \frac{dw_b^i}{dx^i} + \frac{dm_{byj}^i}{dx^i} \right) \tag{28c}$$

$$E_b^i C_S^i \frac{d^4 \theta_{bx}^i}{dx^{i4}} - G_b^i I_t^i \frac{d^2 \theta_{bx}^i}{dx^{i2}} = \sum_{j=1}^{2} \left(m_{bxj}^i - q_{xj}^i \frac{I_p^i}{A_b^i} \frac{d\theta_{bx}^i}{dx^i} \right) \tag{28d}$$

where the linearized expression of the axial stress resultant N_b^i is given as

$$N_b^i = E_b^i A_b^i \frac{du_b^i}{dx^i} \tag{29}$$

Equations (6) and (19) constitute a set of seven coupled and nonlinear partial differential equations including thirteen unknowns, namely u_p, v_p, w_p, u_b^i, v_b^i, w_b^i, θ_{bx}^i, q_{x1}^i, q_{y1}^i, q_{z1}^i, q_{x2}^i, q_{y2}^i, q_{z2}^i. Six additional equations are required, which result from the displacement continuity conditions in the direction of z^i local axes and linear relationships between interface slip and corresponding tractions in the directions of x^i and y^i local axes along the two interface lines of each (*i-th*) plate – beam interface. These conditions can be expressed as

In the direction of x^i local axis:

$$u^i_{p1} - u^i_b = \frac{h_p}{2}\frac{\partial w^i_{p1}}{\partial x} + \frac{h^i_b}{2}\frac{dw^i_b}{dx^i} + \frac{b^i_f}{4}\left(\frac{dv^i_b}{dx^i} + \frac{dw^i_b}{dx^i}\theta^i_{bx}\right) + \frac{d\theta^i_{bx}}{dx^i}\left(\phi^{iP}_S\right)_{f1} + \frac{q^i_{x1}}{k^i_{x1}}$$

along interface line 1 ($f^i_{j=1}$) (30a)

$$u^i_{p2} - u^i_b = \frac{h_p}{2}\frac{\partial w^i_{p2}}{\partial x} + \frac{h^i_b}{2}\frac{dw^i_b}{dx^i} - \frac{b^i_f}{4}\left(\frac{dv^i_b}{dx^i} + \frac{dw^i_b}{dx^i}\theta^i_{bx}\right) + \frac{d\theta^i_{bx}}{dx^i}\left(\phi^{iP}_S\right)_{f2} + \frac{q^i_{x2}}{k^i_{x2}}$$

along interface line 2 ($f^i_{j=2}$) (30b)

In the direction of y^i local axis:

$$v^i_{p1} - v^i_b = \frac{h_p}{2}\frac{\partial w^i_{p1}}{\partial y} + \frac{h^i_b}{2}\theta^i_{bx} + \frac{q^i_{y1}}{k^i_{y1}}$$

along interface line 1 ($f^i_{j=1}$) (31a)

$$v^i_{p2} - v^i_b = \frac{h_p}{2}\frac{\partial w^i_{p2}}{\partial y} + \frac{h^i_b}{2}\theta^i_{bx} + \frac{q^i_{y2}}{k^i_{y2}}$$

along interface line 2 ($f^i_{j=2}$) (31b)

In the direction of z^i local axis:

$$w^i_{p1} - w^i_b = -\frac{b^i_f}{4}\theta^i_{bx}$$

along interface line 1 ($f^i_{j=1}$) (32a)

$$w^i_{p2} - w^i_b = \frac{b^i_f}{4}\theta^i_{bx}$$

along interface line 2 ($f^i_{j=2}$) (32b)

where $\left(\phi^{iP}_S\right)_{fj}$ is the value of the primary warping function with respect to the shear center of the beam cross section (coinciding with its centroid) at the point of the j-th interface line of the i-th plate – beam interface f^i_j and k^i_{xj}, k^i_{yj} are the stiffnesses of the arbitrarily distributed shear connectors along x^i and y^i directions, respectively. In all the aforementioned equations the values of the primary warping function $\phi^{iP}_S(y^i, z^i)$ should be set having the appropriate algebraic sign corresponding to the local beam axes.

In the special case of a small deflection analysis, continuity conditions (30) in the direction of the x^i local axis, after ignoring the nonlinear terms are simplified to

$$u_{p1}^i - u_b^i = \frac{h_p}{2}\frac{\partial w_{p1}^i}{\partial x} + \frac{h_b^i}{2}\frac{dw_b^i}{dx^i} + \frac{b_f^i}{4}\frac{dv_b^i}{dx^i} + \frac{d\theta_{bx}^i}{dx^i}\left(\phi_S^{iP}\right)_{f1} + \frac{q_{x1}^i}{k_{x1}^i}$$

along interface line 1 ($f_{j=1}^i$) (33a)

$$u_{p2}^i - u_b^i = \frac{h_p}{2}\frac{\partial w_{p2}^i}{\partial x} + \frac{h_b^i}{2}\frac{dw_b^i}{dx^i} - \frac{b_f^i}{4}\frac{dv_b^i}{dx^i} + \frac{d\theta_{bx}^i}{dx^i}\left(\phi_S^{iP}\right)_{f2} + \frac{q_{x2}^i}{k_{x2}^i}$$

along interface line 2 ($f_{j=2}^i$) (33b)

while continuity conditions in the directions of y^i and z^i local axes remain the same.

3 Numerical Solution

According to the precedent analysis, the nonlinear flexural analysis of plates stiffened by arbitrarily placed parallel beams subjected to an arbitrary loading reduces in establishing the displacement and force components u_p, v_p, w_p, u_b^i, v_b^i, w_b^i, θ_{bx}^i, q_{x1}^i, q_{y1}^i, q_{z1}^i, q_{x2}^i, q_{y2}^i, q_{z2}^i. These components must satisfy the coupled boundary value problems described by Equations (6), (7), (19) and (21-24) subjected to the continuity conditions (30)-(32). The numerical solution of the aforementioned boundary value problems is accomplished by employing the analog equation method [67], as it is developed in [45].

4 Numerical examples

On the basis of the analytical and numerical procedures presented in the previous sections, a computer program has been written and representative examples have been studied to demonstrate the efficiency and the range of applications of the developed method. In all the examples treated $E = E_b^i = 3.00E7$, $v = v_b^i = 0.20$, while the numerical results have been obtained using *180* constant boundary elements and *324* constant domain rectangular cells.

4.1 Example 1: Small Deflection Analysis

A rectangular plate with dimensions $l_{px} \times l_{py} = 18.0 \times 9.0\ m$ subjected to a uniform load $g = 10kN/m^2$ and stiffened by a rectangular beam of *1.0 m* width eccentrically placed with respect to the center line of the plate (Figure 4) has been studied. The plate is clamped along its small edges, while the other two edges are free according to both its transverse and inplane boundary conditions, while the beam is also clamped at its edges according to its transverse, axial and torsional boundary conditions. The connection between the slab and the beam is accomplished using a linear distribution of shear connectors along the interface (Figure 4). In Table 1 the torsion I_{bx}^t and warping I_{bw}^t constants of the beam cross section and the values of the primary warping function $\left(\phi_S^P\right)_{fj}$ (j=1,2) at the nodes of the two interface lines for various beam heights are presented. In Table 2 the obtained deflections of the stiffened plate analyzed employing either one [44] or two interface lines at its center and at the middle of the free edges A and C (Figure 4) for various beam heights and in Figure 6 the deflection surfaces for $h_b = 1.0m$, for the cases of a partial connection in both directions and a full shear connection are presented as compared with those obtained from finite element (FE) solutions using Bernoulli beam finite elements, eight-noded quadrilateral shell finite elements (parabolic elements) or 8-noded hexahedral solid finite elements (parabolic elements). More specifically, a shell-beam model (modeling the plate with shell elements and the beam with beam ones ignoring torsional warping effects) using rigid offsets for the plate-beam connection, a shell-shell model and a solid one are employed as are shown in Figure 5. It is worth noting here that in the shell-beam model, due to the coupling of various structural elements with rigid offsets, kinematic or static assumptions cannot always be valid at the interface of the slab and the beams [20]. Also, the shell-shell model can be applied only for thin walled rectangular beams, while in the solid FE model a large number of solid FE in the plate must be used in order to avoid shear or membrane locking effects (long or unrealistic computational times) [19]. The discrepancy of the results between the models employing one or two interface lines, which is more pronounced at the high height beams justifies the employment of the proposed improved model, which better describes the actual response of the plate beams system. Moreover, the accuracy of the results and the validity of the proposed model are noteworthy, while the increment of the deflections with the decrement of the connectors' stiffness is easily verified. In Figures 7 and 8 the distributions of the beam deflections w_b, v_b for various beam heights and for the cases of a partial connection in both directions and a full shear connection are presented. Moreover, in Figure 9 the total forces q_x along the axis of the beam of the stiffened plate for the cases of a full and a partial in x direction for $1000k_{xj}^i$ shear connection are presented demonstrating the significant reduction of these forces with the decrement of the connectors' stiffness, while in Figure 10 the total q_y, q_z forces along the axis of the beam for the case of a partial in x direction for $1000k_{xj}^i$ shear connection are also presented. Finally, in order to demonstrate the general character of the proposed model, in Figure 11 the deflection surfaces of the stiffened plate with full shear connection, for the cases of a centrally and an eccentrically placed stiffening beam at the free end with $h_b = 2.0m$ are presented as compared with those obtained from a shell-shell and a solid FE model. The accuracy of the results of the proposed model is verified.

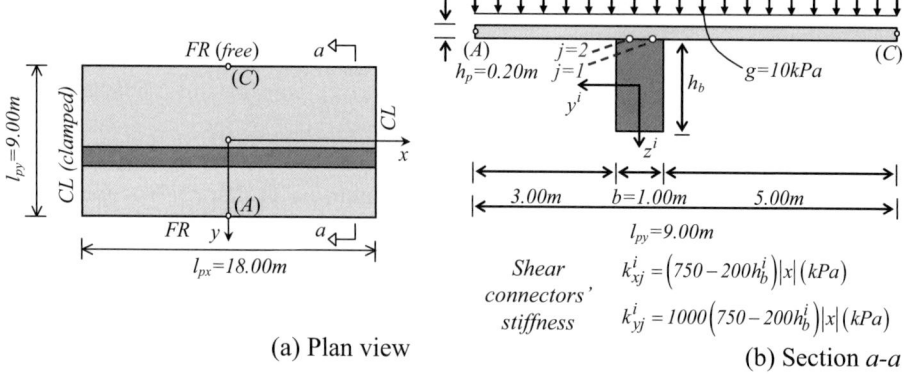

(a) Plan view

(b) Section a-a

Figure 4: Plan view (a) and section a-a (b) of the stiffened plate of Example 1

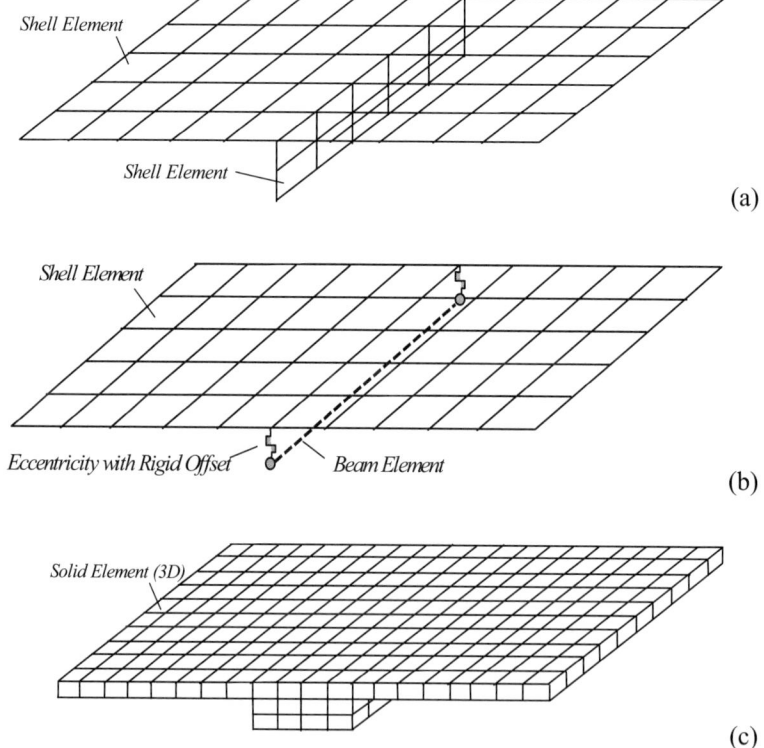

Figure 5: (a) Shell-shell model, (b) Shell-beam model using rigid offsets and (c) solid model for the analysis of the stiffened plate of example 1

h_b (m)	I^1_{bx} (m⁴)	I^1_{bw} (m⁶)	$\left(\phi^P_S\right)_{f1}$ (m²)	$\left(\phi^P_S\right)_{f2}$ (m²)
0.50	2.8584E-02	3.1760E-04	-4.9366E-02	4.9366E-02
1.00	1.4057E-01	1.3441E-04	-3.5037E-02	3.5037E-02
1.50	2.9363E-01	3.7909E-03	1.6372E-02	-1.6372E-02
2.00	4.5733E-01	2.0326E-02	7.5813E-02	-7.5813E-02

Table 1: Torsion, warping constants and primary warping function at the interface lines for various beam heights of Example 1

	Full Connection					Partial Connection
	AEM		FEM			AEM
h_b (m)	2 Interface Lines (Present study)	1 Interface Line [44]	Shell–Shell (Nastran) [72]	Shell–Beam (SAP2000) [73]	Solid (Nastran) [72]	2 Interface Lines (Present study)
Center of the Plate						
0.0	13.1345	13.1345	13.111	13.0872	13.0977	13.1345
0.5	2.5689	2.0568	1.8346	1.8752	1.6994	7.2005
1.0	0.7747	0.9532	0.6245	0.6161	0.4908	1.9125
1.5	0.4315	0.7593	0.3996	0.3864	0.2622	0.7860
2.0	0.3198	0.6994	0.3261	0.3133	0.1862	0.4657
Middle of the free edge A of the plate						
0.0	13.8025	13.8025	13.788	13.7522	13.7719	13.8025
0.5	2.3052	1.0946	1.8550	1.7479	1.6078	7.2293
1.0	0.9661	–0.1812	0.9773	0.9492	0.6963	2.0788
1.5	0.8252	–0.4046	0.8552	0.8406	0.5704	1.1526
2.0	0.8036	–0.4735	0.8249	0.8142	0.5401	0.9328
Middle of the free edge C of the plate						
0.0	13.8025	13.8025	13.788	13.7522	13.7719	13.8025
0.5	5.8435	5.8338	5.1030	5.2138	4.5751	9.3309
1.0	4.0732	5.0324	3.9177	3.9096	3.3441	4.9834
1.5	3.6634	4.8919	3.6701	3.6484	3.0789	3.9538
2.0	3.5147	4.8484	3.5848	3.5598	2.9836	3.6355

Table 2: Deflections w_p (cm) of the stiffened plate of Example 1

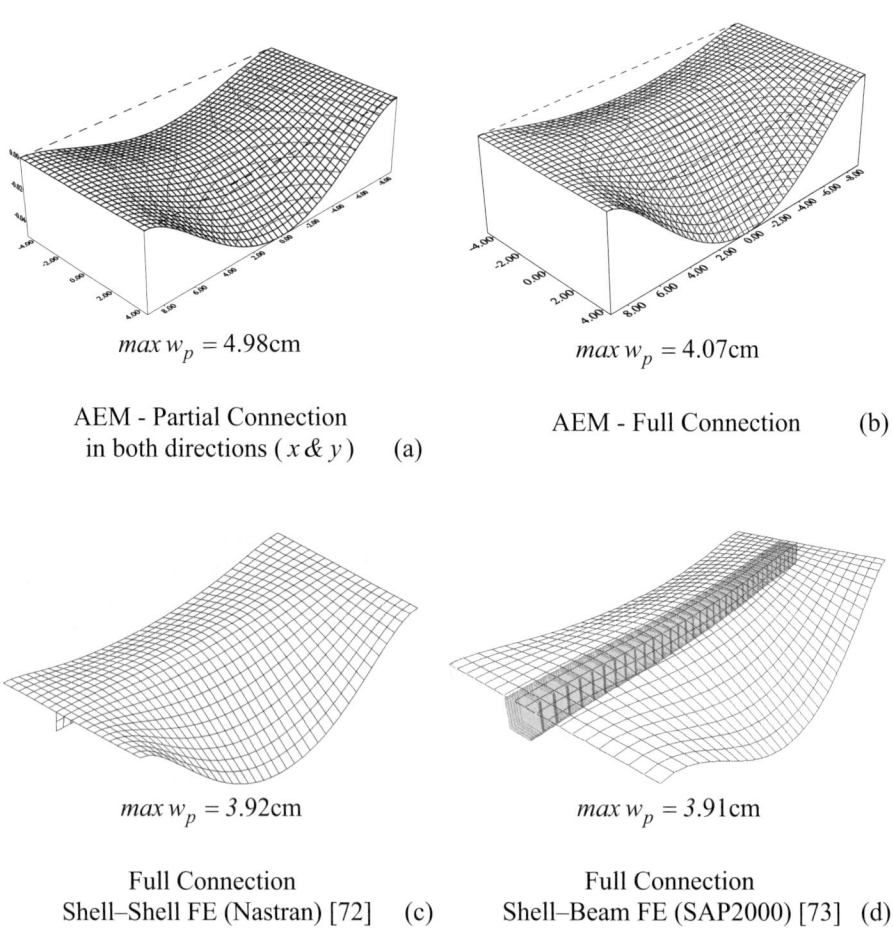

Figure 6: Deflection surfaces and corresponding maximum values of the stiffened plate of Example 1 for $h_b = 1.0m$

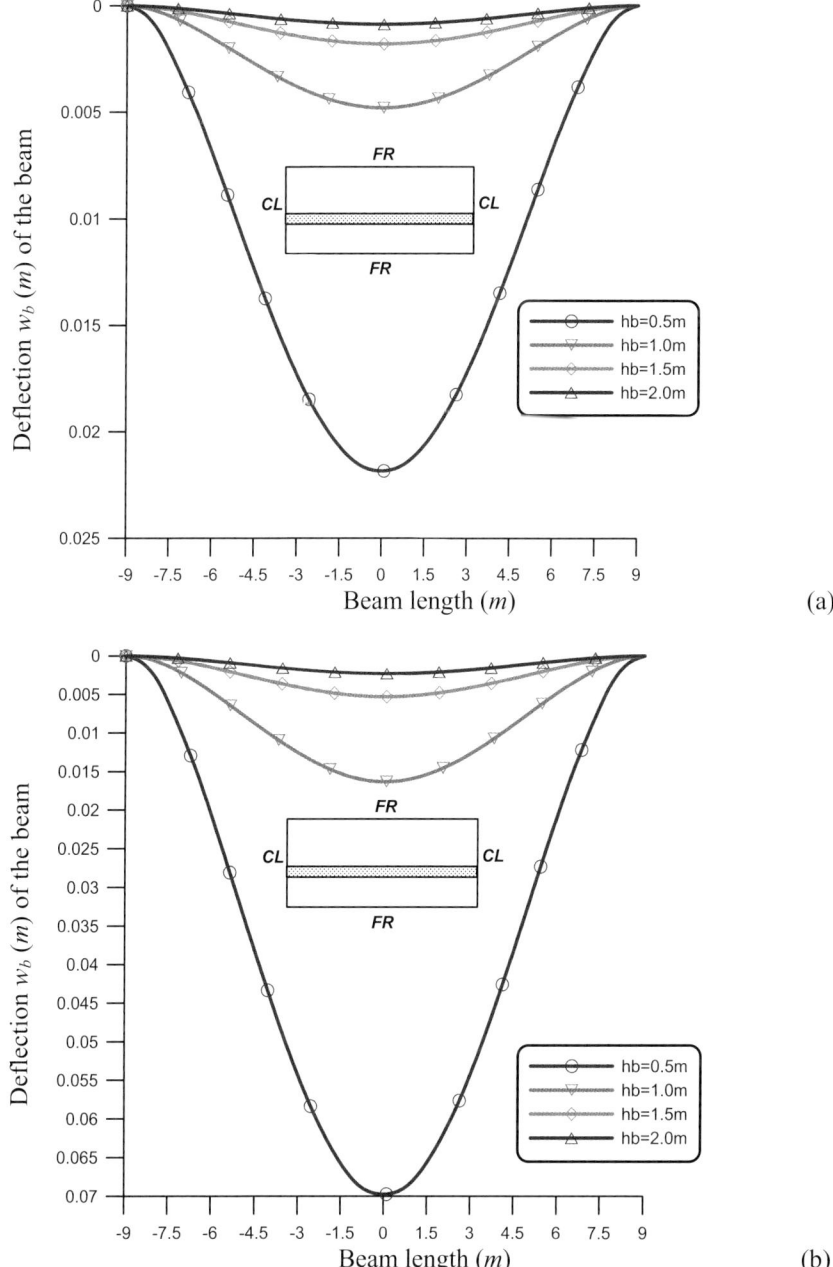

Figure 7: Deflections w_b along the axis of the beam of the stiffened plate of Example 1 for the cases of a full (a) and a partial in both directions (b) shear connection

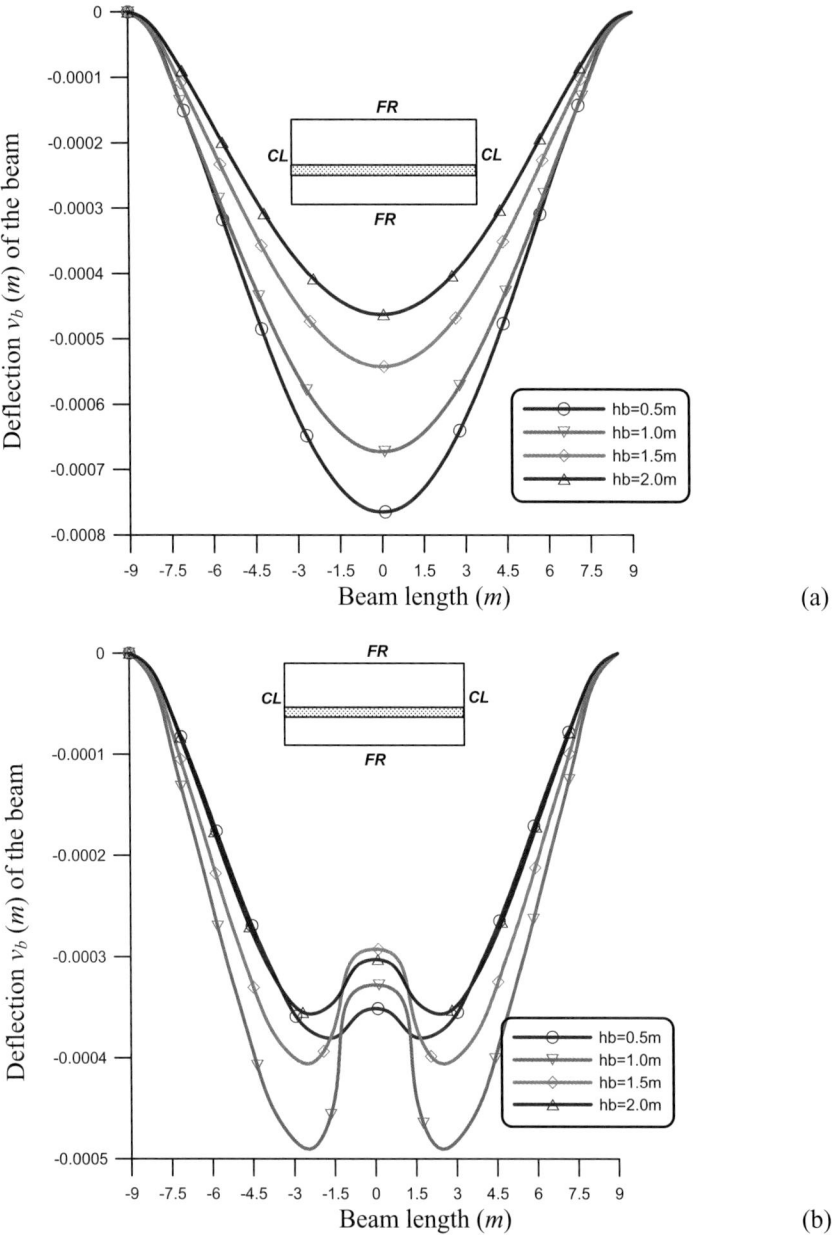

Figure 8: Deflections v_b along the axis of the beam of the stiffened plate of Example 1 for the cases of a full (a) and a partial in both directions (b) shear connection

Recent Developments in the Analysis of Stiffened Plates

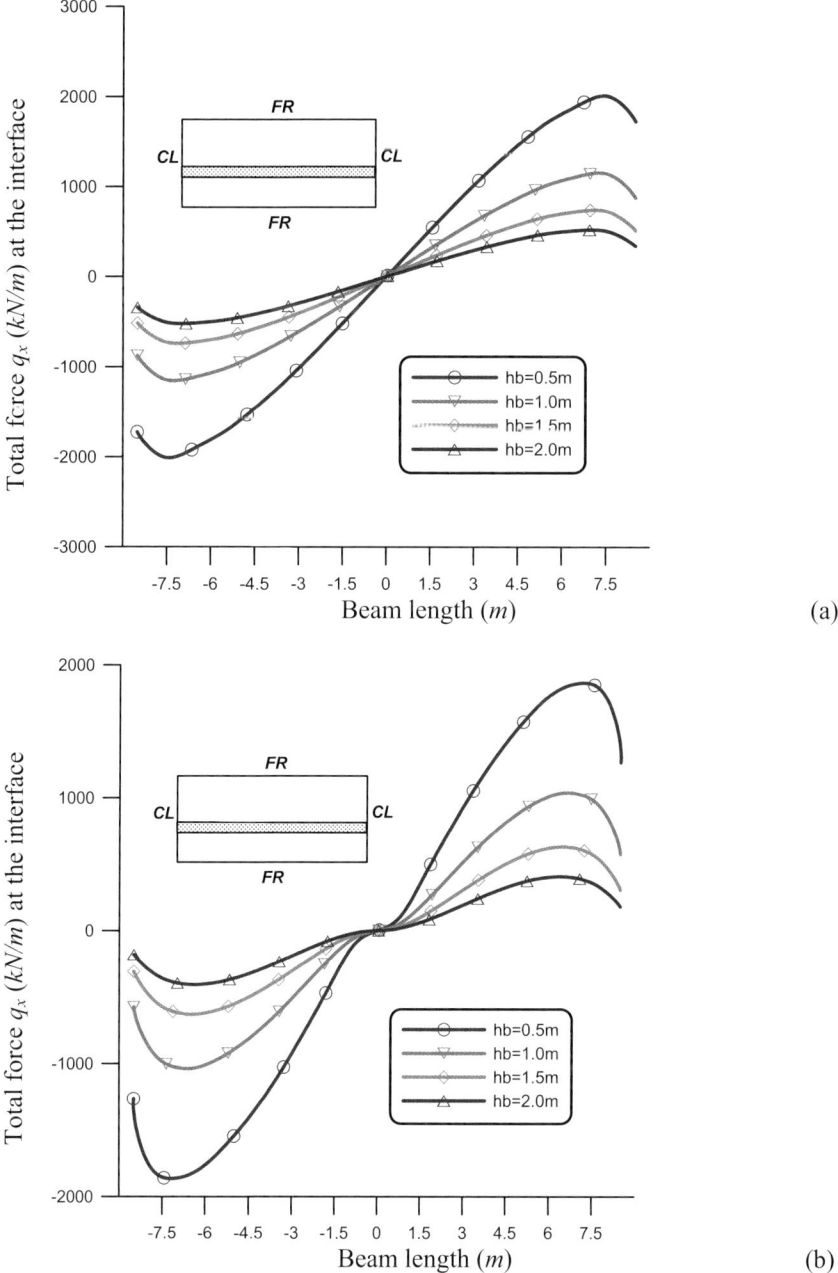

Figure 9: Total forces q_x along the axis of the beam of the stiffened plate of Example 1 for the cases of a full (a) and a partial in x direction for $1000k_{xj}^i$ (b) shear connection

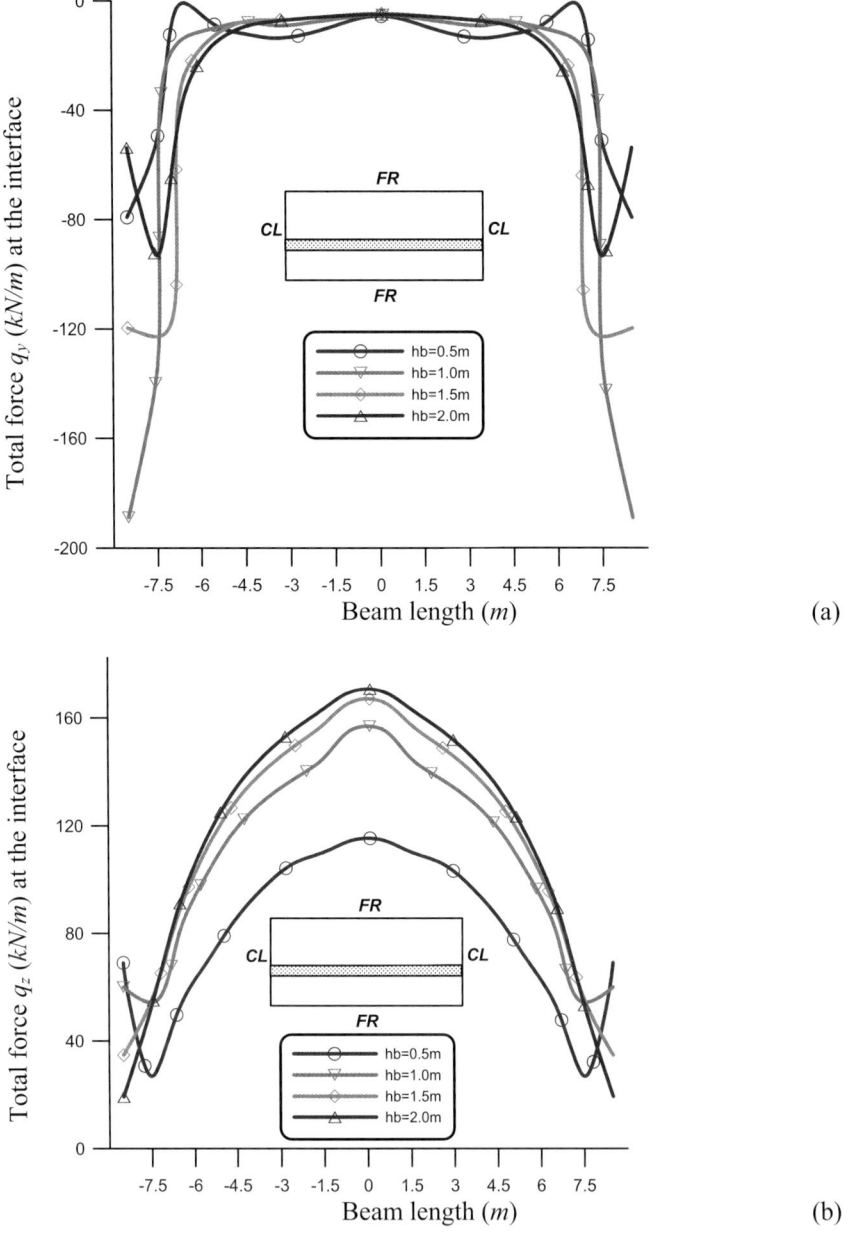

Figure 10: Total forces q_y (a) and q_z (b) along the axis of the beam of the stiffened plate of Example 1 for the case of a partial in x direction for $1000k_{xj}^i$ shear connection

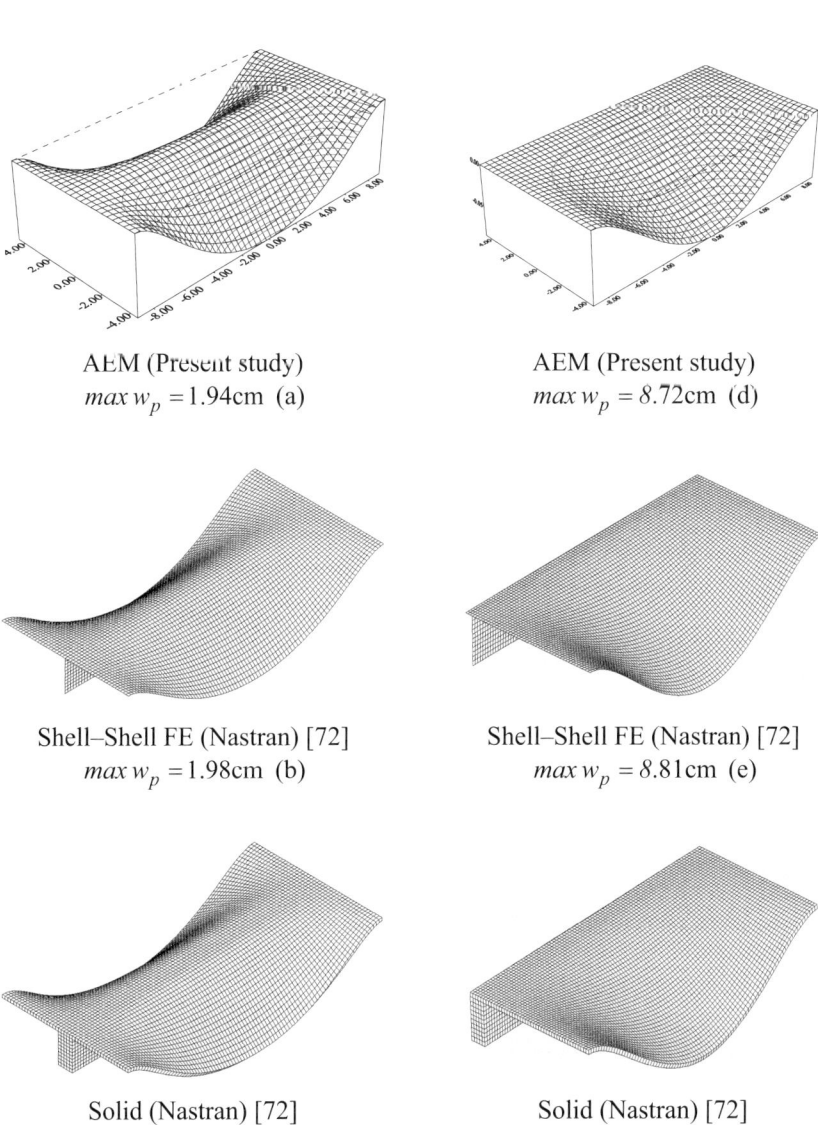

Centrally placed stiffening beam   Eccentrically placed stiffening beam

AEM (Present study) $max\ w_p = 1.94$cm (a)

AEM (Present study) $max\ w_p = 8.72$cm (d)

Shell–Shell FE (Nastran) [72] $max\ w_p = 1.98$cm (b)

Shell–Shell FE (Nastran) [72] $max\ w_p = 8.81$cm (e)

Solid (Nastran) [72] $max\ w_p = 1.48$cm (c)

Solid (Nastran) [72] $max\ w_p = 8.19$cm (f)

Figure 11: Deflection surfaces of the stiffened plate of Example 1 (full shear connection) for the cases of a centrally (a, b, c) and an eccentrically at the free end (d, e, f) placed stiffening beam with $h_b = 2.0m$

4.2 Example 2: Large Deflection Analysis

The stiffened plate of Example 1, having the plate and the stiffening beam firmly bonded together, subjected to a uniformly distributed load $g = 160\, kN/m^2$ has also been studied. In Table 3 the obtained deflections w_p of the stiffened plate at its center and at the middle of the free edges A and C (Figure 4a) for various beam heights are shown as compared with those obtained from FE solutions [72] already described in Example 1. In Figure 12 the contour lines of the deflection w_p of the stiffened plate with $h_b = 2\, m$ and in Figure 13 the load-deflection curve of w_p at the middle of the free edge A of the plate (Figure 4(a)) with $h_b = 0{,}5\, m$ are presented as compared with the aforementioned FE solutions [72]. From both of the aforementioned figures and table, the accuracy of the proposed method is demonstrated. Finally, in Figure 14 the deflections w_b along the beam axis are presented for various beam heights. From the results obtained it is observed that the reduction of the beam height leads to the significant increment of the intensity of the geometrical nonlinear effects on the stiffened plate behaviour.

$h_b\,(m)$	Linear analysis	Nonlinear analysis			
		AEM Present study	Solid (Nastran) [72]	Shell–Shell (Nastran) [72]	Shell–Beam (Nastran) [72]
Center of the plate					
0,50	0,2674	0,2093	0,1980	0,2100	0,2110
1,00	0,0796	0,0707	0,0778	0,0885	0,0581
1,50	0,0450	0,0396	0,0498	0,0622	0,0286
2,00	0,0336	0,0292	0,0352	0,0479	0,0189
Middle of the free edge A of the plate					
0,50	0,7231	0,4477	0,4325	0,4470	0,4529
1,00	0,5401	0,4085	0,3990	0,4310	0,4333
1,50	0,5018	0,3946	0,3875	0,4251	0,4228
2,00	0,4878	0,3890	0,3810	0,4215	0,4187
Middle of the free edge C of the plate					
0,50	0,2670	0,2626	0,2210	0,2680	0,2570
1,00	0,1120	0,1231	0,1230	0,1790	0,1550
1,50	0,0891	0,0952	0,1030	0,1570	0,1400
2,00	0,0833	0,0877	0,0918	0,1450	0,1360

Table 3: Deflections $w_p\,(m)$ of the stiffened plate of Example 2, for various beam heights h_b

Recent Developments in the Analysis of Stiffened Plates

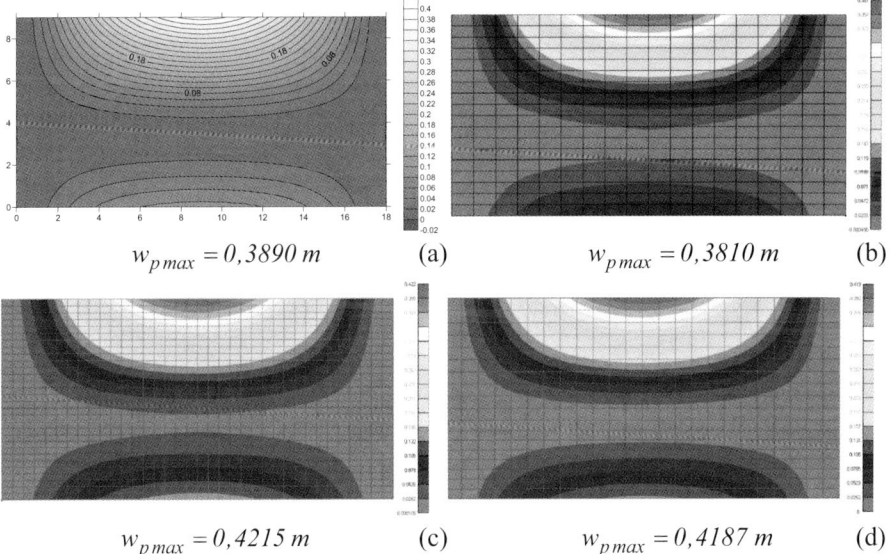

Figure 12: Contour lines of w_p of the stiffened plate of Example 2 for $h_b = 2\ m$, employing the present study (a) and a FE solution using solid elements (b), shell-shell elements (c) or shell-beam elements (d)

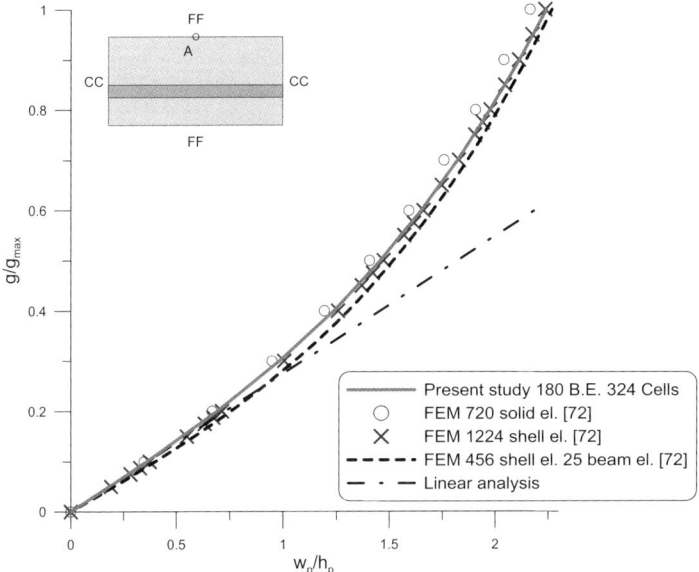

Figure 13: Load-displacement curve of w_p at the middle of the free edge A of the stiffened plate of Example 2 for $h_b = 0,5\ m$

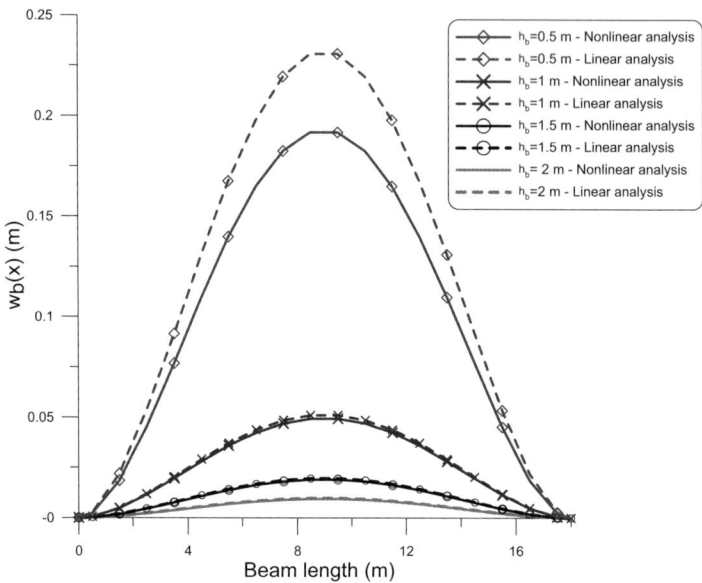

Figure 14: Deflection w_b of the stiffening beam of Example 2 for various beam heights

5 Concluding remarks

The small and large deflection analysis of plates stiffened by arbitrarily placed parallel beams subjected to an arbitrary loading is presented. The proposed model is an improved one, which contrary to previous approaches, takes into account the nonuniform distribution of the interface transverse shear force and the nonuniform torsional response of the beams. The main conclusions that can be drawn from this investigation are:

a. The discrepancy in the results between the models employing one or two interface lines, which is more pronounced at the high height beams justifies the employment of the proposed improved model, which better describes the actual response of the plate beams system.
b. The accuracy of the results and the validity of the proposed model are noteworthy.
c. As was expected, the increment of the deflection with the decrement of the connectors' stiffness is easily verified.
d. The values of the shear interface forces are significantly reduced with the decrement of the shear connectors' stiffness or in other words with the increment of the slip.
e. The discrepancy in the results and especially in the transverse interface forces near the supports necessitates the inclusion of the warping beam behavior in the analysis of a stiffened plate.

f. The utilization of two interface lines for each beam better describes the actual response of the plate beams system since the beam angle of twist is indirectly equated with the corresponding plate slope.

Future works will be devoted to considering material nonlinearities as well as viscoelastic material in the analysis of the stiffened plate.

References

[1] C. Massonet, "Method of Calculations for Bridges with Several Longitudinal Beams Taking into Account their Torsional Resistance", International Association for Bridges and Structural Engineering, 147-182, 1950.
[2] W. Cornelius, "Die Berechnung der Ebener Flächentragwerke mit Hilfe der Theorie der Orthogonal Anisotropen Platten", Der Stahlbau, 2, 21-26, 1952.
[3] R.P. Pama, A.R. Cusens, "Edge Beam Stiffening of Multibeam Bridges", Journal of the Structural Division, ASCE, 93(ST2), 141-161, 1967.
[4] G.H. Powell, D.W. Ogden, "Analysis of Orthotropic Steel Plate Bridge Decks", Journal of the Structural Division, ASCE, 95(ST5), 909-921, 1969.
[5] M. Mukhopadhay, "Stiffened Plates in Bending", Computers & Structures, 541-548, 1994.
[6] A.R. Kukreti, Y. Rajapaksa, "Analysis Procedure for Ribbed and Grid Plate Systems used for Bridge Decks", Journal of Structural Engineering, ASCE, 116(2), 372-391, 1990.
[7] A.R. Kukreti, E. Cheraghi, "Analysis Procedure for Stiffened Plate Systems Using an Energy Approach", Computers and Structures, 46(4), 649-657, 1993.
[8] Z.A Siddiqi, R.A. Kukreti, "Analysis of Eccentrically Stiffened Plates with Mixed Boundary Conditions Using Differential Quadrature Method", Applied Mathematical Modelling, 251-275, 1998.
[9] Y.K. Cheung, "Finite Strip Method in Analysis of Elastic Plates with Two Opposite Sides Simply Supported Ends", Proceedings of the Institution of Civil Engineers, 40, 1-7, 1968.
[10] I.P. King, O.C. Zienkiewicz, "Slab Bridges with Arbitrary Shape and Support Conditions: A General Method of Analysis Based on Finite Element Method", Proceedings of the Institution of Civil Engineers, 40, 9-36, 1968.
[11] M. Mukhopadhay, "Stiffened Plate Plane Stress Elements for the Analysis of Ships' Structures", Computers and Structures, 563-573, 1981.
[12] A. Deb, M. Booton, "Finite Element Models for Stiffened Plates under Transverse Loading", Computers and Structures, 361-372, 1988.
[13] G.S. Palani, N.R. Iyer, T.V.S.R. Apa Rao, "An Efficient Finite Element Model for Static and Vibration Analysis of Eccentrically Stiffened Plates/Shells", Computers and Structures, 651-661, 1992.
[14] S. Peng-Cheng, H. Dade, W. Zongmu, "Static Vibration and Stability Analysis of Stiffened Plates Using B Spline Functions", Computers and Structures, 73-78, 1993.

[15] M.P. Rossow, A.K. Ibrahimkhail, "Constraint Method Analysis of Stiffened Plates", Computers and Structures, 51-60, 1978.
[16] A.H. Sheikh, M. Mukhopadhay, "Analysis of Stiffened Plate with Arbitrary Planform by the General Spline Finite Strip Method", Computers and Structures, 53-67, 1992.
[17] C. Katz, J. Stieda, "Praktische FE-Berechnungen mit Plattenbalken", Bauinformatik, 1, 30-34, 1992.
[18] W. Wunderlich, G. Kiener, W. Ostermann, "Modellierung und Berechnung von Deckenplatten mit Unterzügen", Bauingenieur, 69, 381-390, 1994.
[19] G. Rombach, "Anwendung der Finite-Elemente-Methode im Betonbau", Ernst & Sohn, Berlin, 2000.
[20] F. Hartmann, C. Katz, "Statik mit Finiten Elementen", Springer, Berlin-Heidelberg, 2002.
[21] E.J. Sapountzakis, J.T. Katsikadelis, "Analysis of Plates Reinforced with Beams", Computational Mechanics, 26, 66-74, 2000.
[22] E.J. Sapountzakis, J.T. Katsikadelis, "Dynamic Analysis of Elastic Plates Reinforced with Beams of Doubly-Symmetrical Cross Section", Computational Mechanics, 23, 30-439, 1999.
[23] E.J. Sapountzakis, J.T. Katsikadelis, "Elastic Deformation of Ribbed Plate Systems Under Static, Transverse and Inplane Loading", Computers and Structures, 74, 571-581, 2000.
[24] E.J. Sapountzakis, J.T. Katsikadelis "Analysis of Prestressed Concrete Slab-and-Beam Structures", Computational Mechanics, 27, 492-503, 2001.
[25] E.J. Sapountzakis, J.T. Katsikadelis, "Creep and Shrinkage Effect on Reinforced Concrete Slab-and-Beam Structures", Journal of Engineering Mechanics, ASCE, 128(6), 625-634, 2002.
[26] M. Tanaka, A.N. Bercin, "A Boundary Element Method Applied to the Elastic Bending Problem of Stiffened Plates", Boundary Element Method XIX, 203-212, 1997.
[27] M. Tanaka, T. Matsumoto, S.Oida, "A Boundary Element Method Applied to the Elastostatic Bending Problem of Beam-Stiffened Plates", Engineering Analysis with Boundary Elements, 24, 751-758, 2000.
[28] G.R. Fernandes, W.S. Venturini, "Building Floor Analysis by the Boundary Element Method", Computational Mechanics, 35, 277-291, 2005.
[29] C. Hu, G.A. Hartley, "Elastic Analysis of Thin Plates with Beam Supports", Engineering Analysis with Boundary Elements, 13, 229-238, 1994.
[30] J.B. de Paiva, "Boundary Element Formulation of Building Slabs", Engineering Analysis with Boundary Elements, 17, 105-110, 1996.
[31] F. Hartmann, BE-SLABS 8.0.0, Boundary Element Analysis of Slabs, Student Version, http://www.be-statik.de, 2003.
[32] P.H. Wen, M.H. Aliabadi, A. Young, "Boundary element analysis of shear deformable stiffened plates", Engineering Analysis with Boundary Elements, 26, 511-520, 2002.
[33] L. Oliveira Neto. J.B. Paiva, "A special BEM for elastostatic analysis of building floor slabs on columns", Computers and Structures, 81, 359–372, 2003.

[34] S.F. Ng, M.S. Cheung, T. Xu, "A Combined Boundary Element and Finite Element Solution of Slab and Slab-on-Girder Bridges", Computers and Structures, 37, 1069-1075, 1990.
[35] M.S. Cheung, G. Akhras, W. Li, "Combined Boundary Element / Finite Strip Analysis of Bridges", Journal of Structural Engineering, 120, 716-727, 1994.
[36] N. Gattesco, "Analytical Modelling of Nonlinear Behaviour of Composite Beams With Deformable Connection", Journal of Constructional Steel Research, 52, 195-218, 1999.
[37] N.A. Jasim, "Deflections of Partially Composite Beams With Linear Connector Density", Journal of Constructional Steel Research, 49, 241-254, 1999.
[38] R.P. Johnson, I.N. Molenstra, "Partial Shear Connection in Composite Beams for Buildings", Proceedings of Institution of Civil Engineers, 91, 679-704, 1991.
[39] D.J. Oehlers, G. Sved, "Composite Beams With Limited-Slip-Capacity Shear Connectors", Journal of Structural Engineering, 932-938, 1995.
[40] T.M. Roberts, "Finite Difference Analysis of Composite Beans With Partial Interaction", Computers and Structures, 21(3), 469-473, 1985.
[41] D.J. Oehlers, N.T. Nguyen, M. Ahmed, M.A. Bradford, "Partial Interaction in Composite Steel and Concrete Beams with Full Shear Connection", Journal of Constructional Steel Research, 41(2/3), 235-248, 1997.
[42] E.J. Sapountzakis, J.T. Katsikadelis, "Interface Forces in Composite Steel-Concrete Structures", International Journal of Solids and Structures, 37, 4455-4472, 2000.
[43] E.J. Sapountzakis, J.T. Katsikadelis "A New Model for the Analysis of Composite Steel – Concrete Slab and Beam Structures with Deformable Connection", Computational Mechanics, 31(3-4), 340-349, 2003.
[44] E.J. Sapountzakis, V.G. Mokos, "Analysis of Plates Stiffened by Parallel Beams", International Journal for Numerical Methods in Engineering, 70, 1209-1240, 2007.
[45] E.J. Sapountzakis, V.G. Mokos, "An Improved Model for the Analysis of Plates Stiffened by Parallel Beams with Deformable Connection", Computers and Structures, 86, 2166-2181, 2008.
[46] E.J. Sapountzakis, V.G. Mokos, "An Improved Model for the Dynamic Analysis of Plates Stiffened by Parallel Beams", Engineering Structures, 30, 1720-1733, 2008.
[47] E.J. Sapountzakis, V.G. Mokos, "Shear Deformation Effect in Plates Stiffened by Parallel Beams", Archive of Applied Mechanics, 79, 893-915, 2009.
[48] E.J. Sapountzakis, V.G. Mokos, "Shear Deformation Effect in the Dynamic Analysis of Plates Stiffened by Parallel Beams", Acta Mechanica, 204(3), 249-272, 2009.
[49] G. Prathap, T.K. Varadan, "Large Amplitude Flexural Vibration of Stiffened Plates", Journal of Sound and Vibration, 57(4), 583-593, 1978.
[50] T.K. Varadan, K.A.V. Pandalai, "Large Amplitude Flexural Vibration of Eccentrically Stiffened Plates", Journal of Sound and Vibration, 67(3), 329-340, 1979.

[51[Y. Ueda, T. Yao, "Ultimate strength of compressed stiffened plates and minimum stiffness ratio of their stiffeners", Engineering Structures, 5, 97-107, 1983.

[52] M.R. Khalil, M.D. Olson, D.L. Anderson, "Nonlinear Dynamic Analysis of Stiffened Plates", Computers & Structures, 29(6), 929-941, 1988.

[53] T.S. Koko, M.D. Olson, "Non-Linear Analysis of Stiffened Plates Using Super Elements", International Journal for Numerical Methods in Engineering, 31, 319-343, 1991.

[54] D. Venugopal Rao, A.H. Sheikh, M. Mukhopadhyay, "A finite element large displacement analysis of stiffened plates", Computers & Structures, 47, 987-993, 1993.

[55] M. Fujicubo, P. Kaeding, "New simplified approach to collapse analysis of stiffened plates", Marine Structures, 15, 251-283, 2002.

[56] A.H. Sheikh, M. Mukhopadhyay, "Linear and nonlinear transient vibration analysis of stiffened plate structures", Finite Elements in Analysis and Design, 38, 477-502, 2002.

[57] M. Fujikubo, T. Yao, M.R. Khedmatic, M. Harada, D. Yanagihara, "Estimation of ultimate strength of continuous stiffened panel under combined transverse thrust and lateral pressure Part 1: Continuous plate", Marine Structures, 18, 383-410, 2005.

[58] R. Ojeda, B.G. Prusty, N. Lawrence, G. Thomas, "A new approach for the large deflection finite element analysis of isotropic and composite plates with arbitrary orientated stiffeners", Finite elements in Analysis and Design, 43, 989-1002, 2007.

[59] G.M. Vörös, "Buckling and free vibration analysis of stiffened panels", Thin-Walled Structures, 47, 382–390, 2009.

[60] P.A. Caridis, P.A. Frieze, "Flexural-Torsional Elasto-plastic Buckling in Flat Stiffened Plating Using Dynamic Relaxation. Part 1: Theory", Thin-Walled Structures, 6, 453-481, 1988.

[61] G.J. Turvey, M. Salehi, "Elastic Large Deflection Analysis of Stiffened Annular Sector Plates", International Journal of Mechanical Sciences, 40, 51-70, 1998.

[62] G.J. Turvey, M. Salehi, "Elasto-plastic large deflection response of pressure loaded circular plates stiffened by a single diametral stiffener", Thin-Walled Structures, 46, 991-1002, 2008.

[63] G.J. Turvey, M. Salehi, "Cross-Stiffened Circular Plates: An Elasto-Plastic Large Deflection Analysis", in B.H.V. Topping, J.M. Adam, F.J. Pallarés, R. Bru, M.L. Romero, (Editors), "Proceedings of the Tenth International Conference on Computational Structures Technology", Civil-Comp Press, Stirlingshire, UK, Paper 289, 2010. doi:10.4203/ccp.93.289

[64] D.K. Shin, "Postbucking Behavior of Rectangular Stiffened Plates Considering Buckled Pattern Change", KSCE Journal of Civil Engineering, 3(4), 319-330, 1999.

[65] G.R. Fernandes, W.S. Venturini, "Non-linear boundary element analysis of floor slabs reinforced with rectangular beams", Engineering Analysis with Boundary Elements, 31, 721-737, 2007.

[66] E.J. Sapountzakis, I.C. Dikaros, "Large Deflection Analysis of Plates Stiffened by Parallel Beams", Proc. of the 12h International Conference on Boundary Element and Meshless Techniques, Brasilia, Brazil, 13-15 July, 2011.

[67] J.T. Katsikadelis, "The Analog Equation Method. A Boundary – only Integral Equation Method for Nonlinear Static and Dynamic Problems in General Bodies", Theoretical and Applied Mechanics, 27, 13-38, 2002.

[68] E. Ramm, T.J. Hofmann, "Stabtragwerke, Der Ingenieurbau, Ed.G. Mehlhorn, Band Baustatik/Baudynamik, Ernst & Sohn, Berlin, 1995.

[69] H. Rothert, V. Gensichen, "Nichtlineare Stabstatik", Springer–Verlag, Berlin, 1987.

[70] E.J. Sapountzakis, V.J. Tsipiras, "Nonlinear Elastic Nonuniform Torsion of Bars of Arbitrary Cross Section by BEM", International Journal of Non-Linear Mechanics, 45, 63-74, 2010.

[71] F. Mohri, L. Azrar, M. Potier-Ferry, "Vibration analysis of buckled thin-walled beams with open sections", Journal of Sound and Vibration, 275, 434-446, 2004.

[72] MSC/NASTRAN for Windows, Finite element modeling and postprocessing system. Help System Index,Version 4.0, USA, 1999.

[73] SAP2000, Linear and Nonlinear Static and Dynamic Analysis and Design of Three-Dimensional Structures, Version 9, Computers and Structures, Inc., Berkeley, California, USA, 2004.

Chapter 10

Numerical Modelling and Analysis of Composite Structures subject to Extreme Loading

R.Y. Xiao[1], Z.W. Gong[1], C.S. Chin[1] and C.G. Bailey[2]
[1] Department of Urban Engineering
 Faculty of Engineering, Science and Built Environment
 London South Bank University, United Kingdom
[2] Faculty of Engineering
 Manchester University, United Kingdom

Abstract

This chapter presents the fire resistance behaviour of composite structures and provides in-depth understanding for the performance of a whole structure with the interactions between each of its elements under fire conditions. In this research the finite element model has been developed to simulate the behaviour of composite connections, floors and frames under fire by using the ANSYS software. The simulation presents the results of the time dependent temperature distribution within the structures, the structural responses under the elevated temperature as well as their effects on the fire resistant capacity of the structure. The composite structure is simulated under exposure to the ISO834 standard fire and the modelling results showed agreement with the fire test.

Keywords: composite structure, connections, composite floor, composite frame, finite element method, elevated-temperature, fire.

1 Introduction

A composite structure is constructed by steel columns, steel beams and a composite floor. It has been widely used in modern construction as a result of advantages such as higher strength and stiffness, a shorter construction period and easier handing on site [1]. However, arising from the significant difference in the thermal properties of steel and concrete elements, the composite structure could be quite different when subject to a fire situation compared with that of a bare steel or concrete structure. It will never be possible to prevent fires from occurring thus the aim of fire safety engineers is to reduce the risk of fire to an acceptable level. Based on its highly developed computing capability, the finite element technique can be used to model the response of a composite structure in a fire situation. The research presented here has focused on investigating the behaviour of flush end-plate beam-to-column connections, composite metal decking slab and composite frame in fire conditions using ANSYS software. The simulation presents the results of the time dependent temperature distribution within the structures, the structural response under the elevated temperatures as well as their effect on the fire resistant capacity of the structure.

2 Modelling of flush end-plate connection

Flush end-plate beam-to-column connections are commonly used for steel structures. A typical flush end-plate connection consists of a rectangular steel plate of the same depth as the beam end and is welded to the beam. This assembly is connected to the flange of a column by structural steel bolts. In the traditional steel structure design, the behaviour of beam-to-column connections has been assumed to be either rigid or pinned. However, observations from the full-scale fire tests and the actual damaged structures [2,3] have confirmed from the behaviour of beam-to-column connections used in actual structures exhibit a range of characteristics between these two extreme cases [4]. The evidence from the collapse of the WTC building [5] and a full scale fire tests at Cardington [6] indicated that connections have a significant impact on the survival time of structures. Many experimental tests have been conducted in the last few decades to study the effect of dynamic load and static load on flush end-plate connections, however, very little information on their behaviour in fire conditions had been disseminated.

The strength and stiffness of flush end-plates are significantly decreased when exposed to high temperature in fire conditions [7]. This may affect the overall load carrying capacity of a structure. Generally, reliable results can be obtained by experimental tests to describe the behaviour of beam-to-column connections. However, in many cases, experiments are either not feasible or too expensive to conduct. Additionally, the geometric sizes and mechanical parameters could be very restricted in fire tests. With significant advances in finite element methods, it is possible to simulate the complicated behaviour of composite connections and frames.

The finite element method (FEM), which is a reliable tool for investigating the effect of different parameters, can provide acceptable and accurate results once the model has been validated. The complicated phenomena such as local buckling, lateral buckling, detailed connection modelling and membranes actions can be accounted for in detail. Therefore, finite element modelling of connections in fire provides a good opportunity for wider parametric consideration and to eliminate the limitations associated with experiments.

In the early 1970s, as the application of computers for solving structural problems became evident, the finite element method was used to study the connection behaviour. Morris and Liu, who developed a finite element model to simulate the various types of connections under fire conditions [8-12]. Sherbourne and Bahaari

[13] developed a three-dimensional finite element model for simulating end-plate connections by using brick elements.

Rahamn et al [14], has investigated the behaviour of fin-plate connections in fire by ANSYS. The entire structures were modelled by four types of element: two types of three-dimensional solid elements, pre-tensioning elements, and contact elements. More recently, Xiao and Gong et al [7, 15, 16] created a three-dimensional finite element to investigate composite connections and floors. The results obtained have been well calibrated against test results. Some simplified analytical connections component models have been developed by some researchers [17-21] with the intention of simplifying connection design for fire conditions. The assessment of previous research [22-25] conducted into steel connections in fire conditions contain well documented experimental and analytical data.

2.1 The finite element model description

In this section, a complicated three-dimensional finite element model developed by Xiao et al. will be discussed. The commercial finite element software ANSYS has been used to simulate the behaviour of endplate connections. In order to illustrate the behaviour of steel connections under fire loading, a cruciform bolted flush end-plate beam-to-column steel connection, tested by Al-Jabri et al. [26] has been selected. In total twenty tests (5 Groups) were conducted in that research, however, in this paper only Group 2 has been selected for comparison. The properties of connection components and the level of loading for tests is shown in Table 1.

Spec. No.	Beam Section	Column Section	Bolts Size & Number	Thickness of End-plate	Moment Level (M)	Applied M (kNm)
FB21	356X171UB51	254X254XUC89	8M20	10mm	$0.2M_{cc}$	27
FB22	356X171UB51	254X254XUC89	8M20	10mm	$0.4M_{cc}$	56
FB23	356X171UB51	254X254XUC89	8M20	10mm	$0.6M_{cc}$	82
FB24	356X171UB51	254X254XUC89	8M20	10mm	$0.8M_{cc}$	110

Table 1: Properties of connection components and the level of loading for tests conducted by Al-Jabri et al. [26]

The flush end-plate connection consists of two 356x171UB51 beams connected to a 254x254UC89 column using 10mm thick flush end-plate and eight M20 bolts. The connection details are shown in Figure 1. Using the symmetry, just half of the connection is modelled. Figure 2 shows a three-dimensional finite element mesh.

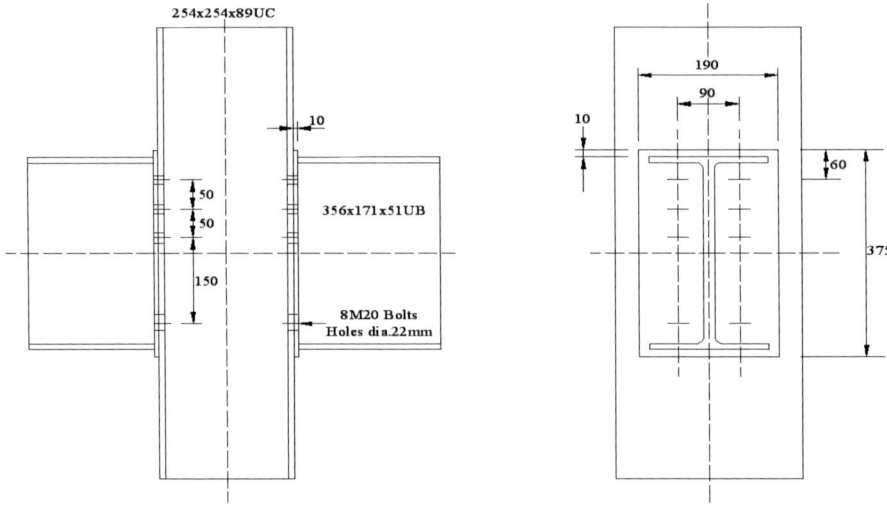

Figure 1: Details of the connection

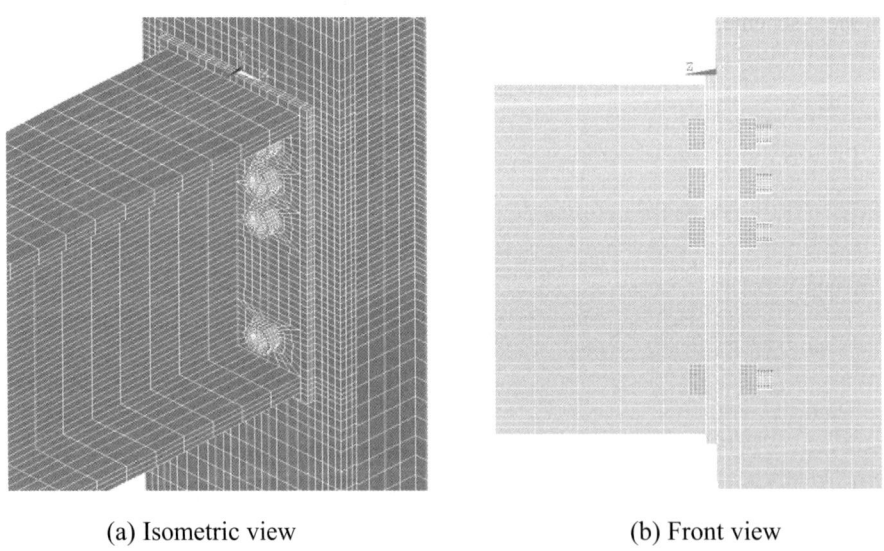

(a) Isometric view (b) Front view

Figure 2: Three-dimensional finite element model of the connection

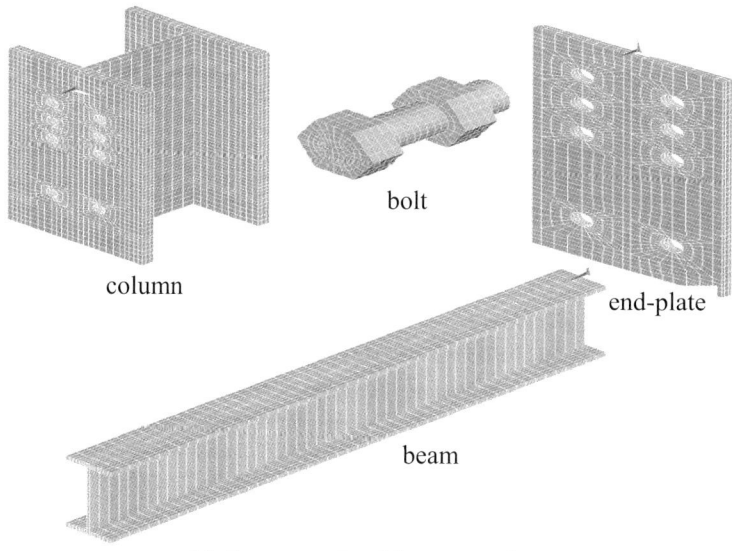

(c) Components of the connection

Figure 2: (continued) Three-dimensional finite element model of the connection

This is a comprehensive model accounting for contact interaction between each component in the connection. The surface to surface contact elements have been used for the contact surface including bolt shank and hole interface, bolt head and end-plate surface, nuts and column flange surface [27]. Figure 3 shows the contact element scheme.

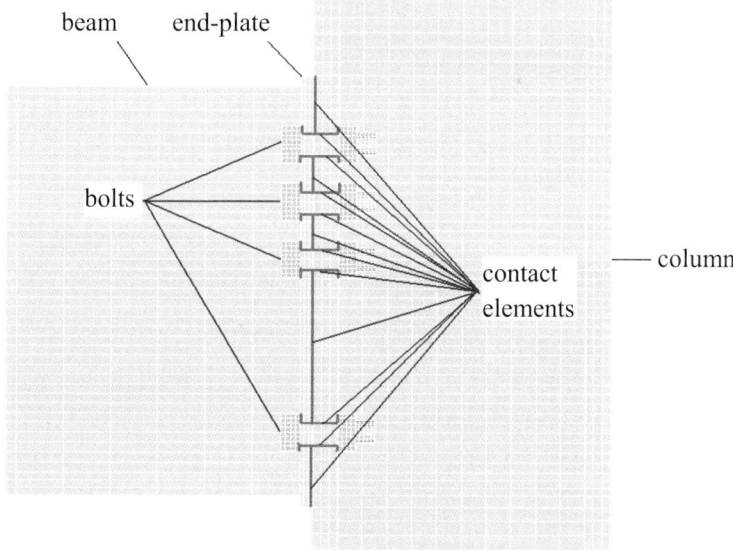

Figure 3: Contact elements distribution

2.2 Material modelling

The material properties for each component of the end-plate connection can be determined from the engineering stress-strain relationship using nonlinear material curves recommended in Eurocode 3 [28]. They may be defined also according to the stress-strain relationship obtained in a standard tensile test of steel. In the connection tests, the steel used was S355 for each connection component. All the bolts used are M20 grade in *2mm* clearance holes. The material properties of the connection tests are summarized in Table2.

Material type	Yield stress [N/mm^2]	Ultimate stress [N/mm^2]	Density [Kg/mm^3]	Young's Modulus [KN/mm^2]
S355	355	545	7850	195
8.8 Bolts	480	600	7850	195

Table 2: Material properties of tests conducted by Al-Jabri *et al.* [26]

The values selected are based on tensile tests conducted on structural steel specimens. However, most material test data have provided engineering stress and strain according to the uniaxial material testing response [29]. True stress-true strain data take into account the section area change. True strain, which is a nonlinear strain measure that is dependent upon the finial length of the model, is used for large strain simulations. Since the model is operated as a large deformation setting in ANSYS, it is necessary to correct material data from engineering stress and strain to true stress and strain. For uniaxial stress-strain data engineering stress and strain can be converted into true stress-strain using the following relationship:

$$\sigma_{true} = \sigma_{eng}(1+\varepsilon_{eng}) \quad (1)$$

$$\varepsilon_{true} = \ln(1+\varepsilon_{eng}) \quad (2)$$

where: σ_{true} is the true stress; σ_{eng} is the engineering stress; ε_{true} is the true strain, and ε_{eng} is the engineering strain.

The true elastic strain (ε_{el}) is known as the true stress (σ_{true}) divided by the Young's modulus (E);

$$\varepsilon_{el,true} = \sigma_{true}/E \quad (3)$$

The true strain is composed of the true elastic strain ($\varepsilon_{el,true}$) and the true plastic strain ($\varepsilon_{pl,true}$);

$$\varepsilon_{true} = \varepsilon_{el,true} + \varepsilon_{pl,true} \quad (4)$$

Therefore, the true plastic strain can be obtained from the following relationship:

$$\varepsilon_{pl,true} = \varepsilon_{true} - \varepsilon_{el,true} = \ln(1+\varepsilon_{eng}) - \sigma_{true}/E \quad (5)$$

Hence, the elastic-plastic material curves, shown in Figures 4 and 5, are used for the flush end-plate connection simulations.

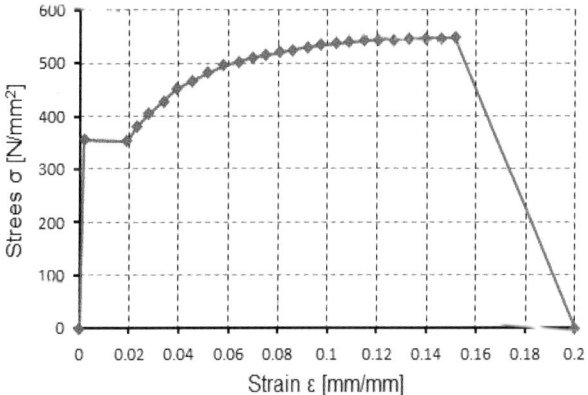

Figure 4: Stress-strain curve for S355 steel

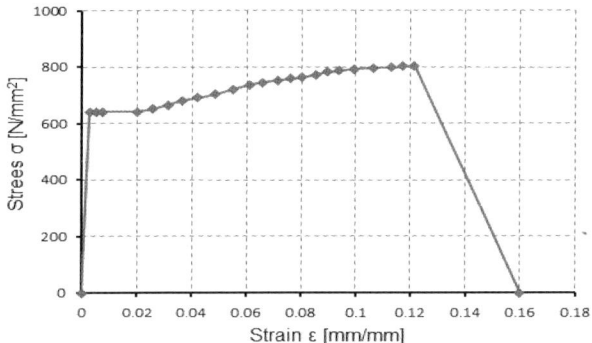

Figure 5: Stress-strain curve for Grade 8.8 bolts

2.3 Fire loading and boundary conditions

The ISO 834 standard temperature-time curve has been considered in this analysis. It is defined as the ability of a structure or its components to keep the bearing capacity during a standard fire exposure, for a specified standard period of time such as *30, 60* or *90* minutes. The standard fire exposure is described by an increasing temperature–time curve of the surrounding air as experienced in typical hydrocarbon fires. Further details about these temperature curves can be found in Eurocode 1 [30]. The ISO-834 standard temperature-time curve can be written in the form:

$$T(t) = 20 + 345\log_{10}(8t+1)$$

where T [$^\circ$C] is the gas temperature in the fire compartment at time t [min]. Figure 6 shows the ISO834 temperature-time curve.

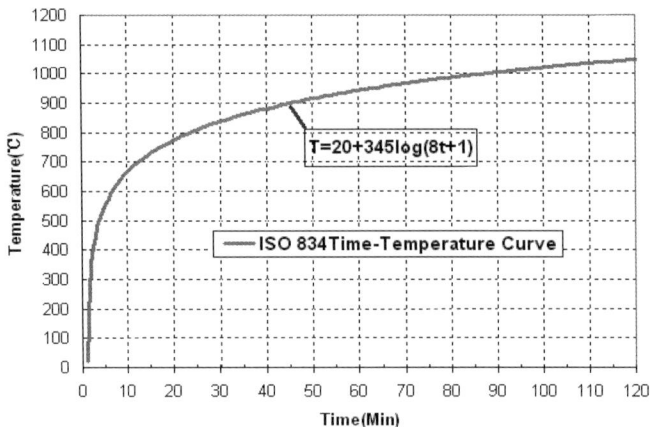

Figure 6: ISO834 Temperature-Time Curve

As the specimen is symmetric, only half of the structure was modelled. Therefore all the nodes along the planes of symmetry are restrained from moving in their corresponding perpendicular direction. The column has been assumed to be fixed at the top and bottom, the beam was allowed to deflect downward only, while horizontal movement was restrained to prevent any possibility of premature failure of the beam by lateral torsional buckling. The connection rotation, φ, can be estimated using the following expression:

$$\varphi = \tan^{-1}(u/L)$$

where u is the vertical deflection of the point along the beam, and L is the distance from the connection centreline to the point where the deflection is taken.

2.4 Model verification

The finite element results were compared with experimental data generated by Al-Jabri. Figure 7 shows the response of the flush end-plate connection under elevated temperature. Figure 8 shows the deflection of the beam member of the flush end-plate connection. It can be seen from Figure 7 that there is a local deformation at the top of the end-plate, where it is subject to the highest tensile stresses; deformation has occurred at the bottom of the end-plate as well, where it is subject to the highest compression stresses. They are accompanied by deformation of the column flange in the tension and compression zones.

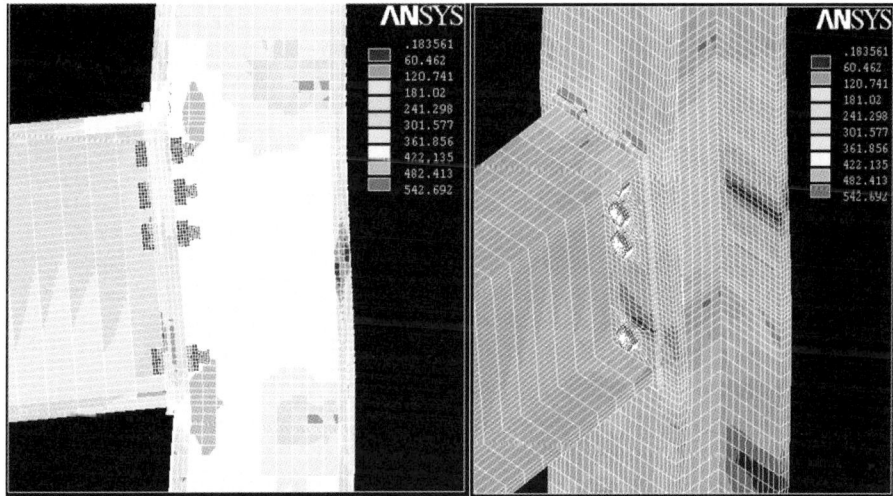

Figure 7: Response of the flush end-plate connection under elevated temperature

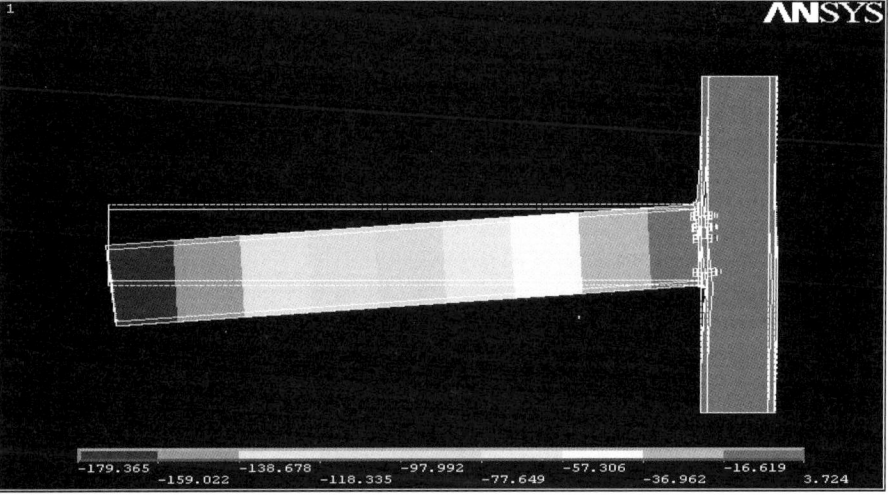

Figure 8: Shows the deformed model of the flush end-plate connection

Four elevated temperature tests were modelled, and a comparison of the temperature-rotation response of the connection at different temperatures is shown in Figure 9. The temperature-rotation response curves of the connection agree very well with tests at the elastic and plastic stages. For all tests, the degradation of the connection characteristics with increasing temperatures has been well predicted by the model. Figure 10 shows a comparison between the moment-rotation curves of the connection at different temperatures. The results also show good agreements between the finite element results and the experimental data.

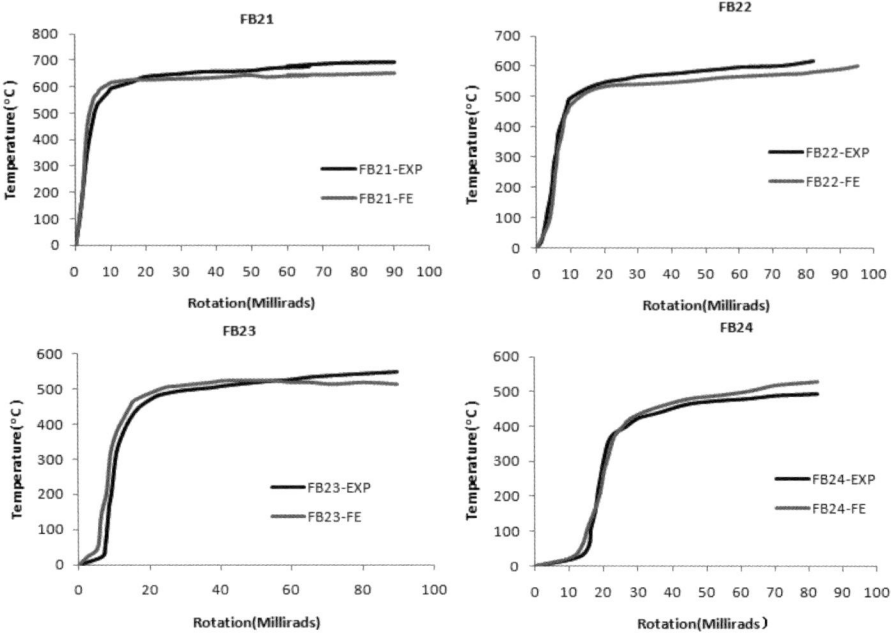

Figure 9: Comparison of finite element and experimental results for the tests

3 Modelling of composite flooring system

3.1 Description of the fire test

The response of a composite floor in fire conditions has been validated by modelling a metal decking floor system using the finite element technique. The results of the fire test conducted by BRANZ (Building Research Association of New Zealand) are used for validation [31].

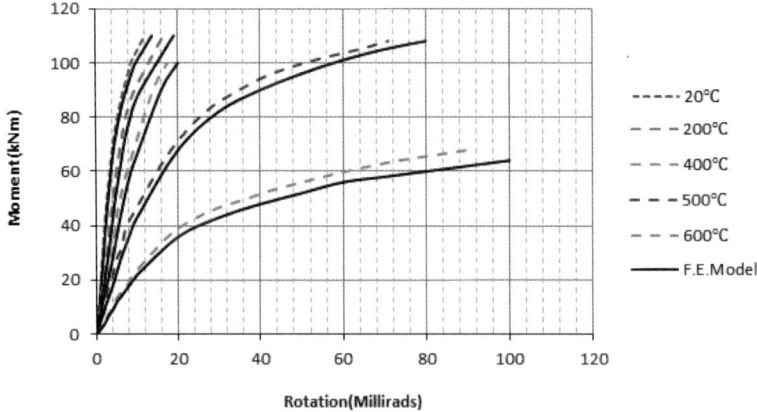

Figure 10: Numerical and experimental moment-rotation-temperature curve

The flat slab used in the test measured *3.3m* wide by *4.3m* long. The details of the slab are shown in Table 3 and the profile of the flat slab is shown in Figure 11. The fracture strains of the steel were determined from tensile tests at ambient conditions, and the measured yield stress of the mesh used in the slab was *565MPa*. The concrete compressive strengths are based on cylinder crushing tests carried out at ambient conditions.

Slab	Thickness (mm)	Reinforcing Steel area (mm^2/m)	Reinforcement Fracture Strain (%)	Concrete Compressive Strength, (MPa)
Decking slab	130	200	2.3	35

Table 3: Configuration of the slab

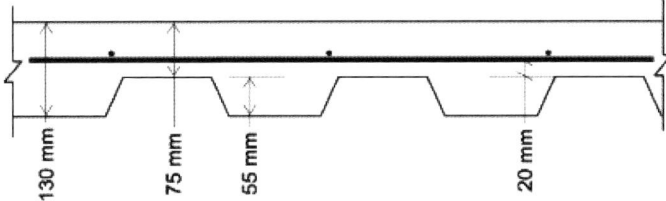

Figure 11: Profile of the composite slab

The slabs were simply supported on rollers over the furnace at all four edges. The edges were horizontally unrestrained. The clear span between the supports in the long and short directions was *4.15* and *3.15m*, respectively. The corners of the slabs were clamped down to prevent curling and to more closely represent the behaviour of a slab under the support conditions of a real building. The slabs were each subjected to ISO fire exposure for 2 hours while carrying a constant live load.

3.2 Validation of the model

A three-dimensional finite element model has been created using ANSYS. Figure 12 shows the finite element model. The metal decking of the concrete slab was modelled with a Shell 57 element. The concrete slab elements are modelled with Solid 70 elements. The element has eight nodes with a single degree of freedom, temperature, at each node. The element is applicable to a three-dimensional, steady-state or transient thermal analysis. The reinforcing bars were modelled with Link 33 elements. All these three elements should be replaced by an equivalent structural element.

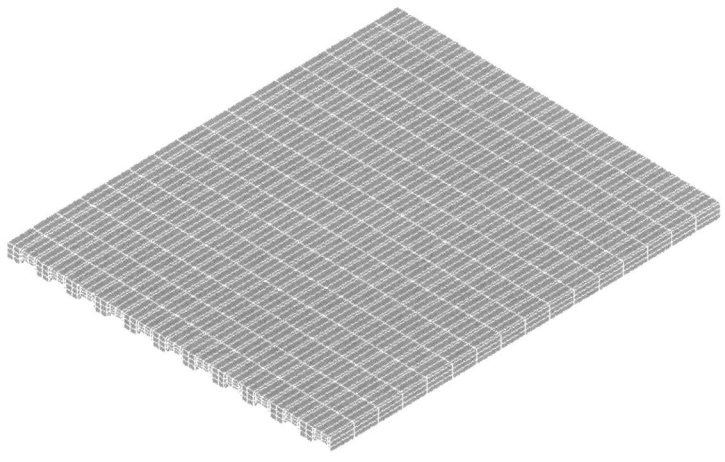

Figure12: The finite element model of concrete slab

In the real fire condition, there are two assumptions being made about the structure. The temperature of inner atmosphere of the model will increase equally. Temperature will be applied on the internal surfaces of the structure. The temperature load will follow the ISO standard fire curve rather than that of the natural fire curve [31]. Fire is usually represented by a temperature-time curve which gives the average temperature reached during a fire in a small size compartment or in the furnaces used for fire resistance tests. International standards are based on the standard fire defined by the heat exposure given by the ISO 834 curve [32].

The finite element results were compared with experimental data generated by BRANZ. Figure 13 shows the temperature distribution in the concrete slab after two hours of fire testing.

For fire, the initial conditions consist of the temperature distribution at the beginning of the analysis (usually the room temperature before the fire); boundary conditions must be defined on every surface of the structure. The rise of temperature in the concrete slab depends on the heat transfer between the fire environment and the structure. According to the second law of thermodynamics, heat is transferred between any two elements, which are at different temperatures. Inside the elements, heat is transferred from point to point only by conduction. Heat is also transferred from the bottom of the slab to the top of the slab; the thermal resources always go up.

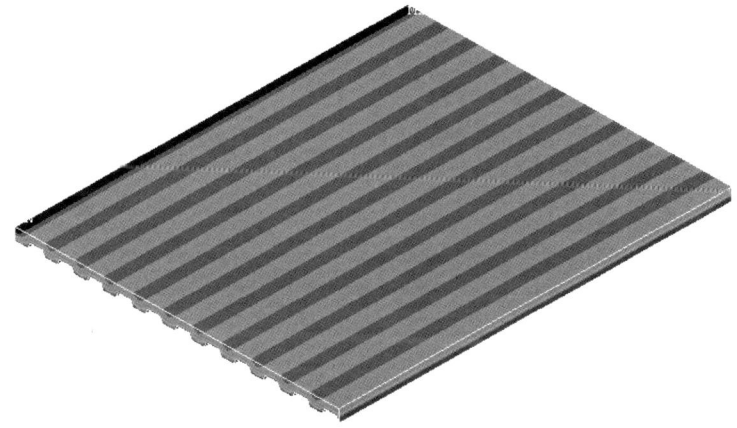

(a) Isometric view

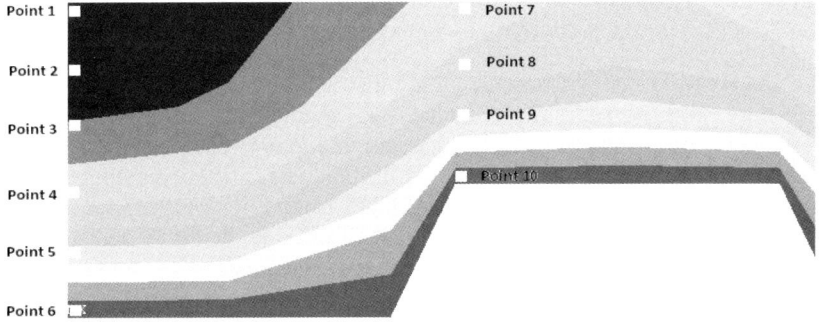

(b) Location of selecting points

Figure 13: Temperature distribution after 2 hours

°C \ Min	1 min	10 min	20 min	40 min	60 min	80 min	100 min	120 min
1	20.3	23.2	36.7	105.9	184.4	252.5	305.4	362.8
2	20.0	26.5	62.4	126.8	205.9	273.4	342.4	385.3
3	19.9	46.5	118.7	192.4	272.5	339.5	407.7	451.2
4	20.2	105.5	185.3	270.4	380.2	452.0	510.3	536.9
5	24.3	25.3	356.9	480.5	594.9	650.1	710.3	752.8
6	130.8	705.9	810.4	889.6	920.8	1006.5	1025.8	1049.0
7	20.1	48.6	86.3	202.6	295.5	380.3	434.7	510.5
8	19.0	88.4	165.8	245.7	338.4	425.7	489.9	552.4
9	23.6	202.5	258.5	405.5	508.6	572.7	638.5	685.7
10	168.5	682.5	792.6	889.3	946.9	999.5	1010.9	1049.0

Table 4: Temperature of 10 points at different time steps

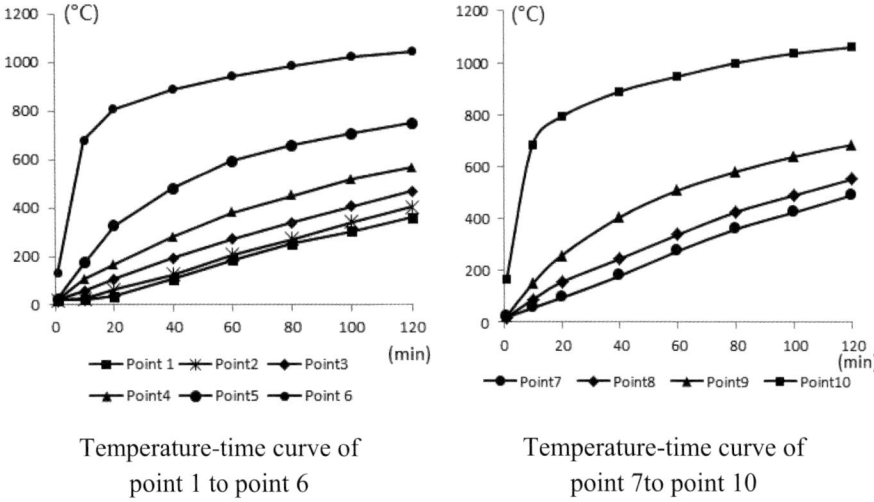

| Temperature-time curve of point 1 to point 6 | Temperature-time curve of point 7 to point 10 |

Figure14: Curves of time- temperature of selected ten points

For the detailed analysis of temperature distribution in the slab components, 10 points were selected from the concrete slab as indicated in Figure 13(b), the corresponding temperatures at times are listed in Table 4. The results are plotted in Figure 14. The variation of temperatures in the concrete calculated with ANSYS and measured during the fire test is shown in Figure 15. The temperatures are plotted at the heated surface and at the unheated surface of the slab. At the heated surface, the ANSYS calculation closely agrees with the experimental temperatures during the first 20min, then the experimental temperatures drop off slightly. At the top (unheated) surface, the ANSYS calculation shows very good agreement with the measured temperature.

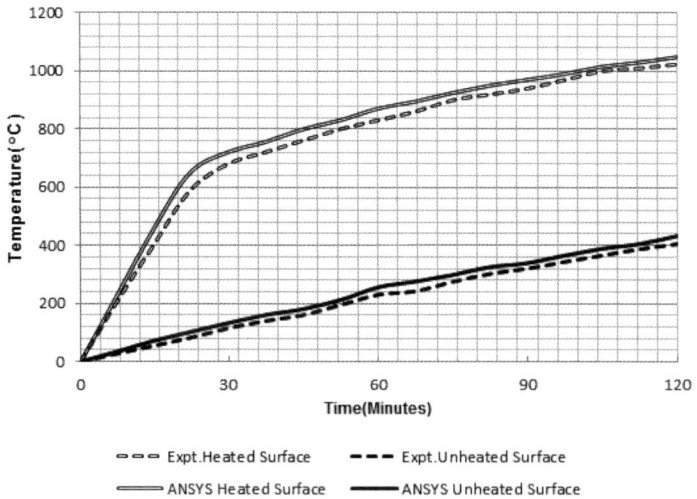

Figure15: Comparison between the test results and the ANSYS results

The results obtained from ANSYS have demonstrated the thermal behaviour of the concrete slab under the fire condition. After two hours, the temperature on the bottom surface of the slab has risen from room temperature to 1049 °C. From the ISO 834 testing curve, after two hours, the gas temperature is 1049 °C; it is the same as the concrete temperature. In the centre depth of the concrete slab (point 4), the temperature rose to 537°C. The temperature on the top face of the concrete slab (point 1) rose to 363°C.

For the thermal-structural analysis, the first step is to conduct the thermal field analysis, to achieve the temperature distribution on the concrete slab, and then one requires to change the element type and material properties from the thermal field to the structure field, based on the results of thermal analysis, to achieve uniform pressure on the top surface of the concrete. Figure 16 shows displacement of the model in the testing time of 7200s (2 hours). Table 5 indicates the development of the maximum vertical displacement UY in the mid-span of the concrete slab.

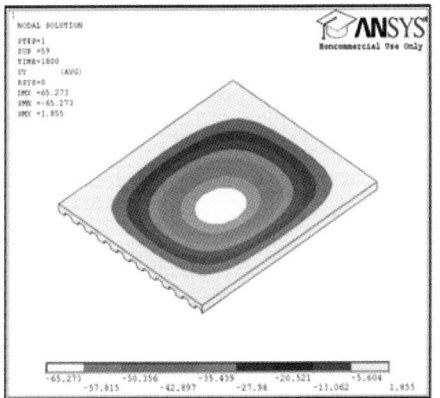
Deflection UY at time=30 minutes

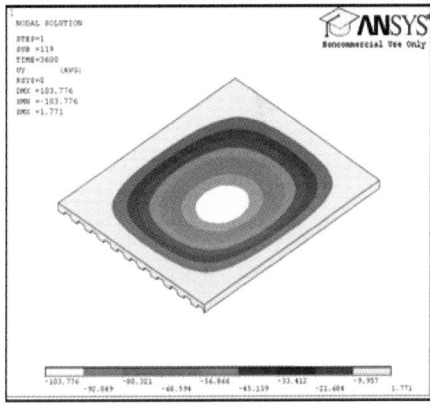
Deflection UY at time=60 minutes

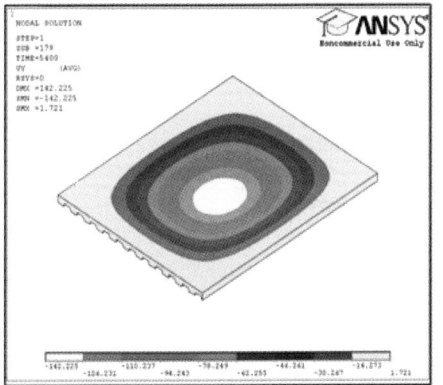
Deflection UY at time=90 minutes

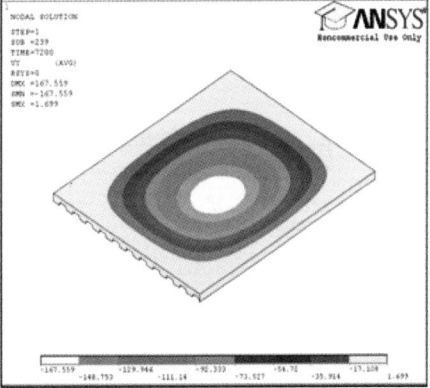
Deflection UY at time=120minutes

Figure16: Vertical displacement UY in the midspan of the concrete slab

Times (minutes)	Maximum deflection (mm)
10	43
30	65
40	83
60	104
80	139
90	142
110	162
120	168

Table 5: The development of the maximum vertical displacement UY

The numerical results from ANSYS were validated by the fire test data for deflection; Figure 17 shows the comparison of these two results, which show good agreement with the fire test data.

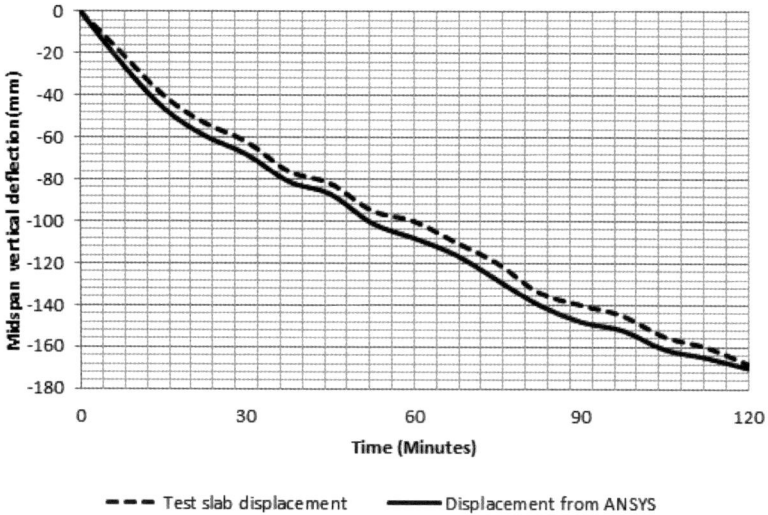

Figure 17: Compare numerical results with fire test results

Figure 16 and Table 5 have clearly assisted understanding of the development of the maximum vertical displacements at time steps of 10mins, 30mins, 40mins, 60mins, 80mins, 90mins, 110mins 120mins. The model is the same as that used in the thermal field analysis.

The mechanical properties of all common building materials degrade with an elevation of temperature [33]. Concrete generally suffers irreversible loss in strength once these materials are heated to temperatures usually in excess of 300°C. At higher temperatures, up to around 500°C at least 60% of the strength is permanently lost [34]. As a result of the strength loss in fire situations, the deflection is also increased with the elevated temperature.

4 Modelling of steel frame in fire situation

Significant developments have been made in analysing the behaviour of steel-framed structures under fire conditions in the last 20 years. During the 1990s a programme of fire tests was completed in the UK at the Building Research Establishment's Cardington Laboratory. The tests were carried out on an eight-storey composite steel-framed building that had been designed and constructed as a typical multi-storey office building. The purpose of the tests was to investigate the behaviour of a real structure under real fire conditions and to collect data that would allow computer programs for the analysis of structures in fire to be verified [35]. In this chapter the finite element models created for connections and floors discussed previously have been used to provide an understanding of the behaviour of the steel frame. The finite element modelling on a full structure allows the displacement and stresses of the structure to be predicted during the fire.

The steel frame model in this computer model is measured at 6m wide by12m long. A three-dimensional thermal analysis technique using an ISO834 temperature curve is utilised [36]. Figure 18 shows the computer model of the frame.

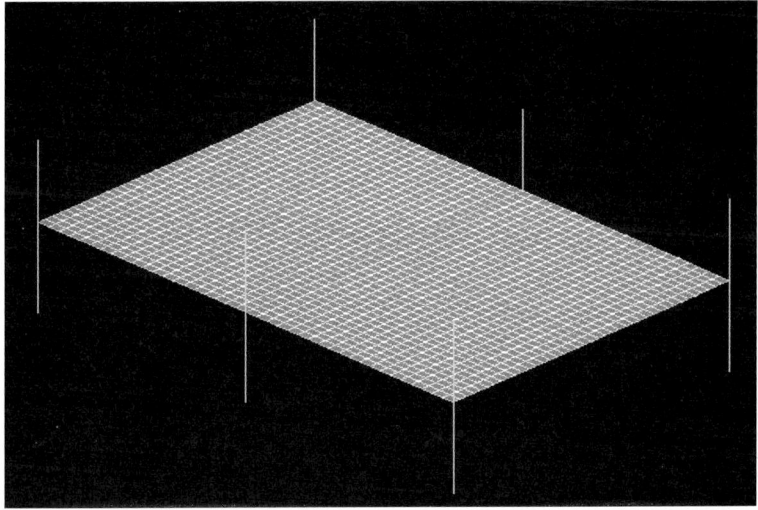

Figure 18: Finite element model of the frame

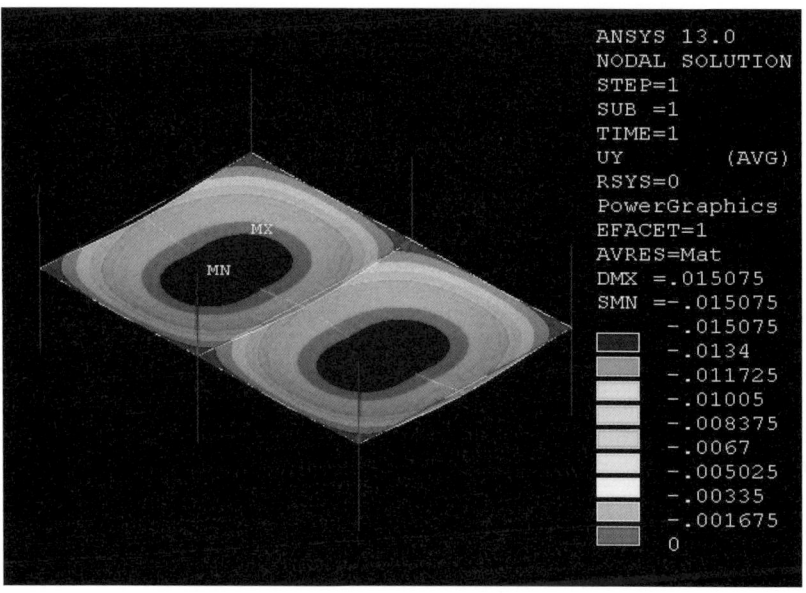

Figure 19: Displacement of the steel frame

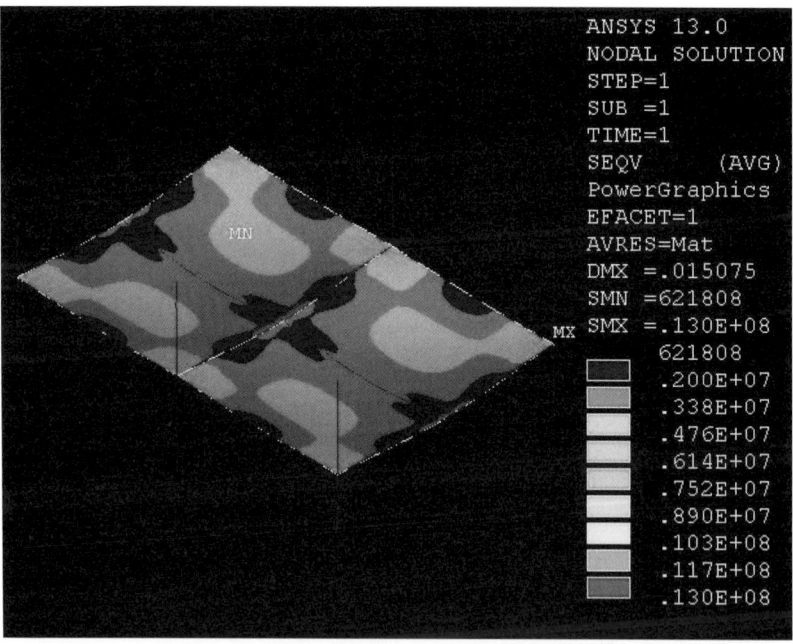

Figure 20: Von-mises stress of the frame

The model takes into consideration the material properties and the effect of the boundary conditions. Figure 19 and Figure 20 show the displacement and the Von-Mises stress of the steel frame under fire conditions respectively. The validated connection model and floor model have worked efficiently in the full structure model.

5 Conclusion

This paper describes the three-dimensional numerical analysis of flush end-plate connections, the composite flooring system and the steel frame under fire conditions using a non-linear finite element program, ANSYS. The observations and conclusions can be drawn from the results of finite element analysis under the fire situation. For the flush end-plate connections, all the components of the connection were modelled by using solid elements, while contact elements have been used for the contact surface to transfer the load from the beam to the column. The finite element results have been compared with experimental data. The results obtained showed good agreement in both the elastic and plastic stages. The degradation of the connection strength and stiffness with the increasing temperatures is well predicted by the model. This demonstrates that the finite element technique is capable of predicting connection response at elevated temperatures. For the composite flooring system, the thermal analysis indicates that the temperature distribution in the same section of the concrete slab is different along the same plane. The greater difference in temperature on the concrete slab surface will cause greater differences in local stresses distribution. After two hours the temperature of the metal decking has been increased to 1049°C, but the temperature towards the top of the concrete slab was just around 400°C. A large vertical displacement was found in the composite decking slab. At the beginning of the analysis, the vertical displacement in the midspan of the concrete slab developed slowly, However, it increased significantly by the end of the two hours fire testing, the vertical displacement in the midspan of the concrete slab increased by over 168mm. The results from three-dimensional analysis of the composite slab using the ANSYS software have shown good agreement with the fire test results. The full finite element models on connections and the composite floor have been successfully implemented to full scale composite frame analysis. This has provided a robust tool for designing composite structure subjected to fire.

References

[1] C.G. Bailey, E. Ellobody, "Whole-building behaviour of bonded post tensioned concrete floor plates exposed to fire", Engineering Structures, 31, 1800-1810, 2009.

[2] K.S. Al-Jabri, T. Lennon, I.W. Burgess, R.J. Plank, "Behaviour of steel and composite beam-column connections in fire", J Construct Steel Res, 46(1–3), paper 180, 1998.

[3]   K.S. Al-Jabri, T. Lennon, I.W. Burgess, R.J. Plank, "The performance of frame connections in fire", ACTA Polytech, 39(5), 65–75, 1999.
[4]   W.F. Chen, E.M. Lui, "Stability design of steel frames", CRC Press, Inc., Boca Raton FL, 1991.
[5]   NIST, "Final Report on the Collapse of the World Trade Center Towers", National Institute of Standards and Technology, USA, 2005.
[6]   G.M. Newman, J.T. Robinson, C.G. Bailey, "Fire safety design: A new approach to multi-storey steel-framed buildings", The Steel Construction Institute, 2004.
[7]   R.Y. Xiao, Z.W. Gong, C.S. Chin, "Nonlinear numerical simulation of steel flush end-plate connections in fire", Proceedings of the 4th International Conference on Steel & Composite Structures, Research Publishing, Sydney, Australia, 597-602, July 2010.
[8]   T.C.H. Liu, L.J. Morris, "Theoretical modeling of steel bolted condition under fire exposure", in Proceedings of the International Conference on Computational Mechanics, 1994.
[9]   T.C.H. Liu, "Moment–rotation–temperature characteristics of steel/composite connections", Journal of Structional Engineering, 125(10), 1188–1197, 1999.
[10] T.C.H. Liu, "Three-dimensional modeling of steel/concrete composite connection behavior in fire", Journal of Constructional Steel Research, 45(1–3), 1998.
[11] T.C.H. Liu, "Effect of connection flexibility on fire resistance of steel beams", Journal of Constructional Steel Research, 45, 99–118, 1998.
[12] T.C.H. Liu, "Finite element modeling of behavior of steel beams and connections in fire", Journal of Constructional Steel Research, 35(3), 181–199, 1996.
[13] A.N. Sherbourne, M.R. Bahaari, "3D Simulation of End-Plate Bolted connections", Journal of Structural Engineering, 120(11), 3122-3136, 1994.
[14] R. Rahman, R. Hawilch, M. Ahamid, "The effect of fire loading on a steel frame and connection", in C.A. Brebbia, W.P. Wilde, (Editors), "High performance structures and materials II", WIT Press, 307-316, 2004.
[15] R.Y. Xiao, Z.W. Gong, C.S. Chin, "Nonlinear analysis of fire resistance of composite flooring system", 9th World Congress on Computational Mechanics and 4th Asian Pacific Congress on Computational Mechanics (WCCM/APCOM 2010), Sydney, Australia, 275, July 2010.
[16] Z.W. Gong, R.Y. Xiao, C.S. Chin, "Fire strength analysis of composite flooring system", Proceedings of the 4th International Conference on Steel & Composite Structures, Research Publishing, Sydney, Australia, 167-172, July 2010.
[17] L. Simoes da Silva, A. Santiago, P. Vila Real, "A component model for the behaviour of steel joints at elevated temperatures", J Constr Steel Res, 57 1169–1195, 2001.
[18] S. Spyrou, J.B. Davison, I.W. Burgess, R.J. Plank, "Experimental and analytical investigation of the 'tension zone' components within a steel joint at elevated temperatures", J Constr Steel Res, 60, 867–896, 2004.

[19] S. Spyrou, J.B. Davison, I.W. Burgess, R.J. Plank, "Experimental and analytical investigation of the 'compression zone' components within a steel joint at elevated temperatures", J Constr Steel Res, 60, 841–865, 2004.
[20] K.S. Al-Jabri, "Component-based model of the behaviour of flexible end-plate connections at elevated temperatures", Compos Struct, 66, 215–221, 2004.
[21] I.W. Burgess, J. El Rimawi, R.J. Plank, "Studies of the behaviour of steel beams in fire", J Const Steel Res, 19, 285–312, 1991.
[22] European Committee for Standardization, "EN 1993-1-2: Eurocode 3: Design of steel structures, Part1.2: Structural fire design", 2005.
[23] R.M. Richard, P.E. Gillett, J.D. Kriegh, B.A. Lewis, "The analysis and design of single plate framing connections", Eng, JAISC, 17(2), 1980.
[24] F. Wald, L. Simoes da Silva, D. Moore, A. Santiago, "Experimental behaviour of steel joints under natural fire", ECCS-AISC Workshop, June 3–4, 2004.
[25] C.G. Bailey, I.W. Burgess, R.J. Plank, "Analyses of the effects of cooling and fire spread on steel-framed buildings", Fire Safety J, 26, 273–293, 1996.
[26] K.S. Al-Jabri, I.W. Burgess, T. Lennon, R.J. Plank, "Moment-rotation-temperature curves for semi-rigid joints", Construct Steel Res, 61, 281–303, 2005.
[27] R.Y. Xiao, Z.W. Gong, C.S. Chin, "Modelling of Flush end-plate steel connections at elevated temperature", Proceedings of The 9th Pacific Structural Steel Conference, Beijing, China, October 2010.
[28] Eurocode 3, "Design of steel structures, Part1.1:Genaral structural rules".
[29] K.J. Bathe, "Finite Element Procedures", Prentice Hall, Englewood Cliffs, New Jersey, 1982.
[30] Eurocode 1, "Actions on structures, Part1.2: Genaral actions-Actions on structures exposed to fire".
[31] L. Lim, C. Wade, "Experimental fire tests of two-way concrete slabs", Fire Engineering Research Report 02/12, University of Canterbury and BRANZ Ltd, New Zealand, 2002.
[32] ISO834, "Fire resistance tests on elements of building construction", International Organisation for Standardisation, 1975.
[33] Z.W. Gong, R.Y. Xiao, C.S. Chin, "Fire strength analysis of composite flooring System", 4th international conference on steel & composite structures, ICSCS'10, Sydney, Australia, July 2010.
[34] C. Zienkiewicz, "The Finite Element Method in Structural and Continuum Mechanics", McGraw-Hill, London, 1967.
[35] SFPE, "Fire exposures to structural elements–engineering guide", Society of Fire Protection Engineers, Bethesda MD, USA, 2004.
[36] B.R. Kirby, "The behaviour of a multi-storey steel framed building subject to fire attack-experimental data", British Steel Swindon Technology Centre, United Kingdom, 1998.

Author Index

Ambrósio, J., 65

Bailey, C.G., 279

Chin, C.S., 279

da S. Vellasco, P.C.G, 187
da Silva, A.T., 187
da Silva, J.G.S., 187
de Andrade, S.A.L., 187
de J. dos Santos, J., 187
de Lima, L.R.O., 187
Dimitrovová, Z., 99

Gantes, C.J., 219
Gattulli, V., 1
Gong, Z.W., 279
Gramegna, F., 35

Kaveh, A., 157

Majorana, C.E., 35

Pomaro, B., 35
Pombo, J., 65
Prete, G., 35

Rahami, H., 157

Salomoni, V.A., 35
Sapountzakis, E.J., 243
Šejnoha, M., 131

Valenta, R., 131
Vorel, J., 131

Xiao, R.Y., 279

Zeman, J., 131

Keyword Index

advanced finite element analysis, 219

beam-to-column joints, 187
bending, 243
boundary element method, 243

canonical forms, 157
circulant, 157
component method, 187
composite floor, 279
composite frame, 279
composite structure, 279
concrete, 131
concrete durability, 35
connections, 279
coupled problem, 35

damage assessment, 1
decomposition, 157
diagonalization, 157
dynamic stiffness matrix, 99

earthquake engineering, 1
eigensolution, 157
elastic stiffened plate, 243
elevated-temperature, 279
equilibrium path, 219
existing buildings, 1
experimental analysis, 187
extended endplate joints, 187

finite element method, 279
fire, 279

geometric nonlinearity, 219
graph products, 157

homogenization, 131

imperfections, 219

localized disturbances, 99

masonry, 131
mastic asphalt, 131
material nonlinearity, 219
modal expansion, 99
moving load, 99
moving mass, 99
multibody systems, 65
multiscale modeling, 131

nonlinear analysis, 243
nonuniform torsion, 243
nuclear facility, 35

radiation damage, 35
radiation heat, 35
radioactive ion beams, 35
railway dynamics, 65
random composites, 131
reinforced plate with beams, 243
ribbed plate, 243

semirigid joints, 187
shielding, 35
slab-and-beam structure, 243
state-space formulation, 99
statistically equivalent periodic unit cell, 131
steel structures, 187, 219
strength, 219
structural modelling, 1

track irregularities, 65

ultimate limit state, 219

vehicle-track interaction, 65

warping, 243
wheel-rail contact, 65